Logic in Asia: Studia Logica Library

Logic in Asia: Studia Logica Library

This book series promotes the advance of scientific research within the field of logic in Asian countries. It strengthens the collaboration between researchers based in Asia with researchers across the international scientific community and offers a platform for presenting the results of their collaborations. One of the most prominent features of contemporary logic is its interdisciplinary character, combining mathematics, philosophy, modern computer science, and even the cognitive and social sciences. The aim of this book series is to provide a forum for current logic research, reflecting this trend in the field's development.

The series accepts books on any topic concerning logic in the broadest sense, i.e., books on contemporary formal logic, its applications and its relations to other disciplines. It accepts monographs and thematically coherent volumes addressing important developments in logic and presenting significant contributions to logical research. In addition, research works on the history of logical ideas, especially on the traditions in China and India, are welcome contributions.

The scope of the book series includes but is not limited to the following:

- Monographs written by researchers in Asian countries.
- Proceedings of conferences held in Asia, or edited by Asian researchers.
- Anthologies edited by researchers in Asia.
- Research works by scholars from other regions of the world, which fit the goal of "Logic in Asia".

The series discourages the submission of manuscripts that contain reprints of previously published material and/or manuscripts that are less than 165 pages/ 90,000 words in length.

Please also visit our webpage: http://tsinghualogic.net/logic-in-asia/background/

Relation with Studia Logica Library

This series is part of the Studia Logica Library, and is also connected to the journal Studia Logica. This connection does not imply any dependence on the Editorial Office of Studia Logica in terms of editorial operations, though the series maintains cooperative ties to the journal.

This book series is also a sister series to Trends in Logic and Outstanding Contributions to Logic.

For inquiries and to submit proposals, authors can contact the editors-in-chief Fenrong Liu at fenrong@tsinghua.edu.cn or Hiroakira Ono at ono@jaist.ac.jp.

More information about this series at http://www.springer.com/series/13080

Mihir Kumar Chakraborty · Soma Dutta

Theory of Graded Consequence

A General Framework for Logics of Uncertainty

Springer

Mihir Kumar Chakraborty
University of Calcutta
Kolkata, India

Jadavpur University
Kolkata, India

Soma Dutta
Department of Mathematics
and Computer Science
University of Warmia and Mazury
in Olsztyn
Olsztyn, Poland

ISSN 2364-4613 ISSN 2364-4621 (electronic)
Logic in Asia: Studia Logica Library
ISBN 978-981-13-8894-1 ISBN 978-981-13-8896-5 (eBook)
https://doi.org/10.1007/978-981-13-8896-5

This Springer imprint is published by the registered company Springer Nature Singapore Pte Ltd.
The registered company address is: 152 Beach Road, #21-01/04 Gateway East, Singapore 189721, Singapore

Dedicated to the memory of
Profesor Lotfi A. Zadeh

Preface

Lotfi A. Zadeh introduced the theory of fuzzy sets through his paper in 1965. Since that time the theory has grown enormously in the theoretical aspects as well as applications. Almost from the beginning, the term 'fuzzy logic' had been of abundant use particularly among the researchers who made significant contribution in the application of the theory. However, after the publication of Pavelka's seminal work Fuzzy logic, I, II and III in 1979, there had been a change; the term 'fuzzy logic' was divided into two senses, fuzzy logic in the broad sense and fuzzy logic in the narrow sense. In the former sense, any use of fuzzy set theory is referred to, while in the latter sense specific issues of the area of knowledge that goes under the title 'logic' are incorporated and discussed. The present book falls broadly within the category of fuzzy logic in narrow sense. This research on the theory of graded consequence (GCT) was initiated by Chakraborty in 1986, and since then a number of articles have been published and several theses are written on this topic.

The present authors have been collaborating in this field since 2004. In fact, it is due to their contributions that the subject has now developed to a considerable extent, and in spite of some initial inhibitions of the community it seems to have gained ground in recent years. The motto of the theory may be summarized in the following remark made by Chakraborty in (Fundamenta Informaticae, 1997):

> *... if vagueness is present at the object-level language and hence multivalence is accepted by providing a semantics for object-level sentences, multivalence cannot be generally denied at the level of meta-concepts like consequence, consistency, tatutologihood etc.*

More precise justifications have been put forward in the Introduction of the book. The term 'graded consequence', representing a consequence relation being graded, seems to appear first in the first published work of Chakraborty in 1988. If one browses through the literature of many-valued and/or fuzzy logics, it may be noticed that apart from a faint reference to degree of consequence in some works, consequence was neither meant to be a graded concept, nor it has been ascribed the adjective 'graded' before. GCT has been the first attempt where all meta-logical notions, including consequence, are systematically treated as a matter of grade. The account for a graded meta-logic has passed through a lot of skepticism for quite many years. But slowly it has started taking a shape. The term 'graded logic' is now in the air, and several researchers are coming up with their proposals for a many-valued notion of consequence, calling them approximate entailment, graded entailment, graded inference, rough reasoning and sometimes with the same name, viz. graded consequence. Though the approaches and treatments towards viewing a graded consequence are quite different, that the consequence may be regarded as a matter of grade has gained a ground. The present treatise may be helpful to get an overview of how the approach towards consequence gradually gets shifted from many-valued logics, fuzzy logics to the theory of graded consequence.

Before giving a short overview of the chapters, we should make it clear that GCT does not present a logical system which has to satisfy a fixed set of properties concerning its object-level connectives. The basic three structural properties of a standard notion of consequence, namely, overlap, dilution and cut, have been incorporated in graded context. Rest depends on the choice of the user of a particular logic. For instance, we have considered some basic properties of negations in the context of graded inconsistency in Chap. 3. They need not be considered as properties of every logic in the framework of GCT. If those specific axioms (GC1) to (GC5) are taken as properties of graded consequence in some system in the framework of GCT, then one can only establish that in such a logic inconsistency and consequence are interderivable under some conditions. If we drop some axioms then it may affect the consequence–inconsistency connection, but it still remains a logic with graded notion of consequence. Similar is the case in the context of the study made in Chap. 4. None of the rules concerning different logical connectives of the object language is considered as a fixed rule. Different logics in the framework of GCT can have different rules, and this solely depends on the object-level and meta-level reasoning.

The chapters of the book are arranged as follows.

Chapter 1 is the introduction of the theme; that is, the objective of this work, namely, to introduce many-valuedness to meta-logical notions like consequence, consistency/inconsistency, tautologihood, etc., involved in a logical discourse, is stated. To arrive at this end, the nodal points that have been passed through are (i) three levels inherent in a logic discourse; (ii) from many-valued logics, fuzzy logics to graded consequence: a brief overview; (iii) a general discussion on uncertainty and vagueness; (iv) notion of consequence in classical logic; and (v) finally, some motivations for lifting many-valuedness to the meta-level notions including consequence.

Chapter 2 introduces the basic technical apparatus. This chapter comprises the basic concepts of the theory of graded consequence including the syntactic, semantic and axiomatic notions of graded consequence, and their interrelations.

Chapter 3 is about the extension of the theory in the presence of a logical connective negation ($\neg$) in the object language. The set of axioms for characterizing a graded consequence relation is extended in the presence of $\neg$. The notion of graded inconsistency is also introduced, and an equivalence between the notions of graded consequence and graded inconsistency is established. In continuation to the study of level cuts of a graded consequence relation, presented in Chap. 2, a few results in the presence of $\neg$ in the language are also presented. Reflection of the newly added axioms in the meta-level algebraic structure because of the presence of $\neg$ in the object language is presented in an extended algebraic structure of a complete residuated lattice. The properties of this structure, called GC($\neg$)-algebra, are studied as well.

Chapter 4 presents a scheme for generating proof theory of a logic of graded consequence. As the theory of graded consequence is not a particular logic, rather a general scheme for logics dealing with multilayered many-valuedness; in this chapter, we shall present the necessary and sufficient conditions for getting specific rules corresponding to logical connectives of the object language. That is, we shall add connectives, say $\#$, in the language and explore the necessary and sufficient conditions to get hold of particular syntactic rules concerning $\#$. This study will help us to present a general scheme for generating different logics based on GCT. We shall show that the interrelation between the object-level algebraic structure and the meta-level algebraic structure determines the proof theory of a logic. That the many-valued logics can be obtained as a special case of this scheme will also be presented.

Chapter 5 deals with different approaches towards retaining the classical equivalence between consequence operator and consequence relation in fuzzy context. As in classical context the notion of consequence can be equivalently presented by consequence operator approach and consequence relation approach, similar attempts are observed in fuzzy context too. Different approaches have come up in order to bridge the gap between Pavelka's notion of fuzzy consequence operator and Chakraborty's notion of graded consequence relation in fuzzy context. In this regard, we shall discuss the notion fuzzy consequence relation by Castro et al., and implicative consequence operator and implicative consequence relation by Rodríguez et al. We shall discuss some of the limitations of those approaches and propose extensions to bridge the gap.

Chapter 6 contains a comparative analysis of the notion of graded consequence with other approaches towards consequence in the context of fuzzy logics. In this regard, we shall first present different systems of fuzzy logics with respect to their proposed notions of consequence, and then analyse the approaches in order to check how faithful those approaches are in incorporating many-valuedness in the notion of consequence.

Chapter 7 deals with a few aspects of decision-making based on the idea of the theory of graded consequence. In the first case, we shall extend the theory in the context of interval semantics and present how different ways of aggregating information collected from different sources/experts/agents incorporate different attitudes of a decision-maker. In the second case, we shall propose an extension of GCT in a distributed network of decision-making among agents (sources of information), each having their own local logics, and the final decision-maker in the sense of Barwise and Seligman. The third will be on dealing with Sorites like paradox, which is usually considered to be one of the testing criteria for any theory of reasoning with vague predicates.

The subject GCT is still new, and there appear many misunderstandings at the first encounter. We would request the readers not to give up because of unfamiliarity with the concepts.

Acknowledgements We extend our thanks to Sanjukta Basu for allowing us to use some relevant portions of her Ph.D. thesis entitled ‘A Study in Logical and Philosophical Implications of Graded Consequence in Many-Valued Systems’.

We express our sincere gratitude to Andrzej Skowron and Mohua Banerjee for giving valuable suggestions not only during the writing of the book but also throughout the development of the theory.

We express our thanks to Giangiacomo Gerla, Lluis Godo, and Francesc Esteva for extending their support by showing positive attitude towards the idea of GCT.

We would also like to thank R. Ramanujam for extending institutional support to one of the authors and general encouragement in the development of the work.

We also thank Purbita Jana for the technical help she has extended to us during the preparation of the manuscript.

Finally, our debt to Calcutta Logic Circle (CLC) and one of its mentors Late Prof. Haragauri Narayan Gupta, a Ph.D. student of Professor Alfred Tarski, has to be acknowledged. CLC has been the main source that inspired both the present authors for a long period of time in pursuing research in logic.

Kolkata, India
Warsaw, Poland
March 2019

Mihir Kumar Chakraborty
Soma Dutta

Contents

Chapter 1
Introduction

Abstract In this chapter, the objective of this work, which is to introduce many-valuedness to meta-logical notions like consequence, consistency/inconsistency, tautologihood, etc. involved in a logical discourse, is stated. To arrive at this end the issues that have been sailed through are (i) three levels inherent in a logic discourse, (ii) from many-valued logics, fuzzy logics to graded consequence: a brief overview, (iii) a general discussion on uncertainty and vagueness, (iv) notion of consequence in classical logic and (v) finally some motivations for lifting many-valuedness to the meta-level.

1.1 Three Levels Inherent in a Logic Discourse

This treatise is on logic. The logical theory that we are going to introduce and elaborate here is called the theory of graded consequence (GCT) which may be placed within the broad category called fuzzy logic. In course of development, however, it will be clear that in spite of similarities, GCT is quite different from the standard fuzzy logics in many aspects. There are, in fact, many different opinions regarding what logic is or should be. So, it is necessary to clarify at the very beginning what is our understanding of the term in the present context. By a logic we shall mean a pair $\langle F, \vdash \rangle$ where F is a formal language (Hunter 1971, p. 4) and $\vdash$ is a binary relation from the power set $P(F)$ of F to F, that is, $\vdash$ is a subset of the Cartesian product $P(F) \times F$. If $(X, \alpha) \in \vdash$ (or in other words $X \vdash \alpha$ holds) where $X \subseteq F$ and $\alpha \in F$, we can read this as 'α follows from X' or 'α is a consequence of X'. The set X of sentences (well-formed formulae) constitutes the premise , and α is called the conclusion from X. We do not consider logic as the study of tautologies (universal truths) only, but of logical consequence , that is,'as certain determinate kinds of structure which embody the notion of logical consequence' (Cleave 1991, p. 12). The consequence relation $\vdash$ has, of course, to satisfy certain basic postulates, namely, reflexivity (overlap) , monotonicity (dilution) , cut (generalized transitivity) and compactness . These will be stated at the end of the introduction. However, these basic principles are also not considered to be sacrosanct, for example, monotonicity is not taken in some logics giving rise to non-monotonic logics . There are logics

M. K. Chakraborty and S. Dutta, *Theory of Graded Consequence*, Logic in Asia: Studia Logica Library, https://doi.org/10.1007/978-981-13-8896-5_1

where the order of the formulae in the premise set counts. That means, $\{\alpha_1, \alpha_2\} \vdash \alpha$ may not automatically imply $\{\alpha_2, \alpha_1\} \vdash \alpha$, thus giving a genre of substructural logics (Ono and Komori 1985; Restall 2000).

Although various kinds of logics had been present in human civilization (the Greek, the Indian, the Chinese and the Arabic), after the rise and development of modern mathematics and due to the inherent troubles within modern mathematics, in particular, set theory, modern logics emerged as the fundamental principles of deductive sciences. But even within this framework various shades have come up. One may recall Carnap's principle of tolerance:

> In logic there are no morals. Everyone is at liberty to build up his own logic, i.e., his own form of language, as he wishes. All that is required of him is that, if he wishes to discuss it he must state his methods clearly and give syntactical rules instead of philosophical arguments.
>
> [cf. Cleave (1991), p. 1]

This principle of Carnap has in a sense bloomed into a reality since the middle of the past century when mainly computer scientists took the baton of logic studies from the hands of philosophers and mathematicians. A plethora of logics appeared in the arena with various purposes to serve but in all cases the fundamental issue was to derive a conclusion from a set of premises. The conclusion as well as the premises are entities of some language. That is, defining the language and the consequence relation has been the main concern of logic. Be it modal logics, many-valued logics, intuitionistic logic, fuzzy logics, paraconsistent logics, non-monotonic logics or substructural logics, in general, the focus of study is on the relation of consequence (syntactic or semantic). Of course, the story goes back to Alfred Tarski who proposed some 'fundamental concepts of the methodology of deductive sciences framed in terms of the most general properties of the consequence relation' (Feferman and Feferman 2004, p. 81). Kurt Gödel and Rudolf Carnap had been highly impressed by Tarski's presentation on the above topic at a meeting of Vienna Circle in 1930. The study of consequence has become so central that a special branch of logic called 'Abstract Algebraic Logic' has gained ground during past two/three decades (Font et al. 2003).

The present study is basically the study of the logical consequence relation and related notions such as consistency, completeness, compactness and the like in many-valued context.

The notion of consequence, however, belongs to the meta-language of a logical discourse. If the pair $\langle F, \vdash \rangle$ is taken as the definition of a logic, the items in F, the sentences (or formal strings of symbols belonging to the alphabet, the well-formed formulae) fall within the object-level entities while the relation $\vdash$, which is a relation between a subset X of F and an element α of F, belongs to the meta-language. $X \vdash \alpha$, if properly written, becomes 'X' $\vdash$ 'α', which is an assertion, viz. the wff named 'α' is a consequence of the set of wffs named 'X'. This assertion speaks something about specific object linguistic items X and α. So, this is a meta-linguistic assertion.

A quotation from Church (1956) would be relevant here.

> In order to set up a formalized language we must of course make use of a language already known to us, ... The device of employing one language in order to talk about another is one for which we shall have frequent occasions not only in setting up formalized languages but also in making theoretical statements as to what can be done in a formalized language, ... Whenever we employ a language in order to talk about some language ... we shall call the latter language the object language, and we shall call the former the meta-language.
>
> [Church (1956), p. 47]

In the study of logic, there always takes place an interplay between items belonging to object level (level 0), meta-level (level 1) and metameta-level (level 2). This may be apparent from the following pictorial entity that belongs to the building blocks of sequent calculus of classical logic:

$$\frac{X, \alpha \vdash \beta}{X \vdash \alpha \supset \beta} \quad \text{(the right rule for implication } \supset\text{).}$$

Here is the intended interpretation. The expressions above and below the horizontal bar say that the formula on the right-hand side of the sign $\vdash$ follows from the set of formulae on the left-hand side. These are assertions about formulas or sets of formulae that are object-level entities (level 0), and hence are meta-level assertions (i.e. level 1). The horizontal bar stands for a relation between the two assertions '$X, \alpha \vdash \beta$' and '$X \vdash \alpha \supset \beta$', and hence the whole picture says '$X, \alpha \vdash \beta$' implies '$X \vdash \alpha \supset \beta$', and is thus a metameta-statement (i.e. level 2). It should be noted that the symbol $\supset$ (implication) giving rise to the string of symbols $\alpha \supset \beta$ of the object level is not a relation, $\supset$ is a logical connective like conjunction ($\wedge$) giving the string $\alpha \supset \beta$ (in natural language interpretation the sentence 'if α then β').

Making a clear distinction of levels is very crucial in the study of graded consequence. But we would like to make note of the fact that logic studies do not usually pay due attention to distinguishing levels due to which, in the opinion of the present authors, there do arise several misconceptions even mistakes. We will discuss this issue in Chap. 6.

Let us recall Carnap's remark on Tarski's talk that has been mentioned before.

> Of special interest to me was his emphasis that certain concepts used in logical investigations e.g., consistency of axioms, the provability of theorems in a deductive system, and the like are to be expressed not in the language of the axioms (later called the object language), but in metamathematical language (later called meta-language).
>
> [Feferman and Feferman (2004), p. 82]

In fact, consistency of a set formulae (not only of axioms) and derivability of a formula from an arbitrary set of formulae (not only from axioms), etc. are concepts belonging to the meta-level that has to be expressed in the meta-language. In order to define these notions a language at the next higher level, the metameta-language is required. Thus, the definition for the notion of consequence is to be a bunch of assertions made in the metameta-language at level 2. We shall observe these distinctions in the subsequent chapters although for the convenience of natural reading they will not be explicitly shown always. For example, the same linguistic pair of words, 'if ... then ...' might be used at the meta and metameta-levels for implication though they might be different having different semantic import (such as in the calculation of the truth values).

We hope that the necessity for and actual presence of three levels of languages in logical discourses have been clarified enough. The question now is how are these levels related with each other. Let us consider the object and meta-levels of classical logic. How much are they dependent on each other? Take, for example, the meta-logical notion of consequence ($\vdash$). Are there any property of $\vdash$ independent of any object language? Of course, there are the so-called structural properties, which are in fact axiomatized in Tarski's consequence operator (see Sect. 1.4). These properties, according to Tarski, have to be present in all deductive methods using consequences from premises irrespective of the particular nature of the object-level language, and logical operators involved in it, or even when no logical connectives are present in it. The trivial example is reflexivity, $\alpha \vdash \alpha$. One is perfectly justified in conceiving of $\alpha \vdash \beta$ (β follows from α) even though $\alpha \supset \beta$ is not available in the object language since $\supset$ may not be present in the object language at all. Even if the implication be present, $\alpha \vdash \beta$ may not yield that $\alpha \supset \beta$ is true or the other way round, since deduction theorem or its converse might not be available in the system. Current development of substructural logics (Ono and Komori 1985; Restall 2000) even questions the availability of the above Tarskian principles, which he thought to be universal, for example, monotonicity or exchange or contraction rule is not always taken.

Having said this, we would like to add that in all interesting logics there do exist certain connections (via the non-structural rules) between consequence relation and object language connectives. These connections determine the nature of the logical connectives, and thus that of the logics. In Chap. 4, there are several examples dealing with such connections and corresponding logics. What we would like to stress upon is that there should not be any 'natural' obligation between the logical and meta-logical connectives but some relations are imposed by the user (or maker) of the logic according to her choice and need. Such a construct gives enormous freedom to the logic-maker which is important from the angle of hierarchical reasoning as well when there are two levels of agents involved, viz. the data collectors giving values to the sentences as information, and the decision-maker aggregating them. We shall discuss this in Chap. 7 in some detail. But our primary objective is philosophical, viz. to make the existence of three levels visible, and to reveal the relative autonomy of these levels along with their interconnections.

As a testimony of the above claim, the case of many-valued logics may be considered. In these logics, the object-level items, viz. the formulae are given a many-valued ($\geq$3) semantics while the meta-level sentences, e.g. 'α is a consequence of X' is two-valued. Although the sentences represented by the formulae may entertain truth values other than the true and false, the meta-statement '$X \vdash \alpha$', which is usually defined in terms of a set of designated values in the truth set shall be either true or false. In Chap. 4, the case of many-valued logics will be analyzed in more detail. Here, in the introduction, we only point out that the sentences used at the object level and meta-level are of distinct nature. The way consequence relation $\vdash$ in many-valued logics is defined, uses natural (set theoretic) language, which is the metameta-language and is two-valued. It should be marked that many-valuedness of the object language formulae does not determine or dictate that the meta-language has to be two-valued; it is the meta-logical notion like consequence that determines its nature itself by the use of designated set, by a sentence of level-2 (see Chap. 4). That this level exists and is different from the level to which $\vdash$ belongs passes usually unnoticed. This fact becomes visible as soon as the assertion '$X \vdash \alpha$' turns out to be many-valued also. In the theory of graded consequence, it is really so. We shall argue for lifting many-valuedness at the meta-level. Before that let us trace back very briefly the history of development of many-valued logics. But whether the predicate $\vdash$ is many-valued or not that two levels of languages are employed, one for $X \vdash \alpha$ for a particular X and α, and the other to give the definition of $\vdash$ must be recognized.

1.2 From Many-Valued Logics, Fuzzy Logics to Graded Consequence: A Brief Overview

A minimal many-valuedness admits three values and thus one gets three-valued logics. The feeling that ascription of only two truth values to all statements has inherent limitations prevailed in Aristotle and the Epicureans in general, in the context of future contingent statements such as 'Tomorrow there will be a war'. Peirce in 1909 also suggested a third value to be admitted.

> I do not say that the principle of excluded middle is downright false but I do say that in every field of thought whatsoever there is an intermediate ground between positive assertion and positive negation which is just as real as they.
>
> [Cleave (1991), p. 253]

More positive initiatives were taken by Vasilev (1924). However, Łukasiewicz (1920) and Post (1921) are considered as the founder architects of the many-valued logic in general, and the three-valued logic in particular. The motivation of Łukasiewicz originated from the philosophical issues; he was strongly in favour of indeterminism. Following Meinong, who advocated the theory of contradictory objects, he was very critical about the law of contradiction and favoured Kotarbinski's view that two-valued logic seemed to interfere with the freedom of human thinking. Post was not led, at least overtly, by any such philosophical motivation. His approach was purely theoretical. Subsequently, a number of three-valued logics had been proposed and curiously the interpretation of the third value had been varied. While in Łukasiewicz's three-valued logic, the third value stands for the situation of indefiniteness (or neutrality), in Bochvar's calculus (where the intension was to overcome logical antinomies) the third value represents nonsense, in Halden's, Aqvist's and Segerberg's calculi it stands for meaningless. Heyting's three-valued logic is designed for rejection of the classical tautology $\neg\neg\alpha \supset \alpha$, Kleene's calculus attempts to cope with the problems involving partial recursive functions, wherefrom the concept of indefiniteness comes in, and Reichenbach's three-valued calculus was proposed to deal with the philosophical and logical problems of quantum theory.

Good surveys on the motivation behind incorporation of the third value are available in Bolc and Borowik (1992), Cleave (1991), Malinowski (1990), Rescher (1969), Resher (1971). The passage from three values to n-values and infinite values seems to have originated from purely theoretical (mathematical) standpoint of generalization. Although generally developed neither from the philosophical motivation nor from the angle of any specific applications, these generalized logical systems have been valued for their mathematical contents only. To give one instance, we mention Chang's celebrated completeness proof of Łukasiewicz infinite-valued logic (Chang 1958, 1959). The algebraic structure emerged from the work, namely, MV-algebra (many-valued algebra) has made a history.

With the introduction of fuzzy sets in 1965 by Zadeh (1965) many-valued logic got a home. Now there arrived actual statements that needed infinitely many truth values, not only discrete but also continuous. Vague predicates like tall, bald, beautiful or wise that had hitherto been banished from the statements worthy of logical discourse, appeared naturally as the meaning of fuzzy sets. Along with the classical sentences 'Socrates is mortal' a sentence like 'Socrates is healthy' entered into the arena of logic. The latter sentence obviously demands a continuous range of truth values between false (0) and the true (1). Instead of the binary 0/1, the whole range $0 \leq x \leq 1$ now turns into the natural truth value set for sentences of the latter kind with vague predicates. The time seemed to have been ripe for serious consideration of vague concepts. Apart from Zadeh, we get the following remark from Körner (1966) who talked about 'inexact sets':

> ... an immediate transition from one class to another is discontinuous unless it is a 'merging' of the two or a 'shading into each other' which presupposes that the two classes not only have neutral candidates but also that some of these are common to the two classes.
>
> [cf. Cleave (1991), p. 247]

The focus of logic was also shifting as regards the study of argumentation, namely, from the normative to the actual—from how a human reasoner *should* argue to how the rational human reasoner *actually* argues. And humans do argue with vague predications. Zadeh developed a model for approximate reasoning with vague data (Zadeh 1975). He suggested that the power of natural language and human reasoning lies in the use of inexact concepts and arguing in an approximate rather than precise manner. Increasing precisification may cost validity and relevance. The characteristic of a vague predicate P, according to Zadeh, is to allow for a seamless passage from P to non-P reminding Körner's words 'shading into each other' in the above quote. In summary, many-valuedness is no more a theoretical, abstract enterprise but a necessity of real life. Second, argumentation, that is, deriving a consequence from a set of premises, is no more a process as hard (definite) as it was considered before but approximate as actual argumentation by a human reasoner where obligatory precisiation, in all contexts, mars the very purposes of language use. Logic instead of dictating norms for validity of argumentation in fixed and certain conditions now delves in investigation and modelling of actual arguments that a rational human being makes in the environment of uncertainty. Besides, what we have proposed in this work and apart from the works that have been referred to in the present text, there are several other approaches to meet basically this shift of logic towards modelling human argumentation procedure in uncertain environment. To mention a few, we refer to the areas of commonsense thinking and architecture for human intelligence (Minsky 1985, 2006; Minsky et al. 2004), non-monotonic reasoning (Brewka 1991; Ghosh and Chakraborty 2004; Makinson 2005), modelling reasoning by paraconsistent logics and inconsistency tolerant approaches (Arruda and Ayda 1989; Bertossi et al. 2005; Cano et al. 1992; Dunn et al. 2009; Lim 1994; Priest and Routley 1989), theory of rough sets and rough logics (Chakraborty and Banerjee 1993; Pawlak 1982; Pawlak and Skowron 2007a, b, c; Skowron and Suraj 2013), evidence theory (Dempster 1967; Shafer 1976), possibility logic (Dubois 2006; Dubois and Prade 2004), some non-crisp approaches to entailment (Bankova et al. 2016; Diaconescu 2014; Muiño David 2010; Muiño and David 2013; She and Ma 2014; Vetterlein 2015; Vetterlein et al. 2016) and computing with words (Zadeh 1978, 2006). Interconnection between GCT and these approaches, particularly with possibility logic, is an important area of research, but not taken up in this treatise.

1.3 Different Shades of Imprecision

Discussion on uncertainty has become a necessity in both theory (philosophical, mathematical and computational) and practice. Future contingent statements already bear uncertainty. This kind of uncertainty gave rise to induction of a third truth value in some three-valued logics as we have mentioned before. In another approach, this uncertainty falls within the domain of probability and may be dealt with probability theory or probability logic. Probability theory (logic) was proposed by Kolmogorov (1956), which has been of great success in theoretical constructs and a wide field of applications. In fact, before the advent of fuzzy set theory probability theory was considered to be the sole formalism to approach situations having uncertainty (Bridges 2016); in other words, uncertainty was regarded as being only of one kind, viz. probabilistic. One may recall the fierce debate that ensued among the probability theorists and fuzzy set theorists during late 60s and early 70s. The plea of probability theorists was that all the cases that fuzzy set theorists had been addressing could be accommodated within probability theory; the other group was denying. However, after all these years, it is now more or less accepted that uncertainties are of several kinds of which one is probabilistic. We mention below some other kinds of uncertainties that would be relevant to the study of graded consequence.

However, use of terminologies is not uniform. As instances, we discuss here views from the computer science community such as Klir et al. (2006) and Godo and Rodriguez (2008). There is ample possibility of more divergences (e.g. uncertainty theory Liu 2010). First, we reproduce below the diagram from Klir et al. (2006).

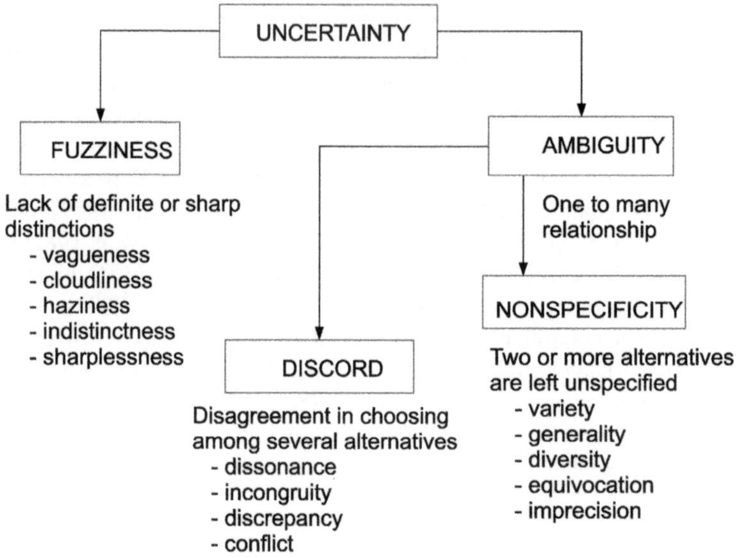

Clearly, here the word 'uncertainty' is used as the general term that includes 'vagueness' as one of the pertinent synonyms of fuzziness. In the existing literature, generally, no distinction is made between vagueness and fuzziness, and all the entities under fuzziness amount to basically the same concept, something with unsharp boundary.

On the other hand, Godo et al. use the word 'imperfect' as the umbrella term, and uses the terms 'vagueness' and 'uncertainty' orthogonally. In their own words

> the modelling of human reasoning usually requires 'imperfect' knowledge to be taken into account in the form of uncertainty, vagueness, truthlikeness, incompleteness, and partial contradiction.
>
> Godo and Rodríguez (2008)

Furthermore, they treat the first three terms orthogonally. While the terms 'imprecision' and 'vagueness' appear in two distinct branches of the tree, mentioned by Klir et al., they mean almost the same in Godo et al. 'Truthlikeness' of Godo and Rodríguez (2008) does not appear in Klir et al. (2006). While 'discord' is considered as one of the main categories in Klir et al. (2006), the same probably in the name of 'partial contradiction' appears in Godo and Rodríguez (2008) but not featured as one of the basics (they consider vagueness, uncertainty and truthlikeness as the basic three orthogonal categories giving rise to imperfection in knowledge and human reasoning.) The point of convergence of both Klir et al. and Godo et al. is in the assignment of a degree from the set [0, 1] with respect to each category, e.g. in Godo et al. degree of vagueness is measured by fuzzy sets, degree of certainty is measured by probability, possibility or evidence theories and degree of truthlikeness is measured by some kind of metric. In Klir et al. (2006), the quantification of the extent of various kinds of uncertainties has been presented in various ways. Although there are scholastic studies regarding the sameness and difference between the notions 'vagueness' and 'fuzziness' (Dubois 2006), use of fuzzy set theory in representing the semantics of vague concepts is widespread.

Refraining ourselves from the semantic nuances, which, we think, is not so important since we are proposing some general method of creating logics in which meta-logical notions are also graded (for which we shall argue later in this introduction), we see that an imprecision is involved in all the above categories. We would like to quote here the much used prophetic remark made by Zadeh, which is known as the 'principle of incompatibility':

> The closer one looks at a 'real world' problem, the fuzzier becomes its solution. Stated informally, the essence of this principle is that as the complexity of a system increases, our ability to make precise and yet significant statements about its behaviour diminishes until a threshold is reached beyond which precision and significance (or relevance) become almost mutually exclusive characteristics.
>
> Zadeh (1973)

The kind of imprecision we shall be dealing with is primarily vagueness or synonymously fuzziness.

> Fuzziness relates not to the uncertainty concerning the membership of a point in a set, but to the graduality of progression from membership to non-membership.
>
> Zadeh (1977)

Such an understanding of fuzziness entails existence of borderline instances. But all cases of existence of the borderline instances of a concept do not necessarily entail seamless (continuous) passage from membership to non-membership. Vagueness is characterized by most of the researchers by the existence of borderline instances (Shapiro 2006; Smith 2008). We shall treat fuzziness in the second sense, i.e. not necessarily seamless, and use a fuzzy set to represent a vague concept. Our membership function will not in general be continuous. We adopt degree-theoretic approach to vagueness and thus our logic is many-valued—the object-level sentences may receive values other than true (1) and false (0).

However, the many-valuedness of the object-level sentences may be associated with a value arising out of any sort of imprecision. It may be noted that logic is not concerned with where the truth values of atomic sentences came from. Rather its task is to suggest ways of computing truth values of complex (non-atomic) sentences and the correct (valid) notion of derivability. The most significant point of the present approach is that the consequence relation will be a fuzzy relation, its degree representing the strength of derivability.

A few general remarks about vagueness may not be out of place here. A vague concept is vague due to existence of borderline cases of the concept. Summarizing from Shapiro (2006) in (2016) Chakraborty has presented the following four categories of borderline instances x of a vague concept P (for instance, red or bald or heap) propounded by various researchers on vagueness.

(a) x is either P or non-P, but it is not known which or even not knowable.
(b) x is actually neither P or non-P.

(c) x is partially P and partially non-P.
(d) Depending on the context (perspective) x is sometimes P and sometimes non-P.

The approach (a) is of epistemic nature and based on ignorance. Philosophers subscribing this view are divided over where this ignorance comes from [cf. Shapiro (2006), p. 2–4]. Some others would like (Godo and Rodríguez 2008) to make a distinction between the two epistemic states, viz. 'not-known' and 'not-knowable' they would put the first kind under 'incomplete information problem' due to lack of information and the second kind under 'interpretation problem' since there is no suitable True/False interpretation possible.

Claims of category (b) are made more due to ontological commitments: the natures of x and P are such that they satisfy the condition (b) (Fine 1975; Keefe 2000; Tye 1994, cf. Shapiro 2006, p. 2–4).

Proponents of approach (c) generally take a degree-theoretic position, the intermediate truth values 'a' between true (1) and false (0) being interpreted as partially true to the degree 'a' and partially false to the degree '$1 - a$' (Machina 1976; Zadeh 1965). Some would like to use the term 'truthlike' (Godo and Rodríguez 2008), if the sentence $P(x)$ is more inclined towards true than false. Then 'a' is a measure of 'truthlikeness degree'.

The category (d) is subscribed by researchers like Raffman (1996), Graff (1996), Gaifman (2001) and Shapiro (2006, p.3). It is not ignorance but variability of the truth values of the sentence $P(x)$ with the context that turns x into a borderline case. One can see its reflection in rough set approach too (Chakraborty 2016).

It should be marked that all the terms pertaining to vagueness apply to sets or concepts or events or phenomena. Various theories associate a measure of applicability of the concept to objects (e.g. outputs, possible decreases and the like). In logic, we are interested in associating imprecision to sentences of a language. Now, this association is accomplished through interpretation (i.e. models) of predicates and logical constants of the language. In the syntax of the language, there remains no ambiguity which surfaces only in interpretation. Imprecision in the object–property relationship of the models gives rise to imprecision in the (truth) values of the sentences. For instance, the sentence 'A is tall' gets the value 0.8 because the object named A belongs to the fuzzy set corresponding to the word 'tall' to the degree 0.8.

In this way, due to various theories various semantic values are assigned to sentences of a given language. This falls within level 0 (object level) activity of any logic. Then comes the level 1 (meta-level) activity of defining the notion of consequence of a sentence, called the conclusion, from a set of sentences, called the premise. Thus, the relation denoted by '$\vdash$' is a predicate of this level which represents 'is a consequence of'. Usually, this predicate is interpreted as a classical two-valued relation whatever might be the interpretation of the object-level sentences. The theory of graded consequence makes this as well as related predicates, e.g. 'is consistent', 'is tautologous', etc multiple valued in general. That means the theory admits imprecision, in general, at the meta-level too. Rationale behind such a lifting to many-valuedness beyond mere mathematical generalization will be taken up in the following subsection.

1.4 Classical Consequence and Motivations for Lifting it to Many-Valuedness

As mentioned at the beginning, logic will be considered as the study of the consequence relation ($\vdash$). In the context of classical logic, the notion of consequence is presented either in terms of the consequence relation ($\vdash$) or in terms of a consequence operator (C). The formal language consists of the set F of all well-formed formulae (wff) over some alphabet. Wffs are finite strings of symbols taken from the alphabet constructed by certain rules of formation. The consequence relation, which is a binary relation between the power set $P(F)$ of the set F and the set of formulae F, should satisfy at least the following conditions:

1. If $\alpha \in X$, then $X \vdash \alpha$ (reflexivity/overlap),
2. If $X \subseteq Y$ and $X \vdash \alpha$, then $Y \vdash \alpha$ (monotonicity/dilution) and
3. If $X \vdash \beta$ for all $\beta \in Y$ and $X \cup Y \vdash \alpha$, then $X \vdash \alpha$ (cut/generalized transitivity).
   Usually, the classical consequence relation is supposed to satisfy the following condition also:
4. If $X \vdash \alpha$ then $X' \vdash \alpha$ for some finite subset X' of X (compactness).

In the Hilbert-type axiomatic presentation of logics, some wffs are taken as axioms and some rules of inference are fixed (usually modus ponens (MP) and generalization, and in case of modal logics, the rule of necessitation). The consequence relation $\vdash$ is then defined which satisfies the conditions (1)–(4). On the other hand, the consequence operator approach (originally proposed by Tarski 1956) takes an operator $C : P(F) \mapsto P(F)$ satisfying the following conditions:

(1′) $X \subseteq C(X)$,
(2′) If $X \subseteq Y$, then $C(X) \subseteq C(Y)$ and
(3′) $C(C(X)) = C(X)$.
Conditions (1) to (3) are equivalent to (1′) to (3′) under the transformations $C(X) = \{\alpha : X \vdash \alpha\}$ and $X \vdash \alpha$ if $\alpha \in C(X)$. Under these translations, (4) is equivalent to
(4′) $C(X) = \cup_{X' \subseteq X, X' finite} C(X')$.

In the current literature, conditions (1)–(3) and equivalently conditions (1′)–(3′) are considered to be the defining criteria of the notion of consequence in logic. Throughout this book we shall adopt this convention.

As stated before, the objective of the theory of graded consequence (GCT) is to introduce many-valuedness to the consequence relation and other related notions such as consistency, tautologihood, etc. This is in the same direction as had been pursued during the origination and development of many-valued logics. While in many-valued logics the object-level formulae are considered to be many-valued, meta-level notions like consequence are taken as two-valued; in GCT, items of both the levels are treated as many-valued. Various reasons may be ascribed for lifting the meta-level notions to many-valued ones.

1. The first is simple generalization. The interest is mathematical, if one can generalize the truth-value set for the formulae from two values to many, even infinitely many, then why not so in the notion of consequence if this generalization is not trivial mathematically? We have observed that assignment of infinitely many values to formulae by Łukasiewicz was simply of mathematical interest. Only after the arrival of fuzzy set theory (Zadeh 1965) such assignment of truth values has obtained a reasonable interpretation.
2. The question has already been raised in some corners about the status of many-valued logics.

   (i) Pelta et al. in (2004) remarked '*Untill now the construction of superficial many-valued logics, that is, logics with an arbitrary number (bigger than two) of truth values but always incorporating a binary consequence relation, has prevailed in investigation of logical many-valuedness*'. These authors did not notice the fact that in 1986 precisely this issue had been addressed by lifting the meta-logical notions to many-valuedness (Chakraborty 1988). In this and another paper, Pelta (1999) mentions about works by Marraud (1998) and Malinowski (1993, 1994) in which consequence relations are considered to be three-valued.

   (ii) In the context of logics for vague sentences, Parikh commented that in dealing with observationality (property of a vague predicate whose impreciseness cannot even be removed theoretically) '*... we seem to have come no closer to observationality by moving from two-valued logic to real valued, fuzzy logic. A possible solution ... is to use continuous valued logic not only for the object language but also for the meta-language*' (Parikh 1983).

   (iii) Harty Field also echos similar attitude, '*If we are to take seriously the idea that vagueness or indeterminacy is a quite widespread phenomenon, then we should consider the possibility that the language in which we discuss the semantics of vague and indeterminate language itself be vague or indeterminate*'. [p. 10, Field (2000) from p. 2 of Andrew Bacon's paper (Bacon 2012)].

   (iv) In Diaconescu (2014), grades to the notion of consequence have been attached in the style of institution theory. In the words of the author, '*... we introduce our entailment-styled concept of graded consequence which may be regarded either as an extension of the main definition of Chakraborty* (1988) *to a multi-signature context or as a generalization of classical two-valued entailment systems from institution theory ... to many-valued truth. We also discuss "graded proofs", which are entailments of the graded entailment systems freely generated by systems of raw graded rules*'.

   (v) Zadeh himself implicitly mentioned (Zadeh 2009) about many-valuedness in meta-logic: '*Fuzzy logic adds to bivalent logic an important capability - a capability to reason precisely with imperfect information. ... In fuzzy logic, results of reasoning are expected to be provably valid, or p-valid for short. Extended fuzzy logic adds an equally important capability - a capability to reason imprecisely with imperfect information. This capability comes into play when precise reasoning is infeasible, excessively costly or unneeded. In extended fuzzy logic, p-validity of results is desirable but not required. What is admissible is a mode of reasoning which is fuzzily valid, or f-valid for short*'.

Thus, all the above-mentioned researchers have entered the same feeling from various angles that there is the need to have grades attached to some of the meta-logical notions.

3. There has been some earlier work of assigning degree to the consequence relation (Basu 2003) that we state below. K. F. Machina, who advocates a degree-theoretic approach towards vagueness (Machina 1976), not only admits degrees of truth (and also falsehood) at the object level but at the meta-level also while defining the notions of validity of argument form, tautologous sentence forms, etc. According to Machina, 'ψ follows from ϕ to degree n' or '$\phi \vdash_n \psi$' means that n is the least upper bound of all numbers m such that $(|\phi| - |\psi|) \leq 1 - m$ for all possible assignments of values to ϕ and ψ, provided that at least some of these assignments result in a positive value for $(|\phi| - |\psi|)$, where $|\phi|$ represents the truth degree of ϕ. It is further stipulated that if $(|\phi| - |\psi|)$ is negative or 0 for all assignments of values to ϕ and ψ, then $n = 1$. Given this definition of '$\phi \vdash_n \psi$', it follows that n is the greatest lower bound of the set of possible values of the formula '$\phi \supset \psi$', where '$\supset$' is the object-level operator for implication defined by

$$\begin{aligned} |p \supset q| &= 1, \text{ when } |q| > |p| \\ &= 1 - |p| + |q|, \text{ when } |q| \leq |p|. \end{aligned}$$

4. In fuzzy logic, the idea has already been present as traced in the following kinds of phrases expressing their motivations: 'many-valued rule of inference' (Pavelka 1979), or 'partially true conclusions from partially true premises (Hájek 1998)', etc. Though fuzzy logical approach cannot appropriately address the issue, a problem which will be discussed later (cf. Chap. 6), the need for such an extension of the notions of consequence has been aired.
5. In many fields concerned with decision-making, the notion of partial consistency has come into existence and methods of drawing conclusion from such a premise have gained ground (Cano et al. 1992; Lim 1994). It is known that according to the classical logic from an inconsistent set any conclusion may be derived. So, which conclusions may be drawn from a partially inconsistent premise become a valid question. If the notion of consistency has a strength (or degree) it is quite likely that derivability should also have a strength (or degree). Exactly this will be the topic of some of the sections in later chapters.
6. Through the previous observations, GCT may be linked with imprecision. If the premise X is imprecise, in particular, vague, its consistency may be assigned with a grade. Now, adopting the classical logic paradigm, one gets that $X \cup \{\neg\alpha\}$ is inconsistent iff α is in consequence relation with X. So, if $X \cup \{\neg\alpha\}$ is inconsistent to a degree 'a', it is natural that α is in consequence relation with X to

degree 'a'. This may be interpreted as that the strength of derivability of α from X is 'a'. If the notion of consistency is graded, the notion of consequence is likely to be graded too.

7. Apart from all these, if we examine the nature of actual (not normative) inferenes made by human brain, we notice that from certain premises the brain often makes inferences not very strongly. The procedure itself might have weakness, tentativeness and vagueness. Cases of medical decisions and judgement by the Juries may offer ample instances (Cibattoni et al. 2013; Sanchez 2006; Vetterlein and Cibattoni 2010). In fact, recently a group of researchers are also advocating a notion of approximate entailment (Esteva et al. 2012). In this regard, though it is quite different from what GCT deals with, one can also consider argumentation theory (Van Eemeran et al. 2004). In argumentation theory, there can be different agents involved in the process of decision-making. One can consider, drawing analogy to our approach, that each agent is an expert. The decision-maker derives conclusions using the information provided by the experts and applying her reasoning, which is unlikely to be the same as that of the experts. In argumentation theory, an element of dialogue usually is included among the agents and the decision-maker. In general, the development of GCT has not considered such a factor of dialogue, but that such a possibility of extension cannot be ruled out would be clear from a prototypical example of local logics of agents and decision-maker, and flow of information among agents discussed in Sect. 7.2 of Chap. 7. Components of a typical argument in the argumentation theory usually include all deduction, abduction and induction. In contrast to that development of GCT is more inclined towards a deductive approach.

In the following chapter, we shall present the formal definition of graded consequence. It will be observed that the standard notion of semantic consequence in the classical logic admits a natural generalization as a graded consequence relation.

The main features of the notion of graded consequence and the issues, on which this book focuses, may be summarized as follows:

(a) GCT is not one particular logic but a general schema for generating logics.
(b) The theory observes strict distinction between the three levels, viz. object, meta and metameta, in the making of logics following the scheme.
(c) The notion of consequence, which is a meta-level relation, is multi-valued in GCT. The other notions in logic studies that belong to the meta-level also such as consistency, inconsistency, tautologihood, etc., in general admit degrees, i.e. these are fuzzy notions too.
(d) While determining the grade of a meta-linguistic notion, following the framework of GCT, we first translate the defining criterion of that very notion to a meta-level sentence, and then compute the value in the respective meta-level algebraic structure. This is missing in many of the existing many-valued and fuzzy logics.
(e) The scheme of obtaining logics in GCT presents a method of constructing logics with specified object-level and meta-level algebras for evaluating, respectively, object-level sentences (wff) and meta-level sentences. The study would yield that a logic building depends on both the algebraic structures for object level

and meta-level and their interrelations. As a result, the standard many-valued logics also can be realized within the schema. This marks its difference from the standard way of viewing fuzzy logics.

(f) GCT not only explains the practice of the existing logics by pointing out the blurring of level distinctions but suggests ways of building new logics. It gives enormous freedom to the logic builder since the object-level value set, its structure and the meta-level structure are not fixed, the builder has the option to choose them.

(g) Because of the features presented in item (e) and (f) the method presented in GCT is extremely user-friendly. The logic may be interpreted as that the consequence relation emerges out of the data supplied by various agents/experts. Agents evaluate the atomic sentences assigning values in the object-level algebra—and that constitutes the initial datum. The decision-maker may use different logic, the meta-level logic, to compute the strength of deriving conclusion.

(h) There is a possibility of extending the idea of item (g) to agents using different kinds of logic (local logics), while the final decision-maker uses a different logic. A simple prototypical case to show this possibility has been added in Chap. 7.

(i) Finally, the method of GCT has been compared with other similar theories like many-valued logics and fuzzy logics. It is claimed that in neither of the logics meta-level notions have been considered multi-valued. This implies that GCT may be more suitable for the users who would wish to incorporate vagueness in concepts like consistency and consequence (or derivability).

References

Arruda, A.I.: Aspects of the historical development of paraconsistent logic. In: Priest, G., Routley, R., Norman, J. (eds.) Paraconsistent Logic: Essays on the Inconsistent, pp. 99–129. Philosophia Verlag, München, Handen, Wien (1989)

Bacon, A.: Non-classical metatheory for non-classical logics. J. Philos. Log. **42**, 335–355 (2012)

Bankova, D., Coecke, B., Lewis, M., Marsden, M.: Graded Entailment for Compositional Distributional Semantics (2016). arXiv:1601.04908

Basu, S.: A study in logical and philosophical implications of graded consequence in many-valued systems. Ph.D. thesis, Jadavpur University (2003)

Bertossi, L., Hunter, A., Schaub, T.: Introduction to inconsistency tolerance. In: Bertossi, L., Hunter, A., Schaub, T. (eds.) Inconsistency Tolerance. LNCS, vol. 3300, pp. 1–14. Springer, Berlin (2005)

Bolc, L., Borowik, P.: Many-valued Logics: Theoretical Foundations. Springer (1992)

Brewka, G.: Nonmonotonic Reasoning: Logical Foundations of Commonsense. Cambridge University Press (1991)

Bridges, W.: Uncertainty: The Soul of Modeling Probability and Statistics. Springer, New York (2016)

Cano, J.E., Moral, S., Verdegay-Lopez, J.F.: Partial inconsistency of probability envelopes. Fuzzy Sets Syst. **52**, 201–216 (1992)

Chakraborty, M.K.: On some issues in the foundation of rough sets: the problem of definitions. Fundam. Inform. (2016)

Chakraborty, M.K.: Use of fuzzy set theory in introducing graded consequence in multiple valued logic. In: Gupta, M.M., Yamakawa, T. (eds.) Fuzzy Logic in Knowledge-Based Systems, Decision and Control, pp. 247–257. Elsevier Science Publishers, (B.V) North Holland (1988)

Chakraborty, M.K., Banerjee, M.: Rough logic with rough quantifiers. Bull. Pol. Acad. Sc. Math. **41**(4), 305–315 (1993)

Chang, C.C.: A new proof of the completeness of the Lukasiewicz axioms. Transactions 8874–8880 (1959)

Chang, C.C.: Algebraic analysis of many-valued logics. Trans. Am. Math. Soc. **88**, 476–490 (1958)

Church, A.: Introduction to Mathematical Logic, vol. 1. Princeton University Press, N.J. (1956)

Cibattoni, A., et al.: Formal approaches to rule-based systems in medicine: the case of CADIAG-2. Int. J. Approx. Reason. **54**, 132–148 (2013)

Cleave, J.P.: A Study of Logics. Clarendon Press, Oxford (1991)

Dempster, A.P.: Upper and lower probabilities induced by a multivalued mapping. Ann. Math. Stat. **38**(2), 325–339 (1967)

Diaconescu, R.: Graded consequence: an institution theoretic study. Soft Comput. **18**, 1247–1267 (2014)

Dubois, D.: Possibility theory and statistical reasoning. Comput. Stat. Data Anal. **51**(1), 47–69 (2006)

Dubois, D., Prade, H.: Possibilistic logic, a retrospective and prospective view. Fuzzy Sets Syst. **144**(1), 3–23 (2004)

Dunn, P.E., et al.: Inconsistency tolerance in wieghted argument systems. In: Decker, K.S., Sichman, J.S., Sierra, C., Castelfranchi, C. (eds.) Proceedings of 8th International Conference on Autonomous Agents and Multiagent Systems (AAMAS2009), pp. 851–858. Budapest, Hungary (2009)

Esteva, F., Rodriguez, R., Godo, L., Vetterlein, T.: Logics for approximate and strong entailments. Fuzzy Sets Syst. **197**, 59–70 (2012)

Feferman, A.B., Feferman, S.: Alfred Tarski: Life and Logic. Cambridge University Press (2004)

Field, H.: Indeterminacy, degree of belief, and excluded middle. Nous **34**(1), 1–30 (2000)

Fine, K.: Vagueness, truth and logic. Synthese **30**, 265–300 (1975)

Font, J.M., Jansana, R., Pigozzi, D.: A survey of abstract algebraic logic. Stud. Log. **74**, 13–97 (2003)

Gaifman, H.: Vagueness, tolerance and contextual logic. Synthese. **174**(1), 5–46 (2001)

Ghosh, S., Chakraborty, M.K.: Non-monotonic proof systems: algebraic foundations. Fundam. Inform. **59**, 39–65 (2004)

Godo, L., Rodríguez, R.: Logical approaches to fuzzy similarity-based reasoning: an overview. In: Giacomo Della, R. et al. (eds.) Preferences and Similarities, vol. 504, pp. 75–128. CISM International Centre for Mechanical Sciences (2008)

Graff, D.: Shifting sands: an interest-relative theory of vagueness. Philos. Top. **28**, 45–81 (2000)

Hájek, P.: Metamathematics of Fuzzy Logic. Kluwer Academic Publishers, Dordrecht (1998)

Hunter, G.: Metalogic: An Introduction to the Metatheory of Standard First-order Logic. Macmillan Student Edition (1971)

Keefe, R.: Theories of Vagueness. Cambridge University Press (2000)

Klir, G.J., Yuan, B.: Fuzzy Sets and Fuzzy Logic: Theory and Applications. Prentice Hall of India Private Limited, New Delhi (2006)

Kolmogorov, A.N.: Foundations of the theory of probability, 2nd English edition, Translation edited by, Nathan Morrison, Chelsea Publishing Co. New York (1956)

Lim, J.: Consistent belief reasoning in the presence of inconsistency, theoretical aspects of reasoning about knowledge. In: Fagin, R. (ed.) Proceedings of the 5th TARK 1994, pp. 80–94 (1994)

Liu, B.: Uncertainty Theory: A Branch of Mathematics for Modeling Human Uncertainty. Springer, Berlin (2010)

Machina, K.F.: Truth belief and vagueness. J. Philos. Log. **5**, 47–78 (1976)

Makinson, D.: Bridges from Classical to Nonmonotonic Logic. College Publications (2005)

Malinowski, G.: Q-consequence operation. Rep. Math. Log. **24**, 49–59 (1990)

Malinowski, G.: Many-Valued Logics. Clarendon Press, London (1993)
Malinowski, G.: Inferential many-valuedness. In: Wolenski, J. (ed.) Philosophical Logic in Poland, pp. 75–84. Kluwer, Amsterdam (1994)
Marraud, H.: Razonamiento approximado y grados de consecuencia. Endoxa **10**, 55–70 (1998)
Minsky, M.: A framework for representation knowledge. In: Brachman, R.J., Levesque, H.J. (eds.) Reading in Knowledge Representation, pp. 246–262. Morgan Kaufman, Los Altos (1985)
Minsky, M.: Emotion Machine: Common Sense Thinking, Artificial Intelligence and the Future of the Mind. Simon & Schuster, New York (2006)
Minsky, M., Singh, P., Sloman, A.: The St. Thomas common sense symposium: designing architecture for human level intelligence. AI Mag. **25**, 113–124 (2004)
Muino, D.P.: A graded inference approach based on infinite-valued Lukasiewicz semantics. In: Proceedings of The International Symposium on Multiple-Valued Logic (2010)
Muiño, David P.: A consequence relation for graded inference within the frame of infinite-valued Lukasiewicz logic. Fundam. Inform. **123**, 77–95 (2013)
Ono, H., Komori, Y.: Logics without the contraction rule. J. Symb. Log. **50**, 169–201 (1985)
Parikh, R.: The problem of vague predicates. In: Cohen, R.S., Wartofsky, M. (eds.) Language, Logic, and Method, pp. 241–261. D. Ridel Publishing Company (1983)
Pavelka, J.: On fuzzy logic I, II, III Zeitscher for Math. Logik und Grundlagen d. Math. **25**, 45–52, 119–134, 447–464 (1979)
Pawlak, Z.: Rough sets. Int. J. Comp. Inf. Sci. **11**, 341–356 (1982)
Pawlak, Z., Skowron, A.: Rudiments of rough sets. Inf. Sci. **177**(1), 3–27 (2007a)
Pawlak, Z., Skowron, A.: Rough sets: some extensions. Inf. Sci. **177**(1), 28–40 (2007b)
Pawlak, Z., Skowron, A.: Rough sets and boolean reasoning. Inf. Sci. **177**(1), 41–73 (2007c)
Pelta, C.: Deep many-valuedness. Log. Anal. **167–168**, 361–371 (1999)
Pelta, C.: Wide sets, deep many-valuedness and sorites arguments. Mathw. Soft Comput. **11**, 5–11 (2004)
Priest, G., Routley, R.: A preliminary history of paraconsistent and dialethic approaches. In: Priest, G., Routley, R., Norman, J. (eds.) Paraconsistent Logic: Essays on the Inconsistent, p. 375. Philosophia Verlag, München, Hamden, Wien (1989)
Raffman, D.: Vagueness and context relativity. Philos. Stud. **81**, 175–192 (1996)
Rescher, N.: A Survey of Many-Valued Logic. McGraw-Hill Book Company, New York (1969)
Resher, N.: Many-Valued Logics. McGraw-Hill, New York (1971)
Restall, G.: An Introduction to Substructural Logics. Routledge (2000)
Sanchez, E. (ed.): Fuzzy Logic and the Semantic Web. Elsevier (2006)
Shafer, G.: A Mathematical Theory of Evidence. Princeton University Press (1976)
Shapiro, S.: Vagueness in Context. Clarendon Press (2006)
She, Y., Ma, L.: On the rough consistency measures of logic theories and approximate reasoning in rough logic. Int. J. Approx. Reason. **55**(1), 486–499 (2014)
Skowron, A., Suraj, Z.: Rough sets and intelligent systems. Professor Zdzisław Pawlak in memoriam. Series Intelligent Systems Reference Library, vol. 42–43. Springer, Berlin (2013)
Smith, J.J.N.: Vagueness and Degrees of Truth. Oxford University Press, Oxford (2008)
Tarski, A.: Methodology of deductive sciences. Logics, Semantics, Metamathematics, pp. 60–109 (1956)
Tye, M.: Sorites paradoxes and the semantics of vagueness. In: Tomberlin, J.E. (ed.) Philosophical Perspectives 8: Logic and Language, pp. 189–206. Ridgeview Publishing Co., Atascadero (1994)
Van Eemeran, F.H., Grootendorst, R.: A Systematic Theory of Argumentation. Philosophy, vol. 12. Syndicate of the University of Cambridge (2004)
Vetterlein, T., Cibattoni, A.: On the (fuzzy) logical content of CADIAG-2. Fuzzy Sets Syst. 1941–1958 (2010)
Vetterlein, T.: Logic of approximate entailment in quasimetric spaces. Int. J. Approx. Reason. **64**, 39–53 (2015)
Vetterlein, T., Esteva, F., Godo, L.: Logics for approximate entailment in ordered universes of discourse. Int. J. Approx. Reason. **71**, 50–63 (2016)

Zadeh, L.A.: Fuzzy set theory—a perspective. In: Gupta, M.M., Saridi, G.N., Gaines, B.R. (eds.) Fuzzy Automata and Decision Process, North Holland, New York, pp. 3–4 (1977)

Zadeh, L.A.: Outline of a new approach to the analysis of complex systems an decision process. IEEE Trans. Syst. Man Cybern. SMC **1**, 28–44 (1973)

Zadeh, L.A.: Fuzzy sets. Inf. Control **8**, 338–353 (1965)

Zadeh, L.A.: Fuzzy logic and approximate reasoning. Synthese **30**, 407–428 (1975)

Zadeh, L.A.: PRUF-a meaning representation language for natural languages. Int. J. Man Mach. Stud. **10**, 395–460 (1978)

Zadeh, L.A.: Generalized theory of uncertainty (GTU)-principal concepts and ideas. Comput. Stat. Data Anal. **51**, 15–46 (2006)

Zadeh, L.A.: Toward extended fuzzy logic - a first step. Fuzzy Sets Syst. **160**, 3175–3181 (2009)

Chapter 2
Basics of the Theory of Graded Consequence

Abstract This chapter comprises the basic concepts of the theory. It includes the syntactic, semantic and axiomatic notions of graded consequence, and their interrelations.

2.1 Characterization of Graded Consequence Relation

Before entering the main concepts, it is necessary to have an acquaintance with some preliminary notions.

Fuzzy set theory was introduced by Zadeh (1965). Here, the basic idea is that of partial belongingness; an object may belong to a (fuzzy) set to some degree which is usually taken to be a value in the unit interval [0, 1]. An object 'a' belongs to a fuzzy set 'A' always to a degree. Thus, the sentence $a \in A$ can not only be true (that is of degree 1) or false (that is of degree 0), but may have any intermediate value. Formally speaking, a fuzzy subset A on the universe X is a mapping from X to [0, 1]. The set $\{x \in X : A(x) > 0\}$ is called the support of A, and is denoted by $Supp(A)$. Similarly, a fuzzy binary relation R from X to Y is a mapping from the Cartesian product $X \times Y$ to [0, 1]. If $R(x, y) = a \in [0, 1]$, then this is interpreted as 'x is R-related to y to the degree a'. So, $Supp(R) = \{(x, y) : R(x, y) > 0\}$. The value set [0, 1] is generalized to an arbitrary lattice $(L, \wedge, \vee)$ when the fuzzy sets are called L-fuzzy sets (Goguen 1967) or L-sets. Accordingly, one gets L-fuzzy relation R or L-relation.

Our main interest would be in a fuzzy binary relation where the value set is a special kind of lattice, called the residuated lattice[1] (Galatos et al. 2007; Hájek 1998; Ono 2003).

Definition 2.1 A residuated lattice is an algebraic structure $(L, \wedge, \vee, *, \rightarrow, 0, 1)$ such that

- $(L, \wedge, \vee, 0, 1)$ is a bounded lattice,
- $(L, *, 1)$ is a commutative monoid,

[1] At the end of this chapter, an appendix for introducing a background for residuated lattices has been added.

M. K. Chakraborty and S. Dutta, *Theory of Graded Consequence*, Logic in Asia: Studia Logica Library, https://doi.org/10.1007/978-981-13-8896-5_2

- $*$ is monotonic with respect to both the arguments, i.e. for $a \leq b, a * c \leq b * c$, and
- $(*, \rightarrow)$ forms an adjoint pair, i.e. $x * y \leq z$ iff $x \leq y \rightarrow z$ for any $x, y, z \in L$.

A complete residuated lattice is a residuated lattice such that the lattice structure is complete.

This definition of residuated lattice corroborates with that in Hájek (1998). The notion, however, dates back to 1939 (Dillworth and Ward 1939). Ever since the seminal work of Ono (2003) by a residuated lattice is meant a structure where the product operation $*$ is not assumed to be commutative, and hence instead of one residuum there are two residua, left and right. With respect to this definition ours is a special case, namely, bounded, integral, commutative residuated lattice. However, throughout this book by residuated lattice, we mean this special case of bounded, integral, commutative and residuated lattice.

We list below some of the standard properties of a complete residuated lattice, which would be used frequently in the sequel. These properties are available in standard books (Hájek 1998; Ono 2003).

(r1) $1 \rightarrow a = a$ for any $a \in L$.
(r2) $a \leq b, c \leq d$ imply $a * c \leq b * d$.
(r3) $a \rightarrow b = 1$ if $a \leq b$.
(r4) If $a \rightarrow b = 1$, then $a \leq b$.
(r5) $a \leq b$ implies $a \rightarrow c \geq b \rightarrow c$.
(r6) $a \rightarrow b \geq ((a \wedge c) \rightarrow b) * (a \rightarrow c)$.
(r7) $a \rightarrow \inf_i b_i = \inf_i (a \rightarrow b_i)$.
(r8) $a * b \leq c$ iff $b \leq a \rightarrow c$ iff $a \leq b \rightarrow c$.
(r9) $\sup_i b_i * a = \sup_i (b_i * a)$.
(r10) $(a \rightarrow b) * (b \rightarrow c) \leq (a \rightarrow c)$.
(r11) $\inf_i a_i * \inf_i b_i \leq \inf_i (a_i * b_i)$.

Any complete lattice with the lattice meet $(\wedge)$ as the monoidal composition $*$ and corresponding residuum $a \rightarrow b = \sup\{z \in L : a \wedge z \leq b\}$ forms a complete residuated lattice. This algebraic structure is known as complete pseudo-Boolean algebra or complete Heyting algebra. The set [0, 1] with the usual ordering is of course a residuated lattice with respect to different kinds of adjoint pair. Here are a few examples which will be used in the sequel.

Example 2.1 1. A Łukasiewicz algebra $([0, 1], \wedge, \vee, *_Ł, \rightarrow_Ł, 0, 1)$, also known as MV-algebra, is an example of complete residuated lattice. The pair $(*_Ł, \rightarrow_Ł)$, defined as $x *_Ł y = \max(0, x + y - 1)$ and $x \rightarrow_Ł y = \min(1, 1 - x + y)$, is the adjoint pair of Definition 2.1.

2. A Gödel algebra $([0, 1], \wedge, \vee, \rightarrow_G, 0, 1)$, where the monoidal operation is the lattice meet itself, is an example of a complete residuated lattice. The residuum $(\rightarrow_G)$ corresponding to $\wedge$ is given by

$$x \rightarrow_G y = 1 \text{ if } x \leq y$$
$$= y, \text{ otherwise.}$$

3. A product algebra ($[0, 1], \wedge, \vee, *_P, \rightarrow_P, 0, 1$), where $*_P$ is the standard product operation, is an example of a complete residuated lattice. The definitions for $*_P$ are the same as the product operation on $[0, 1]$, and

$$x \rightarrow_P y = 1 \text{ if } x \leq y,$$
$$= \frac{y}{x}, \text{ otherwise.}$$

($*_P, \rightarrow_P$) forms the adjoint pair.
4. The set $L = \{0, a, b, 1\}$ along with the adjoint pair $(\wedge, \rightarrow)$, defined as below, is

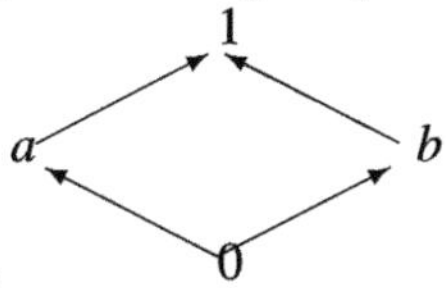

a complete residuated lattice.

$\rightarrow$	0	b	a	1
0	1	1	1	1
b	a	1	a	1
a	b	b	1	1
1	0	b	a	1

A graded consequence relation (Chakraborty 1988, 1995) $\mid\!\sim$ is a generalization of the consequence relation of classical logic. This is a fuzzy binary relation from $P(F)$ to F where F is the set of well-formed formulae constructed over some logical alphabet. For the general theory GCT, however, the language constituted from F may be taken as an arbitrary non-empty set. But as more specific instances, F can be considered as the set of wffs of any logic which is proposed in practice with standard logical operators and quantifiers. Elements of F will be denoted by Greek letters $\alpha, \beta, \gamma, \ldots$ etc., and subsets of F by English capital letters $X, Y, Z \ldots$, etc. As discussed in the introduction, both the items, viz. the elements of F and $P(F)$, are to be considered as entities of the lowest level, that is, the object level or level-0. The value set for the fuzzy relation $\mid\!\sim$ is taken to be a complete residuated lattice L endowed with the structure $(L, \wedge, \vee, *_m, \rightarrow_m, 0, 1)$. Thus $\mid\!\sim$ is a binary L-fuzzy relation, in other words, a mapping from $P(F) \times F$ to L. If $\mid\!\sim (X, \alpha) = a$, we say that '$a$' is the degree or grade to which the wff α is a consequence of the set of wffs X, and 'a' is written as $gr(X \mid\!\sim \alpha)$. The definition of graded consequence relation is as follows.

Definition 2.2 A binary L-fuzzy relation $\mid\!\sim$ from $P(F)$ to F is said to be a graded consequence relation if the following conditions hold:

(GC1) $gr(X \mid\!\sim \alpha) = 1$ for $\alpha \in X$.
(GC2) If $X \subseteq Y$, then $gr(X \mid\!\sim \alpha) \leq gr(Y \mid\!\sim \alpha)$.
(GC3) $\inf_{\beta \in Y} gr(X \mid\!\sim \beta) *_m gr(X \cup Y \mid\!\sim \alpha) \leq gr(X \mid\!\sim \alpha)$.

It can be noticed that (GC1), (GC2) and (GC3) are, respectively, the generalized versions of the conditions for overlap/reflexivity, dilution/monotonicity, and cut of classical consequence relation [cf. Chap. 1].

The algebraic structure $(L, \wedge, \vee, *_m, \rightarrow_m, 0, 1)$ is considered to be the structure for interpreting the meta-linguistic sentences; more specifically, the meta-linguistic connectives 'and' and 'if-then', respectively, would be interpreted by the monoidal composition $*_m$ and its residuum $\rightarrow_m$ in the residuated lattice. The meta-linguistic quantifier 'for all' would be interpreted by the 'infimum' operator of the complete residuated lattice. The expression for (GC3) is a direct translation of the defining criterion 'for all $\beta \in Y$, $X \vdash \beta$ and $X \cup Y \vdash \alpha$ imply $X \vdash \alpha$' of the condition cut (cf. Chap. 1) in many-valued context.

It is to be noted that as a special case of (GC3) when Y is considered to be a singleton set $\{\beta\}$ we get $gr(X \mid\!\sim \beta) * gr(X \cup \{\beta\} \mid\!\sim \alpha) \leq gr(X \mid\!\sim \alpha)$. This is the generalized version for the notion of cut for an element of the two-valued classical logic (Shoesmith and Smiley 1978). The graded counterpart of cut for finite set is the same as the version for cut with the restriction of Y being taken as a finite set.

2.2 Semantic Consequence Relation Generalized

In this section, graded counterpart of semantic consequence relation of classical logic is defined. Since this notion is based on the semantics of the wffs, we will need a (truth) value set (along with operations for respective connectives as per requirement) for the object level. Besides, the defining sentence (II) below being a meta-linguistic statement, to determine its (truth) value, a meta-level algebraic structure over the same value set will be required. Thus, we need two algebraic structures on the same value set, one for the object level and the other for the meta-level.

In two-valued classical logic, the semantic consequence relation $\models$ between a set X of wffs and a wff α is defined as follows.

$X \models \alpha$ holds if and only if for all valuations T whenever $T(x)$ is true for all $x \in X$, $T(\alpha)$ is also true; a valuation is a mapping from the set of wffs to the value set $\{0, 1\}$ where 1 represents 'true' and 0 represents 'false'.

In fact, the value set, from where the meta-linguistic sentence $X \models \alpha$ is getting the value, is an algebra containing operators corresponding to all the logical connectives present in the meta-language. Now each valuation T can be identified with the set of wffs which are mapped into 1 under T. After this identification, the above definition of $\models$ turns out to be the following:

$X \models \alpha$ if and only if for all valuations T, if $X \subseteq T$ then $\alpha \in T$. (I)

It may be easily observed that the relation $\models$ defined as above satisfies the conditions (1), (2) and (3) of classical consequence relation (cf. Chap. 1). However, if instead of taking all possible valuations T, one fixes a subcollection $\{T_i\}_{i\in I}$ and defines $\models$ by using (I), the relation turns out to be a classical consequence also. This is, in fact, done by Shoesmith and Smiley in the context of multiple-conclusion logic (Shoesmith and Smiley 1978) where $\models$ is defined between two sets X and Y ($X \models Y$). We shall, however, restrict only to single-conclusion logic, meaning thereby that Y is a singleton set consisting of a single formula. Thus, the notion of semantic consequence relation is relativized with respect to $\{T_i\}_{i\in I}$ of valuations and (I) reduces to

$X \models_{\{T_i\}_{i\in I}} \alpha$ iff for all valuations $T_i \in \{T_i\}_{i\in I}$, if $X \subseteq T_i$ then $\alpha \in T_i$. (II)

The valuation functions T_i's are now generalized to many-valued range, that is, the object-level formulae are many-valued in general. If instead of the set $\{0, 1\}$ as the value set, the unit interval $[0, 1]$ is taken, T_i's are standard fuzzy subsets of the set F of wffs, $T_i(\alpha)$ being the membership degree of α to the fuzzy set T_i. We shall, however, take a more general value set, an algebra, for the range of the valuation functions T_i's. The operators in the algebra are determined according to the logical connectives (e.g. conjunction, disjunction, negation, etc.) present in the language of the logic. Let L be the base set of this algebra. Now, what could be the interpretation of the definiendum in (II), that is, how to calculate the (truth) value of the sentence

"for all valuations $T_i \in \{T_i\}_{i\in I}$, if $X \subseteq T_i$ then $\alpha \in T_i$." (A)

This is a meta-level assertion, and for its (truth) value determination we need a meta-level truth set. The truth set should be endowed with operators for meta-level logical connectives; that is, we need a meta-level algebra. The underlying base set for this algebra is taken to be the same set L as is the base set for object-level algebra. The value of (A) will be determined in the set L but with respect to a different set of operators generally. For example, we shall need binary operators $*_m$ and $\to_m$ for meta-linguistic conjunction and implication, respectively. We will also need some operator for computation of the (universal) quantifier, 'for all'. So, L is now endowed with the operators $*_m$, $\to_m$ giving an algebraic structure, which is, in general, different from the base structure for the object language. For reasons, that will unfold gradually, the meta-level structure is taken to be a residuated lattice with respect to $*_m$ and $\to_m$. Thus, on the same base set L, there are two algebraic structures: one for calculating the values of object-level formulae, and the other for calculating values of the sentences belonging to the meta-level. Whatever be the former one, the latter structure is taken to be a complete residuated lattice where $\to_m$ gives the residuum with respect to $*_m$.

So, in the present context, the semantic consequence relation given by (II), which is a graded notion, is generalized. Based on a collection of many-valued valuations $\{T_i\}_{i\in I}$, which are functions from the set F to L, the notion of graded semantic consequence, denoted by $\approx_{\{T_i\}_{i\in I}}$, is an L-fuzzy binary relation from $P(F)$ to F defined as follows.

Definition 2.3 Given a collection $\{T_i\}_{i\in I}$ of fuzzy sets over the set F of wffs, for any set X of wffs and a wff α

$$gr(X \approx_{\{T_i\}_{i\in I}} \alpha) = gr(\text{for all } T_i \text{ if } X \subseteq T_i \text{ then } \alpha \in T_i)$$

The right-hand side of the above inequality is the (truth) value of the meta-linguistic sentence (A) that involves a meta-linguistic connective implication, i.e. $\rightarrow$, whose algebraic interpretation is given by an implication function $\rightarrow_m$. The value of the right-hand expression within bracket is calculated as below:

$$\begin{aligned} gr(X \approx_{\{T_i\}_{i\in I}} \alpha) &= gr(\forall_{T_i}(X \subseteq T_i \rightarrow \alpha \in T_i)) \\ &= gr(\forall_{T_i}\{\forall_x(x \in X \rightarrow x \in T_i) \rightarrow \alpha \in T_i\}) \\ &= \inf_{i\in I}\{(\inf_x(X(x) \rightarrow_m T_i(x)) \rightarrow_m T_i(\alpha)\} \\ &= \inf_{i\in I}\{\inf_{x\in X} T_i(x) \rightarrow_m T_i(\alpha)\}. \end{aligned}$$

It can be easily noticed that the above simplification makes use of the two properties, viz. $1 \rightarrow_m a = a$ and $a \rightarrow_m b = 1$ when $a \leq b$.

So, Definition 2.3 turns out to be

$gr(X \approx_{\{T_i\}_{i\in I}} \alpha) = \inf_{i\in I}\{\inf_{x\in X} T_i(x) \rightarrow_m T_i(\alpha)\}$ ...(B)

For all future references, we shall take (B) as the definition of graded semantic consequence relation. For concrete examples, readers are referred to as Examples 2.2, 2.3 and 2.4.

Below we shall present the representation theorems bridging the connection between the general notion of graded consequence relation given in Definition 2.2, and one of its semantic counterparts given by Definition 2.3.

Theorem 2.1 (Representation Theorems, Chakraborty 1995) *Let* $(L, \wedge, \vee, *_m, \rightarrow_m, 0, 1)$ *be a complete residuated lattice.*

(1) Given a collection of L-fuzzy sets $\{T_i\}_{i\in I}$ *over the set of formulae, the graded semantic consequence* $\approx_{\{T_i\}_{i\in I}}$, *given by Definition 2.3, is a graded consequence relation, and*

(2) Given a graded consequence relation $\mid\!\sim$, *there exists a collection of L-fuzzy sets* $\{T_i\}_{i\in I}$ *over the set of formulae such that* $\mid\!\sim$ *coincides with* $\approx_{\{T_i\}_{i\in I}}$.

Proof (1) For $\alpha \in X$, $\inf_{x\in X} T_i(x) \leq T_i(\alpha)$, i.e. by (r3), $\inf_{x\in X} T_i(x) \rightarrow_m T_i(\alpha) = 1$, for any T_i. Hence $gr(X \approx_{\{T_i\}_{i\in I}} \alpha) = 1$.

Let $X \subseteq Y$. Then $\inf_{x\in X} T_i(x) \geq \inf_{x\in Y} T_i(x)$. Hence as $\rightarrow_m$ is antitone on the first variable (r5), we have $\inf_{x\in X} T_i(x) \rightarrow_m T_i(\alpha) \leq \inf_{x\in Y} T_i(x) \rightarrow_m T_i(\alpha)$ for any T_i. That is, $gr(X \approx_{\{T_i\}_{i\in I}} \alpha) \leq gr(Y \approx_{\{T_i\}_{i\in I}} \alpha)$.

For each T_i,

$\inf_{x\in X} T_i(x) \rightarrow_m T_i(\beta)$

$\geq [(\inf_{x\in X} T_i(x) \wedge \inf_{y\in Y} T_i(y)) \rightarrow_m T_i(\beta)] *_m [\inf_{x\in X} T_i(x) \rightarrow_m \inf_{y\in Y} T_i(y)]$ (by r6)

$= [(\inf_{x\in X} T_i(x) \wedge \inf_{y\in Y} T_i(y)) \rightarrow_m T_i(\beta)] *_m \inf_{y\in Y}[\inf_{x\in X} T_i(x) \rightarrow_m T_i(y)]$ (by r7)

Hence, $\inf_{i\in I}[\inf_{x\in X} T_i(x) \rightarrow_m T_i(\beta)]$

$\geq \inf_{i\in I}[\{(\inf_{x\in X} T_i(x) \wedge \inf_{y\in Y} T_i(y)) \rightarrow_m T_i(\beta)\} *_m \inf_{y\in Y}\{\inf_{x\in X} T_i(x) \rightarrow_m T_i(y)\}]$

$= \inf_{i\in I}[(\inf_{x\in X\cup Y} T_i(x) \rightarrow_m T_i(\beta)] *_m \inf_{y\in Y} \inf_{i\in I}[\inf_{x\in X} T_i(x) \rightarrow_m T_i(y)]$ (by r11)

So, $gr(X \approx_{\{T_i\}_{i\in I}} \beta) \geq gr(X \cup Y \approx_{\{T_i\}_{i\in I}} \beta) *_m \inf_{\gamma\in Y} gr(X \approx_{\{T_i\}_{i\in I}} \gamma)$.

Hence, $\approx_{\{T_i\}_{i\in I}}$ is a graded consequence relation.

(2) Let us consider $\{T_X\}_{X\in P(F)}$ to be the collection of fuzzy sets over formulae, and that for each $Y \in P(F)$, $T_Y(\alpha)$ is defined in terms of $|\!\sim$, the given graded consequence relation, in the following way: $T_Y(\alpha) = gr(Y |\!\sim \alpha)$.
We want to prove $gr(X |\!\sim \beta) = \inf_{Y\in P(F)}[\inf_{x\in X} T_Y(x) \to T_Y(\beta)]$.
For any $Y \subseteq F$, (GC3) ensures $\inf_{\alpha\in X} gr(Y |\!\sim \alpha) *_m gr(Y \cup X |\!\sim \beta) \leq gr(Y |\!\sim \beta)$.
That is, $gr(Y \cup X |\!\sim \beta) \leq \inf_{\alpha\in X} gr(Y |\!\sim \alpha) \to_m gr(Y |\!\sim \beta)$ (by r8).
So, $\inf_{Y\in P(F)}[gr(Y \cup X |\!\sim \beta)] \leq \inf_{Y\in P(F)}[\inf_{\alpha\in X} gr(Y |\!\sim \alpha) \to_m gr(Y |\!\sim \beta)]$. ...(a)
Now, as $X \in P(F)$, considering Y to be X, we have $gr(X |\!\sim \beta) = gr(X \cup X |\!\sim \beta)$, and by (GC2) $gr(X |\!\sim \beta) \leq gr(X \cup Y |\!\sim \beta)$ for any $Y \in P(F)$.
So, $\inf_{Y\in P(F)}[gr(Y \cup X |\!\sim \beta)] = gr(X |\!\sim \beta)$ ...(b).
Also, $\inf_{Y\in P(F)}[\inf_{\alpha\in X} gr(Y |\!\sim \alpha) \to_m gr(Y |\!\sim \beta)]$
$\leq \inf_{X\subseteq Y\subseteq F}[\inf_{\alpha\in X} gr(Y |\!\sim \alpha) \to_m gr(Y |\!\sim \beta)]$. ...(c)
Now, as $X \subseteq Y$, by (GC2) $gr(X |\!\sim \alpha) \leq gr(Y |\!\sim \alpha)$.
That is, $1 = \inf_{\alpha\in X} gr(X |\!\sim \alpha) \leq \inf_{\alpha\in X} gr(Y |\!\sim \alpha)$.
Hence (c) becomes, $\inf_{Y\in P(F)}[\inf_{\alpha\in X} gr(Y |\!\sim \alpha) \to_m gr(Y |\!\sim \beta)]$
$\leq \inf_{X\subseteq Y}[1 \to_m gr(Y |\!\sim \beta)] = \inf_{X\subseteq Y} gr(Y |\!\sim \beta)$ (by r1)
$= gr(X |\!\sim \beta)$. ...(d)
Hence, combining (a), (b) and (d), we have
$gr(X |\!\sim \beta) \leq \inf_{Y\in P(F)}[\inf_{\alpha\in X} gr(Y |\!\sim \alpha) \to_m gr(Y |\!\sim \beta)] \leq gr(X |\!\sim \beta)$.
That is, $gr(X |\!\sim \beta) = \inf_{Y\in P(F)}[\inf_{\alpha\in X} gr(Y |\!\sim \alpha) \to_m gr(Y |\!\sim \beta)]$
$= \inf_{Y\in P(F)}[\inf_{\alpha\in X} T_Y(\alpha) \to_m T_Y(\beta)]$. □

Note 2.1 1. From the proof of Theorem 2.1 (1), we can notice that (r3), (r5), (r6), (r7) and (r11) are used. The property (r11) can be obtained using property (r2) and the definition of infimum. So, basically the value set L with properties (r2), (r3), (r5), (r6) and (r7), apart from (r1) which is used in defining graded inclusion between two fuzzy sets, ensure the first part of the representation theorem. Residuation is needed in order to prove the second part.

2. It is to be noted here that the notion of semantic graded consequence, defined in Definition 2.3, gives one way of constructing a graded consequence relation. There could be different constructions to generate a graded consequence relation as per Definition 2.2. We will touch upon this issue in some of the later chapters (Chaps. 5 and 7) of this book.
3. In Definition 2.3, the use of the infimum operation 'inf' in aggregating the values given by the T_i's, which can be counted as experts'/agents' opinions, is a straightforward translation of the universal quantifier ∀. The operation 'infimum' has some drawbacks as it pulls down the value to the greatest lower bound given a set of varied values. In practice, some other aggregation operation might be more realistic. In this book, in general, we have avoided such endeavour in order to not to deviate much from the basic definition of the notion of semantic consequence. But in Chap. 7, a few more realistic ways of aggregating experts' opinion are presented. Moreover, no connection among the T_i's is also assumed in general. But in Sect. 7.2 of Chap. 7 we have extended GCT further considering a possibil-

ity of information flow between an expert T_i and a decision-maker. This might be developed further by introducing flow of information among T_i's as well.
4. At this point, we would like to mention that before the notion of graded consequence introduced in Chakraborty (1995), in Chakraborty (1988) a notion of α-consequence was defined where instead of the present condition of reflexivity the condition of α-reflexivity was considered; it was assumed that for any formula $x \in X, gr(X \mid\!\sim x) \geq \alpha$ for some $\alpha \in [0, 1]$. In Castro et al. (1994), made an extensive study comparing the notion of α-consequence with a notion of consequence in fuzzy context, developed by them.

Definition 2.4 A binary fuzzy relation between $P(F)$ and $F, \mid\!\sim$, is said to be a compact relation if $gr(X \mid\!\sim \alpha) \leq \sup_{X' \subseteq X, X' \; finite} gr(X' \mid\!\sim \alpha)$.

Proposition 2.1 *If $\mid\!\sim$ is a compact relation satisfying (GC2), then cut for finite set implies (GC3).*

Proof Let $\mid\!\sim$ be a compact relation satisfying (GC2). Then combining (GC2) and compactness $gr(X \cup Z \mid\!\sim \alpha) = \sup_{X' \subseteq X,\; Z' \subseteq Z,\; X', Z' \; finite} gr(X' \cup Z' \mid\!\sim \alpha)$. ...(a)
By (GC2), $gr(X' \cup Z' \mid\!\sim \alpha) \leq gr(X \cup Z' \mid\!\sim \alpha)$ for any $X' \subseteq X$ ...(b)
Hence, $\sup_{X' \subseteq X,\; Z' \subseteq Z} gr(X' \cup Z' \mid\!\sim \alpha) \leq \sup_{Z' \subseteq Z} gr(X \cup Z' \mid\!\sim \alpha)$ where X', Z' are finite subsets of X and Z, respectively. ...(c)
By cut for finite set, $gr(X \mid\!\sim \alpha) \geq gr(X \cup Z' \mid\!\sim \alpha) *_m \inf_{\beta \in Z'} gr(X \mid\!\sim \beta)$
$\geq gr(X \cup Z' \mid\!\sim \alpha) *_m \inf_{\beta \in Z} gr(X \mid\!\sim \beta)$ for $Z' \subseteq Z, Z'$ finite.
So, $gr(X \mid\!\sim \alpha) \geq \sup_{Z' \subseteq Z} gr(X \cup Z' \mid\!\sim \alpha) *_m \inf_{\beta \in Z} gr(X \mid\!\sim \beta)$ for finite Z' (by r9)
$\geq \sup_{X' \subseteq X, Z' \subseteq Z} gr(X' \cup Z' \mid\!\sim \alpha) *_m \inf_{\beta \in Z} gr(X \mid\!\sim \beta)$
where X', Z' are finite [by (c)]
$= gr(X \cup Z \mid\!\sim \alpha) *_m \inf_{\beta \in Z} gr(X \mid\!\sim \beta)$ [by (a)] □

Proposition 2.2 *If the meta-level algebraic structure $(L, \wedge, \vee, \rightarrow_m, 0, 1)$ is a complete pseudo-Boolean algebra, and $\mid\!\sim$ satisfies (GC2), then the graded version of cut for an element satisfies cut for finite sets, and conversely.*

Proof Let Y be a finite set. We shall apply induction on the number of elements of Y. Let $|Y| = 1$. Then it is straightforward from the cut for an element.
Induction hypothesis: Let the result hold for any set Y with less than n elements.
Now let us consider $Y = \{\beta_1, \beta_2, \ldots, \beta_n\}$.
$inf_{\beta \in Y} gr(X \mid\!\sim \beta) \wedge gr(X \cup Y \mid\!\sim \alpha)$
$= inf_{\beta \in Y \setminus \{\beta_n\}} gr(X \mid\!\sim \beta) \wedge gr(X \cup Y \mid\!\sim \alpha) \wedge gr(X \mid\!\sim \beta_n)$
$\leq \inf_{\beta \in Z} gr(X \mid\!\sim \beta) \wedge gr(X \cup Z \cup \{\beta_n\} \mid\!\sim \alpha) \wedge gr(X \mid\!\sim \beta_n)$ [assume $Y \setminus \{\beta_n\} = Z$]
$\leq \inf_{\beta \in Z} gr(X \cup \{\beta_n\} \mid\!\sim \beta) \wedge gr(X \cup Z \cup \{\beta_n\} \mid\!\sim \alpha) \wedge gr(X \mid\!\sim \beta_n)$ [by (GC2)]
$\leq gr(X \cup \{\beta_n\} \mid\!\sim \alpha) \wedge gr(X \mid\!\sim \beta_n)$ [by induction hypothesis]
$\leq gr(X \mid\!\sim \alpha)$ [by cut for an element]
The converse is immediate by assuming Y to be a singleton set. □

Proposition 2.3 *If $\{\mid\!\sim_i\}_{i \in I}$ is a collection of graded consequence relations, then $\mid\!\sim = \bigcap_{i \in I} \mid\!\sim_i$ is a graded consequence relation.*

In a complete Heyting algebra, i.e. a complete residuated lattice where the monoidal product is the lattice meet itself, some additional results may be obtained.

Definition 2.5 For any fuzzy relation $\mid\!\sim$, another fuzzy relation $\mid\!\sim'$ is defined as follows. $gr(X \mid\!\sim' \alpha) = gr(X \mid\!\sim \alpha) \wedge \sup_{X' \subseteq X,\ X' finite} gr(X' \mid\!\sim \alpha)$.

Proposition 2.4 *For any fuzzy relation* $\mid\!\sim$, $\mid\!\sim'$ *defined as above, is the maximal compact subrelation contained in* $\mid\!\sim$.

Proposition 2.5 *If* $\mid\!\sim$ *is a graded consequence relation, then so is* $\mid\!\sim'$.

Proposition 2.6 *Any fuzzy relation* $\mid\!\sim$ *can be extended to a minimal graded consequence relation* $\mid\!\sim^c$, *which is the closure of* $\mid\!\sim$ *under (GC1), (GC2) and (GC3).*

Let $(\mid\!\sim^c)'$ be the maximal compact subrelation of $\mid\!\sim^c$ (by Proposition 2.4). Then $(\mid\!\sim^c)'$ is a graded consequence relation by Proposition 2.5. Generally, $(\mid\!\sim^c)'$ can be smaller than $\mid\!\sim$. But when $\mid\!\sim$ is compact by Proposition 2.4, $\mid\!\sim \subseteq (\mid\!\sim^c)' \subseteq \mid\!\sim^c$. As $\mid\!\sim^c$ is the smallest extension of $\mid\!\sim$ satisfying (GC1) to (GC3), and $(\mid\!\sim^c)'$ is also one such extension of $\mid\!\sim$, we have $(\mid\!\sim^c)' = \mid\!\sim^c$. Thus, we get the following proposition.

Proposition 2.7 *If* $\mid\!\sim$ *is compact, its closure* $\mid\!\sim^c$ *is also compact.*

2.3 A Generalization of Hilbert-Type Axiom System

The standard Hilbert-type axiom system for classical logic consists of specifying a set $\mathscr{A}_x$ of wffs as axioms, and a finite set of rules of inference. By the derivation of a formula α from a set X of wffs is meant a finite sequence $\alpha_1, \alpha_2, \ldots \alpha_n$ of wffs such that (i) $\alpha_n = \alpha$, (ii) any $\alpha_i, i = 1, 2, \ldots n$, is either in $\mathscr{A}_x$ or in X or is obtained from a subset of $\{\alpha_1, \alpha_2, \ldots, \alpha_{i-1}\}$ by the application of some of the rules of inference. We then write $\langle \alpha_1, \alpha_2, \ldots, \alpha_n \rangle D(X, \alpha)$, and read as 'the sequence $\langle \alpha_1, \alpha_2, \ldots, \alpha_n \rangle$ is a derivation of α from X'. There can be more than one derivations of a wff α from a set X of wffs. If there exists at least one derivation of α from X, one can write $X \vdash \alpha$ since '$\vdash$' defined thus turns out to be a classical consequence relation, which is also compact. All these are purely syntactic activities. It may be noted that many-valued logics, modal logics or intuitionistic logics, follow the same methodology as regards derivation. We want to propose a generalization of this notion in the context of GCT.

In GCT, the notion of derivation is itself graded. The grade arises due to assignment of grades to the axioms as well as rules of inference. A rule of inference in the classical context is a crisp binary relation between a set of wffs and a formula. For instance, the rule Modus Ponens (MP) relates the set of formulae $\{\alpha, \alpha \supset \beta\}$ to β for all choices of α, β. In the case of GCT, these rules of inference are fuzzy binary relations. Thus any GCT-rule R is a mapping from $P^{fin}(F) \times F$, where $P^{fin}(F)$ is the set of all finite subsets of F, to some value set, which is taken to be the same lattice structure as is used to calculate the values of meta-level sentences. In the present context, this is a complete residuated lattice as has been discussed before.

For a general theory, it is not required to specify the types of a fuzzy rule. Any collection of fuzzy relations may be taken. But for the construction of useful logics, some specific kinds of rules (like MP) might be necessary. So, we define a finitary fuzzy rule as a fuzzy relation R on $P^{fin}(F) \times F$ such that $Supp(R)$ consists of pairs of the form $(\{\alpha_1, \alpha_2, \ldots, \alpha_{m-1}\}, \alpha_m)$ for some formulae $\alpha_1, \alpha_2, \ldots, \alpha_{m-1}, \alpha_m \in F$. For example, the fuzzy modus ponens rule, denoted as *FuzzMP* (or simply R_{MP}), has $Supp(FuzzMP) = \{(\{\alpha, \alpha \supset \beta\}, \beta) : \alpha, \beta \in F\}$. The only difference from classical MP rule lies in that the degree $FuzzMP(\{\alpha, \alpha \supset \beta\}, \beta)$, though > 0 is not necessarily 1 for all choices of α, β. How would the value for different choices of α, β be calculated will be discussed later in Sect. 2.3.1. Depending on these values, a value $|R|$, called the strength of the rules, will be calculated. In the classical case $|R| = 1$ for all rules R.

Also the axioms in GCT constitute a fuzzy subset $\mathscr{A}_x$ of the set of wffs. Although any fuzzy subset of wffs may be taken as axioms, we will prefer a special way of degree assignment as will be discussed below in Sect. 2.3.1.

Let X be a (crisp) set of wffs and α be a wff. By a derivation of α from X, in GCT, is meant a sequence $\langle \alpha_1, \alpha_2, \ldots, \alpha_n \rangle$ of wffs and a value associated with the triple $(\langle \alpha_1, \alpha_2, \ldots, \alpha_n \rangle, X, \alpha)$ in a way that will be formally defined in Definition 2.7. In fact, $\alpha_n = \alpha$ and each α_i will be tagged with a value $|\alpha_i|$, called the value of the ith step, and the value $|\alpha_1| *_m |\alpha_2| *_m \ldots *_m |\alpha_n|$ will be associated with the triple. The values $|\alpha_i|$, though considered to be of syntactic nature and give a measure of a particular step of a derivation, are linked with the semantic context. In this sense, the syntax in the Hilbert-type axiomatic system in GCT is not totally delinked with the semantics.

In classical logic, although formally any set of wffs may be chosen as axioms and any relation as rule of inference, in 'useful' logics the selection is made keeping an eye on the semantics.[2] This means associating truth values to wffs by use of operators on truth set, choosing an appropriate subset of tautologies as axioms, and a set of rules that are sound. Generally, axioms and rules are so chosen that the logical system becomes complete. However, the logic system remains equivalent if the set of all tautologies are taken as the set of axioms.

2.3.1 Graded Consequence by Fuzzy Axioms and Fuzzy Rules

In the case of GCT, Hilbert-type axiomatic theory may be introduced in a way, based on the semantics, as follows. In what follows no specific object language is required.

Let $\{T_i\}_{i \in I}$ be a set of valuations of the basic (object language) formulae. The set $\{T_i\}_{i \in I}$ is also called the semantic context of the logic. Each wff α gets a value $\inf_i T_i(\alpha)$. Let by $[\alpha]$ we denote all wffs having the same syntactic form as that of α. That is, if α is a formula of the form $\beta \supset \gamma$ then formulas with all possible

[2] Like the abstract theory of syntax, formal logic without a powerful procedural semantics cannot deal with meaningful situations—Minsky (1992).

substitutions for β and γ are of the same syntactic form as α. Then each formula α may be associated with another value $\inf_{\alpha' \in [\alpha]} \inf_i T_i(\alpha')$. This value may be called the tautologihood degree of α, denoted by $Taut(\alpha)$, relative to the context $\{T_i\}_{i \in I}$. To fix the fuzzy set $\mathscr{A}_x$ of axioms, we consider some non-atomic formulae having the tautologihood degree greater than 0. A threshold a (>0) may also be chosen (but not essential). The set $\mathscr{A}$ of such formulae is fixed as the support set of $\mathscr{A}_x$. This threshold may be taken in particular as the top element 1 of the value set. Thus, a set of wffs is obtained, and by associating with them the corresponding tautologihood degrees, and 0 to the rest of formulae a fuzzy subset $\mathscr{A}_x$ of axioms is obtained. The fuzzy set $\mathscr{A}_x$ of axioms is thus defined by $\mathscr{A}_x(\alpha) = Taut(\alpha)$ if $\alpha \in \mathscr{A}$, and 0 otherwise. The value $\mathscr{A}_x(\alpha)$ is called the axiomhood degree of α with respect to $\{T_i\}_{i \in I}$. It may be possible that for some α, $\mathscr{A}_x(\alpha) = 0$ but $Taut(\alpha) \neq 0$.

Next to fix the graded inference rules, we proceed as follows. As mentioned before, a finite set of fuzzy relations constitutes the set of rules. Now, for each fuzzy rule R, we take the support set $Supp(R)$. Then the value $|R|$associated with R is

$$\inf_{(\{\alpha_1,\alpha_2,\ldots,\alpha_{n-1}\},\alpha_n) \in Supp(R)} gr(\{\alpha_1, \alpha_2, \ldots, \alpha_{n-1}\} \approx\!\!\!|_{\{T_i\}_{i \in I}} \alpha_n).$$

It should be noted that neither the syntactic form of the formulae nor the number n has to be fixed in a particular R. But as a particular case and usually for each rule R, n is taken to be fixed called the arity of the rule and $Supp(R)$ contains pairs of a definite pattern. As an example, in case of the rule $FuzzMP$ we have

$$|FuzzMP| = \inf_{\alpha,\beta} gr(\{\alpha, \alpha \supset \beta\} \approx\!\!\!|_{\{T_i\}_{i \in I}} \beta).$$

By $\mathscr{R}$ we shall denote a set $\{Supp(R)$: where R is a fuzzy rule$\}$. The pair $(\mathscr{A}, \mathscr{R})$ is called the syntatic base of the Hilbert-type axiom system in GCT.

Definition 2.6 Given a syntactic base $(\mathscr{A}, \mathscr{R})$, a derivation chain for a formula α from a set of formulae X is a sequence of formulae $\langle \alpha_1, \alpha_2, \ldots, \alpha_n \rangle$ such that (i) $\alpha_n = \alpha$, and (ii) each α_i, $1 \leq i \leq n$, is either a member of X, or an element from $\mathscr{A}$, or $(\{\alpha_{i_1}, \alpha_{i_2}, \ldots, \alpha_{i_j}\}, \alpha_i) \in Supp(R)$, for some rule R and $1 \leq i_1, i_2, \ldots, i_j < i$.

The definition is exactly the same as that of any standard logical system. However, the difference can be observed in the following.

Definition 2.7 Given a derivation chain $\langle \alpha_1, \alpha_2, \ldots, \alpha_n \rangle$ of α from X, the value (strength) of the derivation, which is denoted as $val(\langle \alpha_1, \alpha_2, \ldots, \alpha_n \rangle D(X, \alpha))$, is given by $|\alpha_1| *_m |\alpha_2| *_m \ldots *_m |\alpha_n|$, where for each i,

$$\begin{aligned} |\alpha_i| &= 1 \text{ if } \alpha_i \in X \\ &= \mathscr{A}_x(\alpha_i) \text{ if } \alpha_i \in \mathscr{A} \setminus X, \text{ and} \\ &= |R| \text{ if } (\{\alpha_{i_1}, \alpha_{i_2}, \ldots, \alpha_{i_j}\}, \alpha_i) \in Supp(R) \text{ for } 1 \leq i_1, i_2 \ldots, i_j < i. \end{aligned}$$

$\mathscr{A}_x(\alpha_i)$ and $|R|$ are, respectively, the axiomhood degree of α_i and the degree of the rule R with respect to a prefixed semantic base $\{T_i\}_{i\in I}$.

If $\langle\alpha_1, \alpha_2, \ldots, \alpha_n\rangle$ does not satisfy the criterion of Definition 2.6 for being a derivation chain of α from X, then $val(\langle\alpha_1, \alpha_2, \ldots, \alpha_n\rangle X D\alpha)$ is 0.

Definition 2.8 Given a syntactic base $(\mathscr{A}, \mathscr{R})$, the grade to which a formula α is an axiomatic consequence of a set of formulae X, denoted by $gr(X \mid\!\sim_{Ax} \alpha)$, is given by, $gr(X \mid\!\sim_{Ax} \alpha) = \sup_{\langle\alpha_1,\alpha_2,\ldots,\alpha_n\rangle} val(\langle\alpha_1, \alpha_2, \ldots, \alpha_n\rangle D(X, \alpha))$.

Thus, $gr(X \mid\!\sim_{Ax} \alpha)$ is the supremum of the values of all derivation sequences $\langle\alpha_1, \alpha_2, \ldots, \alpha_n\rangle$ of α from X. Considering the supremum over the derivations generalizes the fact that in classical logic $X \vdash \alpha$ iff there exists a derivation of α from X.

Proposition 2.8 $\mid\!\sim_{Ax}$ *is a compact relation.*

Proof $gr(X \mid\!\sim_{Ax} \alpha) = \sup_{\langle\alpha_1,\alpha_2,\ldots,\alpha_n\rangle} val(\langle\alpha_1, \alpha_2, \ldots, \alpha_n\rangle D(X, \alpha))$.
As $\langle\alpha_1, \alpha_2, \ldots, \alpha_n\rangle$ is a finite sequence such that some of the α_is are taken from X, let $X' = \{\alpha_{i_1}, \alpha_{i_2}, \ldots, \alpha_{i_m}\} \subseteq X$, where $1 \leq i_1, \ldots, i_m \leq n$.
Then, $val(\langle\alpha_1, \alpha_2, \ldots, \alpha_n\rangle D(X, \alpha)) = val(\langle\alpha_1, \alpha_2, \ldots, \alpha_n\rangle D(X', \alpha))$ ...(i)
So, for each $\langle\alpha_1, \alpha_2, \ldots, \alpha_n\rangle$ there is a finite $X'(\subseteq X)$ such that (i) holds.
$\sup_{\langle\alpha_1,\alpha_2,\ldots,\alpha_n\rangle} val(\langle\alpha_1, \alpha_2, \ldots, \alpha_n\rangle D(X, \alpha))$
$= \sup_{\langle\alpha_1,\alpha_2,\ldots,\alpha_n\rangle} val(\langle\alpha_1, \alpha_2, \ldots, \alpha_n\rangle D(X', \alpha))$ for some $X' \subseteq X$ and X' is finite.
$\leq \sup_{X'\subseteq X,\ X':finite}[\sup_{\langle\alpha_1,\alpha_2,\ldots,\alpha_n\rangle} val(\langle\alpha_1, \alpha_2, \ldots, \alpha_n\rangle D(X', \alpha))]$
$= \sup_{X'\subseteq X,\ X':finite}\, gr(X' \mid\!\sim_{Ax} \alpha)$.
So, $gr(X \mid\!\sim_{Ax} \alpha) \leq \sup_{X'\subseteq X,\ X':finite}\, gr(X' \mid\!\sim_{Ax} \alpha)$. □

Theorem 2.2 $\mid\!\sim_{Ax}$ *satisfies (GC1), (GC2), and cut for an element.*

Proof That $\mid\!\sim_{Ax}$ satisfies (GC1) and (GC2) is simple to prove. An outline of the proof that $\mid\!\sim_{Ax}$ satisfies cut for an element,
i.e. $gr(X \mid\!\sim_{Ax} \beta) *_m gr(X \cup \{\beta\} \mid\!\sim_{Ax} \alpha) \leq gr(X \mid\!\sim_{Ax} \alpha)$, is as follows.
Let us consider the two cases—(i) $gr(X \mid\!\sim_{Ax} \beta) = 0$ and (ii) $gr(X \mid\!\sim_{Ax} \beta) \neq 0$.
For (i) the inequality holds trivially. So, let (ii), i.e. $gr(X \mid\!\sim_{Ax} \beta) \neq 0$. Then there must be a sequence $\langle\alpha_1, \alpha_2, \ldots, \alpha_n\rangle$ such that $val(\langle\alpha_1, \alpha_2, \ldots, \alpha_n\rangle D(X, \beta)) > 0$.
Let us consider different derivation sequences for obtaining α from $X \cup \{\beta\}$. Let $\{a_j\}_{j\in J}$ and $\{a_k\}_{k\in K}$, respectively, be the two collections of values of the derivations that derive α from X using β, and not using β. So,
$gr(X \cup \{\beta\} \mid\!\sim_{Ax} \alpha) *_m gr(X \mid\!\sim_{Ax} \beta)$
$= \sup_{\langle\beta_1,\beta_2,\ldots,\beta_m\rangle} val(\langle\beta_1, \beta_2, \ldots, \beta_m\rangle D(X \cup \{\beta\}, \alpha) *_m gr(X \mid\!\sim_{Ax} \beta)$
$= (\sup_{j\in J} a_j \vee \sup_{k\in K} a_k) *_m gr(X \mid\!\sim_{Ax} \beta)$
$= (\sup_{j\in J} a_j *_m gr(X \mid\!\sim_{Ax} \beta)) \vee (\sup_{k\in K} a_k *_m gr(X \mid\!\sim_{Ax} \beta))$
$\leq (\sup_{j\in J} a_j *_m gr(X \mid\!\sim_{Ax} \beta)) \vee gr(X \mid\!\sim_{Ax} \alpha)$ (by definition of $\{a_k\}_{k\in K}$) ...(a)
Let for each a_j, $j \in J$, there be a respective derivation chain $\langle\alpha_{j_1}, \alpha_{j_2}, \ldots, \alpha_{j_t}\rangle$ for obtaining α from $X \cup \{\beta\}$ where β occurs at some of the steps. Without loss of generality, let β appear exactly once in the derivation sequence, and $\alpha_{j_i} = \beta$. So, $|\alpha_{j_1}| *_m \ldots *_m |\alpha_{j_{i-1}}| *_m |\beta| *_m \ldots *_m |\alpha_{j_t}| = |\alpha_{j_1}| *_m \ldots *_m |\alpha_{j_{i-1}}| *_m |1| *_m \ldots *_m |\alpha_{j_t}|$ as $|\beta| = 1$.

Now $(\sup_{j\in J} a_j *_m gr(X \mid\!\sim_{Ax} \beta))$
$= \sup_{\langle\alpha_{j_1},\ldots,\alpha_{j_t}\rangle}(|\alpha_{j_1}| *_m \ldots *_m |\beta| *_m \ldots *_m |\alpha_{j_t}|) *_m \sup_{\langle\alpha_1,\ldots,\alpha_n\rangle} val(\langle\alpha_1,\ldots,\alpha_n\rangle D(X,\beta))$
$= \sup_{\langle\alpha_{j_1},\ldots,\alpha_{j_t}\rangle} \sup_{\langle\alpha_1,\ldots,\alpha_n\rangle}(|\alpha_{j_1}| *_m \ldots |\beta| \ldots *_m |\alpha_{j_t}| *_m val(\langle\alpha_1,\ldots,\alpha_n\rangle D(X,\beta))$ (by r9)
$= \sup_{\langle\alpha_{j_1},\ldots,\langle\alpha_1,\ldots,\alpha_n\rangle,\ldots\alpha_{j_t}\rangle}(|\alpha_{j_1}| *_m \ldots *_m val(\langle\alpha_1,\ldots,\alpha_n\rangle D(X,\beta)) *_m \ldots |\alpha_{j_t}|)$
$\leq gr(X \mid\!\sim_{Ax} \alpha)$ as replacing β in $\langle\alpha_{j_1},\ldots,\beta,\ldots,\alpha_{j_t}\rangle$ by $\langle\alpha_1,\ldots,\alpha_n\rangle$ gives rise to a derivation chain $\langle\alpha_{j_1},\ldots,\alpha_1,\ldots,\alpha_n,\ldots\alpha_{j_t}\rangle$ for α from X. Hence (a) becomes, $gr(X \cup \{\beta\} \mid\!\sim_{Ax} \alpha) *_m gr(X \mid\!\sim_{Ax} \beta) \leq gr(X \mid\!\sim_{Ax} \alpha)$. □

Corollary 2.1 *If the meta-level algebraic structure $(L, \wedge, \vee, \rightarrow_m, 0, 1)$ is a complete pseudo-Boolean algebra, $\mid\!\sim_{Ax}$ is a graded consequence relation.*

Proof The proof follows from Theorem 2.2 and Propositions 2.1, 2.2 and 2.8. □

Note 2.2 From Theorem 2.2 and Corollary 2.1, we can notice that with respect to any arbitrary complete residuated lattice $\mid\!\sim_{Ax}$ becomes a restrictive kind of graded consequence relation which satisfies graded version of cut for an element. In particular, when the meta-level algebraic structure is considered to be a complete pseudo-Boolean algebra, $\mid\!\sim_{Ax}$ becomes a graded consequence relation in the sense of Definition 2.2.

2.3.2 *Soundness and Examples of Hilbert-Type Axiomatic Consequence in GCT*

Definition 2.9 A graded consequence relation $\mid\!\sim$ is said to be sound with respect to another graded consequence relation $\approx\!\mid$ if and only if for any $X \subseteq F$, and $\alpha \in F$, $gr(X \mid\!\sim \alpha) \leq gr(X \mid\!\approx \alpha)$.

Theorem 2.3 *Given the syntactic base $(\mathscr{A}, \mathscr{R})$ with respect to the semantic context $\{T_i\}_{i\in I}$, the graded consequence relation $\mid\!\sim_{Ax}$ is sound with respect to the graded consequence relation $\mid\!\approx_{\{T_i\}_{i\in I}}$.*

Proof It will be shown that for every derivation chain $\langle\alpha_1,\alpha_2,\ldots,\alpha_m\rangle$ of α from X, $val(\langle\alpha_1,\alpha_2,\ldots,\alpha_m\rangle D(X,\alpha)) \leq gr(X \mid\!\approx_{\{T_i\}_{i\in I}} \alpha)$, and this will be shown by using induction on the number of steps of the derivation.
Let $\alpha_1 = \alpha$. If $\alpha \in X$, then following Definitions 2.7 and 2.8, $gr(X \mid\!\sim_{Ax} \alpha) = 1$, and as $\mid\!\approx_{\{T_i\}_{i\in I}}$ is a graded consequence relation $gr(X \mid\!\approx_{\{T_i\}_{i\in I}} \alpha) = 1$ as well.
If $\alpha \in \mathscr{A} - X$, then $val(\langle\alpha\rangle D(X,\alpha)) = \mathscr{A}_x(\alpha) = \inf_{\alpha'\in[\alpha]} \inf_i T_i(\alpha') \leq \inf_i T_i(\alpha)$
$$= gr(\phi \mid\!\approx_{\{T_i\}_{i\in I}} \alpha)$$
$$\leq gr(X \mid\!\approx_{\{T_i\}_{i\in I}} \alpha).$$
Let $val(\langle\alpha_1,\alpha_2,\ldots,\alpha_{m-1}\rangle D(X,\alpha)) \leq gr(X \mid\!\approx_{\{T_i\}_{i\in I}} \alpha)$ hold for any number of steps $\leq m-1$. Let us consider the derivation chain $\langle\alpha_1,\alpha_2,\ldots,\alpha_m\rangle$ for obtaining α from X.
As the other cases are simple, we would only discuss the case where α_m (=α) is obtained because of $(\{\alpha_{m_1},\ldots,\alpha_{m_s}\},\alpha_m) \in Supp(R)$, such that $1 \leq m_1,\ldots,m_s < m$.

So, $val(\langle\alpha_1, \alpha_2, \ldots, \alpha_m\rangle D(X, \alpha)) = val(\langle\alpha_1, \alpha_2, \ldots, \alpha_{m-1}\rangle D(X, \alpha_{m-1}) *_m |R| \ldots$(a)
Now $val(\langle\alpha_1, \alpha_2, \ldots, \alpha_{m-1}\rangle D(X, \alpha_{m-1})) \leq val(\langle\alpha_1, \ldots, \alpha_{m_j}\rangle D(X, \alpha_{m_j}))$ for each $m_j \in \{m_1, \ldots, m_s\}$.
Hence, $val(\langle\alpha_1, \alpha_2, \ldots, \alpha_{m-1}\rangle D(X, \alpha_{m-1})) \leq \wedge_{j=1}^{s} val(\langle\alpha_1, \ldots, \alpha_{m_j}\rangle D(X, \alpha_{m_j}))$.
So, (a) becomes

$val(\langle\alpha_1, \alpha_2, \ldots, \alpha_m\rangle D(X, \alpha)) \leq \wedge_{j=1}^{s} val(\langle\alpha_1, \ldots, \alpha_{m_j}\rangle D(X, \alpha_{m_j})) *_m |R|$
$\leq \wedge_{j=1}^{s} gr(X \approx_{\{T_i\}_{i\in I}} \alpha_{m_j}) *_m |R|$ (by induction hypothesis)
$\leq \wedge_{j=1}^{s} gr(X \approx_{\{T_i\}_{i\in I}} \alpha_{m_j}) *_m \inf_{i\in I}\{\wedge_{j=1}^{s} T_i(\alpha_{m_j}) \rightarrow_m T_i(\alpha_m)\}$ (by replacing $|R|$)
$= \wedge_{j=1}^{s}[\inf_{i\in I}\{\inf_{x\in X} T_i(x) \rightarrow_m T_i(\alpha_{m_j})\}] *_m \inf_{i\in I}\{\wedge_{j=1}^{s} T_i(\alpha_{m_j}) \rightarrow_m T_i(\alpha_m)\}$
$= [\inf_{i\in I}\{\inf_{x\in X} T_i(x) \rightarrow_m \wedge_{j=1}^{s} T_i(\alpha_{m_j})\}] *_m \inf_{i\in I}\{\wedge_{j=1}^{s} T_i(\alpha_{m_j}) \rightarrow_m T_i(\alpha_m)\}$ (by r7)
$\leq \inf_{i\in I}[\{\inf_{x\in X} T_i(x) \rightarrow_m \wedge_{j=1}^{s} T_i(\alpha_{m_j})\} *_m \{\wedge_{j=1}^{s} T_i(\alpha_{m_j}) \rightarrow_m T_i(\alpha_m)\}]$ (by r11)
$\leq \inf_{i\in I}\{\inf_{x\in X} T_i(x) \rightarrow_m T_i(\alpha_m)\}$ (by r10)
$= gr(X \approx_{\{T_i\}_{i\in I}} \alpha)$. □

We make two remarks at this point.

Remark 2.1 From the heuristic standpoint, a derivation in GCT turns out to be an ordinary derivation of classical logic with a grade attached to it.

Remark 2.2 Because of the strength of a rule R, a value is transmitted to the conclusion of the derivation which is the same as $|\alpha_1| *_m |\alpha_2| *_m \ldots *_m |\alpha_n|$. This value may not be the actual value of the conclusion α_n with respect to any valuation T_i. Such as in the classical context, the value transmitted to the conclusion β through a sequence $\langle\alpha, \beta \supset \alpha, \beta\rangle$ with respect to the premise $\{\alpha, \beta \supset \alpha\}$ is 0 since

α	$\|\alpha\| = 1$	premise
$\beta \supset \alpha$	$\|\beta \supset \alpha\| = 1$	premise
β	$\|R\| = 0$	as the rule is invalid in classical logic

and as a consequence the value transmitted to β is $1 *_m 1 *_m 0 = 0$. But β may be taken to be a tautology.

Let us take a few exemplary cases to see how does a derivation look like.

Example 2.2 Let us consider the value set $L = \{0, b, a, 1\}$ with the linear ordering $0 \leq b \leq a \leq 1$, and the operations of the meta-level algebraic structure $(L, \wedge, \vee, *_m, \rightarrow_m, 0, 1)$ are given as follows:

$*_m$	0	b	a	1
0	0	0	0	0
b	0	0	b	b
a	0	b	a	a
1	0	b	a	1

$\rightarrow_m$	0	b	a	1
0	1	1	1	1
b	b	1	1	1
a	0	b	1	1
1	0	b	a	1

Let the object language (F) consist of propositional variables and the compound formulae involving $\supset$, the connective for implication. The table for $\rightarrow_o$, the algebraic operation corresponding to $\supset$ is given below:

$\rightarrow_o$	0	b	a	1
0	1	1	1	1
b	a	1	1	1
a	b	a	1	1
1	0	b	a	1

Let us now consider the semantic context as any arbitrary collection of fuzzy sets $\{T_i\}_{i\in I}$ over F. Let the syntactic base be $(\mathscr{A}, \mathscr{R})$ where the set of axioms $\mathscr{A}$ contains all possible instances of the forms $\{\alpha \supset (\beta \supset \alpha), (\alpha \supset (\beta \supset \gamma)) \supset (\alpha \supset (\beta \supset \gamma))\}$, and $\mathscr{R}$ consists of the only rule modus ponens (MP). It can be checked that for all possible values of α, β, $T_i(\alpha \supset (\beta \supset \alpha)) = 1$, as $x \rightarrow_o y \geq y$ holds.
So, $Taut(\alpha \supset (\beta \supset \alpha)) = 1$. That $Taut((\alpha \supset \beta) \supset \gamma) \supset (\alpha \supset (\beta \supset \gamma))) = 1$ also can be checked. We now need to check

$$\begin{aligned} |FuzzMP| &= \inf_{\alpha,\beta} gr(\{\alpha, \alpha \supset \beta\} \approx_{\{T_i\}_{i\in I}} \beta) \\ &= \inf_{\alpha,\beta} gr(\{\alpha, \alpha \supset \beta\} \approx_{\{T_i\}_{i\in I}} \beta). \end{aligned}$$

The following table shows the value of $gr(\{\alpha, \alpha \supset \beta\} \approx_{\{T_i\}_{i\in I}} \beta)$ for all possible cases:

$T_i(\alpha)$	$T_i(\beta)$	$(T_i(\alpha) \wedge T_i(\alpha \supset \beta)) \rightarrow_m T_i(\beta)$
0	0	1
	b	1
	a	1
	1	1
b	0	b
	b	1
	a	1
	1	1
a	0	b
	b	b
	a	1
	1	1
1	0	1
	b	1
	a	1
	1	1

This indicates that $|MP| = b$.

Let $X = \{\alpha_1\}$. Then we can write $\langle \alpha_1 \supset (\alpha_3 \supset \alpha_1), \alpha_1, \alpha_3 \supset \alpha_1 \rangle XD(\alpha_3 \supset \alpha_1)$ where the derivation sequence is justified as follows:

- $\alpha_1 \supset (\alpha_3 \supset \alpha_1)$ (Axiom)
- α_1 (Premise)
- $\alpha_3 \supset \alpha_1$ (MP)

Hence $val(\langle \alpha_1 \supset (\alpha_3 \supset \alpha_1), \alpha_1, \alpha_3 \supset \alpha_1 \rangle X D(\alpha_3 \supset \alpha_1))$
$= |\alpha_1 \supset (\alpha_3 \supset \alpha_1)| *_m |\alpha_1| *_m |\alpha_3 \supset \alpha_1| = 1 *_m 1 *_m b = b$
and we can conclude $gr(\{\alpha_1\} \mid\!\sim_{Ax} \alpha_3 \supset \alpha_1) \geq b$.

It is to be noted that for evaluating $gr(X \mid\!\sim_{Ax} \alpha_3 \supset \alpha_1)$ we need to depend on the semantics as the degree of the rule and axiomhood degree of the axioms are determined from the semantic base. But unlike $gr(\{\alpha_1\} \approx\!\!|_{\{T_i\}_{i\in I}} \alpha_3 \supset \alpha_1)$, we are not at all here concerned about the values that the particular formulas α_1, α_3 are getting under the T_i's. As an instance, we can consider α_1 at the second step of the derivation chain. In the derivation though we are considering a step value for α_1, which is 1 as α_1 is taken from the premise, we do not bother about the value of the formula under the T_i's.

Let us now consider one more derivation from a premise set $Y = \{\alpha_1, \alpha_2\}$.

- α_1 (Premise)
- $\alpha_1 \supset ((\alpha_2 \supset \alpha_3) \supset \alpha_1)$ (Axiom)
- $(\alpha_2 \supset \alpha_3) \supset \alpha_1$ (MP)
- $((\alpha_2 \supset \alpha_3) \supset \alpha_1) \supset (\alpha_2 \supset (\alpha_3 \supset \alpha_1))$ (Axiom)
- $\alpha_2 \supset (\alpha_3 \supset \alpha_1)$ (MP)
- α_2 (Premise)
- $\alpha_3 \supset \alpha_1$ (MP)

So, renaming the formulas appeared at step 1 to step 7, respectively, as $\beta_1, \beta_2, \ldots, \beta_7$, we have $val(\langle \beta_1, \beta_2, \ldots, \beta_7 \rangle\ Y\ D\,(\alpha_3 \supset \alpha_1)) = 1 *_m 1 *_m b *_m 1 *_m b *_m 1 *_m b = 0$.

The previous derivation gives $val(\langle \alpha_1 \supset (\alpha_3 \supset \alpha_1), \alpha_1, \alpha_3 \supset \alpha_1 \rangle X D(\alpha_3 \supset \alpha_1))$ $= b$, and hence by (GC2) $val(\langle \alpha_1 \supset (\alpha_3 \supset \alpha_1), \alpha_1, \alpha_3 \supset \alpha_1 \rangle Y D(\alpha_3 \supset \alpha_1)) \geq b$. So with respect to the earlier derivation $\alpha_3 \supset \alpha_1$ can be derived from Y at least to the grade b, whereas with respect to the later derivation grade of the derivation of $\alpha_3 \supset \alpha_1$ from Y is 0. This exhibits that if use of a weak rule, in the sense of having grade less than 1, increases in a derivation, then the strength of the derivation of a formula from a set of formulae decreases.

Generally, an argument for deriving a proposition can be considered good compared to another argument if the former takes less number of essential steps required for reaching to the conclusion. The above two derivations with their respective grades, in some sense, may be considered as a way of differentiating which argument is better than the other.

Example 2.3 In the context of the above example, let us consider a different meta-level algebraic structure. Let the table for $*_m$ and $\rightarrow_m$ be given as follows, while the table for $\rightarrow_o$ is that of Example 2.2.

$*_m$	0	b	a	1
0	0	0	0	0
b	0	b	b	b
a	0	b	b	a
1	0	b	a	1

$\rightarrow_m$	0	b	a	1
0	1	1	1	1
b	0	1	1	1
a	0	a	1	1
1	0	b	a	1

Let us consider a collection of fuzzy sets $\{T_i\}_{i\in I}$ such that for any $i \in I$, $T_i(p) \neq 0$ for any propositional variable p. Then from the table of $\rightarrow_o$ it can be shown that no compound formula would get the value 0, and hence following the same method we can show that $|FuzzMP|$ or simply $|MP| = a$.

So, with respect to the present context for the same derivations, as mentioned in Example 2.2, we would have $val(\langle\alpha_1 \supset (\alpha_3 \supset \alpha_1), \alpha_1, \alpha_3 \supset \alpha_1\rangle Y D(\alpha_3 \supset \alpha_1)) = a$, and $val(\langle\beta_1, \beta_2, \ldots, \beta_7\rangle\ Y\ D\ (\alpha_3 \supset \alpha_1)) = b$.

Thus, this example shows that if we change our meta-level reasoning base, and modify the collection of T_i's, then the strength of the same derivation may also change significantly.

Example 2.4 Let us consider the same meta-level algebraic structure, i.e. $*_m$ and $\rightarrow_m$, as considered in Example 2.3. The object language contains only one connective $\supset$, and corresponding operator $\rightarrow_o$ is also the same one as has been taken in Examples 2.2 and 2.3. The syntactic base $(\mathscr{A}, \mathscr{R})$ is such that $\mathscr{A}$ contains the axioms $\alpha \supset (\beta \supset \alpha)$, $((\alpha \supset \beta) \supset \gamma) \supset (\alpha \supset (\beta \supset \gamma))$, as in the earlier examples, and $(\alpha \supset (\beta \supset \gamma) \supset ((\alpha \supset \beta) \supset (\alpha \supset \gamma))\}$; $\mathscr{R}$ consists of only rule MP. Here also we consider the same collection of fuzzy sets $\{T_i\}_{i\in I}$ over formulae such that for any $i \in I$, $T_i(p) \neq 0$ for any propositional variable p. Then as discussed in Example 2.3, both of the first two axioms have tautologihood degrees 1, and the grade of the rule MP is a. Now, it can be checked that the tautologihood degree of the third axiom $((\alpha \supset \beta) \supset \gamma) \supset (\alpha \supset (\beta \supset \gamma))$ is a.

Let us now consider $Z = \{\alpha_1, \alpha_3 \supset \alpha_2\}$. Then we have the following derivation:

- α_1 (Premise)
- $\alpha_1 \supset ((\alpha_3 \supset \alpha_2) \supset \alpha_1)$ (Axiom 1)
- $(\alpha_3 \supset \alpha_2) \supset \alpha_1$ (MP)
- $((\alpha_3 \supset \alpha_2) \supset \alpha_1) \supset (\alpha_3 \supset (\alpha_2 \supset \alpha_1))$ (Axiom 2)
- $(\alpha_3 \supset (\alpha_2 \supset \alpha_1))$ (MP)
- $(\alpha_3 \supset (\alpha_2 \supset \alpha_1)) \supset ((\alpha_3 \supset \alpha_2) \supset (\alpha_3 \supset \alpha_1))$ (Axiom 3)
- $((\alpha_3 \supset \alpha_2) \supset (\alpha_3 \supset \alpha_1))$ (MP)
- $\alpha_3 \supset \alpha_2$ (Premise)
- $\alpha_3 \supset \alpha_1$ (MP)

Let us rename the above formulas appeared in the derivation chain sequentially as $\beta_1, \beta_2, \ldots, \beta_9$. Then $val(\langle\beta_1, \ldots, \beta_9\rangle\ ZD(\alpha_3 \supset \alpha_1))$

$$= 1 *_m 1 *_m a *_m 1 *_m a *_m a *_m a *_m 1 *_m a = b.$$

2.4 Consistency, Inconsistency and Equivalence

In this section, we shall see how other meta-logical notions, like consistency of a set of wffs, inconsistency of a set of wffs and equivalence between two sets of wffs, are treated in the context of graded consequence (Chakraborty and Basu 1997, 1999).

2.4.1 Consistency and Inconsistency

Notion of inconsistency, in graded context, is a generalization of classical notion of absolute inconsistency, which in classical context is equivalent to the notion of negation inconsistency.

Definition 2.10 For any set $X \subseteq F$, (absolute) inconsistency degree of X, denoted as $INCONS_{abs}(X)$, is dependent on the notion of graded consequence $\mid\!\sim$, and is given by $INCONS_{abs}(X) = \inf_\alpha gr(X \mid\!\sim \alpha)$.

The above definition is based upon the fuzzy version of the meta-logical statement that 'X is inconsistent if and only if for all wff α, $X \vdash \alpha$'.

Development of the meta-logical notions, proposed in this chapter, is independent of the specification of the object language. One can think of defining the notion of consistency as negation of the notion of absolute inconsistency. In that case, the meta-language needs to be endowed with a connective for negation. We, for the time being, are not bringing in any additional connective, other than the ones already involved in the meta-language.[3] The notion of consistency, thus, would be defined from semantic angle, where a set of formulae X is consistent if and only if X has a model.

Definition 2.11 The degree of consistency of a set X of formulae, relative to a semantic base $\{T_i\}_{i\in I}$ of fuzzy sets of formulae, is given as $CD(X) = \sup_{i\in I}(\inf_{\alpha\in X} T_i(\alpha))$.

Theorem 2.4 *For $X \subseteq Y$ (i) $INCONS_{abs}(X) \leq INCONS_{abs}(Y)$, (ii) $CD(Y) \leq CD(X)$.*

Proof (i) We know $INCONS_{abs}(X) = \inf_\alpha gr(X \mid\!\sim \alpha)$. By virtue of Theorem 2.1, given the graded consequence relation $\mid\!\sim$, there exists a collection of fuzzy sets $\{T_i\}_{i\in I}$ such that $INCONS_{abs}(X) = \inf_\alpha \inf_i\{\inf_{x\in X} T_i(x) \rightarrow_m T_i(\alpha)\}$. As $X \subseteq$

[3] In Chap. 7 in order to modify a notion of implicative consequence relation, developed by Rodríguez et al. (2003), in the light of GCT, we bring in the connective negation in the meta-language.

Y, $\inf_{x\in X} T_i(x) \geq \inf_{x\in Y} T_i(x)$. By (r5) $\inf_{x\in X} T_i(x) \to_m T_i(\alpha) \leq \inf_{x\in Y} T_i(x) \to_m T_i(\alpha)$ holds for any T_i. Hence, $INCONS_{abs}(X) \leq INCONS_{abs}(Y)$.
(ii) $CD(Y) \leq CD(X)$ is also straightforward from $\inf_{x\in X} T_i(x) \geq \inf_{x\in Y} T_i(x)$. □

Proposition 2.9 *If $gr(X \mid\sim \alpha) = 1$, then $CD(X) = CD(X \cup \{\alpha\})$.*

Proof By Theorem 2.1, there is a collection $\{T_i\}_{i\in I}$ such that $gr(X \approx_{\{T_i\}_{i\in I}} \alpha) = 1$. That is, for any $i \in I$, $\inf_{x\in X} T_i(x) \leq T_i(\alpha)$. Hence, $\sup_{i\in I} \inf_{x\in X} T_i(x) \leq \sup_{i\in I} T_i(\alpha)$, i.e. $\sup_{i\in I} \inf_{x\in X} T_i(x) \leq \sup_{i\in I} \inf_{x\in X} T_i(x) \vee \sup_{i\in I} T_i(\alpha) = \sup_{i\in I} \inf_{x\in X\cup\{\alpha\}} T_i(x)$.
So, combining this with Theorem 2.4, we have $CD(X \cup \{\alpha\}) = CD(X)$. □

Proposition 2.10 *If $INCONS_{abs}(X) = 1$, then $CD(X) \leq \inf_\beta CD(\{\beta\})$.*

Proof As $INCONS_{abs}(X) = 1$, for all β, for all $i \in I$, $\inf_{x\in X} T_i(x) \leq T_i(\beta)$, i.e. $\sup_i \inf_{x\in X} T_i(x) \leq \sup_i T_i(\beta)$. That is, $CD(X) \leq CD(\{\beta\})$ for any β. Hence the result. □

Proposition 2.11 *Given a graded consequence relation $\mid\sim$, and corresponding collection of fuzzy sets $\{T_i\}_{i\in I}$, $CD(X) \to_m \inf_\alpha \inf_{i\in I} T_i(\alpha) \leq INCONS_{abs}(X)$.*

Proof $\inf_{x\in X} T_i(x) \leq \sup_i \inf_{x\in X} T_i(x) = CD(X)$.
So, for any $i \in I$, by (r5) $CD(X) \to_m T_i(\alpha) \leq \inf_{x\in X} T_i(x) \to_m T_i(\alpha)$.
i.e. $\inf_i[CD(X) \to_m T_i(\alpha)] \leq \inf_i[\inf_{x\in X} T_i(x) \to_m T_i(\alpha)]$. Using (r7), we have $CD(X) \to_m \inf_i T_i(\alpha) \leq \inf_i[\inf_{x\in X} T_i(x) \to_m T_i(\alpha)]$. That is, for any $\alpha \in F$, $CD(X) \to_m \inf_i T_i(\alpha) \leq INCONS_{abs}(X)$. Hence, again with the use of (r7), we can show $CD(X) \to_m \inf_\alpha \inf_i T_i(\alpha) \leq INCONS_{abs}(X)$. □

Corollary 2.2 *$CD(X) = 0$ implies $INCONS_{abs}(X) = 1$.*

Proposition 2.12 *If in particular L is a complete Heyting algebra, i.e. $*_m = \wedge$, then $CD(X) \leq gr(X \mid\sim \alpha)$ implies $gr(X \mid\sim \alpha) = 1$.*

Proof Let $gr(X \mid\sim \alpha) \geq a$ for some $\alpha \in F$, and $a \geq CD(X)$. Then by Theorem 2.1, $\inf_i\{\inf_{x\in X} T_i(x) \to_m T_i(\alpha)\} \geq a$. That is, for each i, $\inf_{x\in X} T_i(x) \to_m T_i(\alpha) \geq a$. So, $a \wedge \inf_{x\in X} T_i(x) \leq \inf_{x\in X} T_i(x) \wedge \inf_{x\in X} T_i(x) \to_m T_i(\alpha) \leq T_i(\alpha)$ as $b \wedge (b \to_m c) \leq c$ holds in a Heyting algebra. Also by assumption $CD(X) = \sup_i \inf_{x\in X} T_i(x) \leq a$. So, for each i, $\inf_{x\in X} T_i(x) \leq a$. Then $\inf_{x\in X} T_i(x) = \inf_{x\in X} T_i(x) \wedge a \leq T_i(\alpha)$ holds for each i. That is, $\inf_i\{\inf_{x\in X} T_i(x) \to_m T_i(\alpha)\} = 1$. Hence the result. □

2.4.2 Equivalence

Classically, two sets of formulae are considered to be equivalent if both of them have the same set of consequences (Tarski 1956). In the present context, the notion of equivalence between two sets of formulae is captured by the following definition.

Definition 2.12 Given two sets $X, Y \subseteq F$, X is said to be equivalent to Y, denoted as $X \equiv Y$, if and only if $gr(X \mid\!\sim \alpha) = gr(Y \mid\!\sim \alpha)$ for any α.

Proposition 2.13 *$X \cup Y \equiv X$ if and only if $gr(X \mid\!\sim \beta) = 1$ for any $\beta \in Y$.*

Proof Let $X \cup Y \equiv X$, i.e. $gr(X \cup Y \mid\!\sim \alpha) = gr(X \mid\!\sim \alpha)$ for any α.
For any $\beta \in Y$, $gr(X \cup Y \mid\!\sim \beta) = 1$. Hence $gr(X \mid\!\sim \beta) = 1$ for any $\beta \in Y$.
Conversely, let $gr(X \mid\!\sim \beta) = 1$ for any $\beta \in Y$.
By (GC2), $gr(X \mid\!\sim \alpha) \leq gr(X \cup Y \mid\!\sim \alpha)$. Also, because of the assumption and (GC3), we have $gr(X \mid\!\sim \alpha) \geq gr(X \cup Y \mid\!\sim \alpha) *_m \inf_{\beta \in Y} gr(X \mid\!\sim \beta)$, which gives $gr(X \mid\!\sim \alpha) \geq gr(X \cup Y \mid\!\sim \alpha)$. Hence the result. □

Proposition 2.14 *$X \equiv Y$ if and only if $gr(X \mid\!\sim \alpha) = 1$ for all $\alpha \in Y$ and $gr(Y \mid\!\sim \beta) = 1$ for all $\beta \in X$.*

Proof The 'only if' part follows directly from the definition of $\equiv$ and (GC1). For the converse let, $gr(X \mid\!\sim \alpha) = 1$ for all $\alpha \in Y$ and $gr(Y \mid\!\sim \beta) = 1$ for all $\beta \in X$. So, by Proposition 2.13 we have $X \cup Y \equiv X$ and $Y \cup X \equiv Y$. Hence $X \equiv Y$. □

Proposition 2.15 *For $X, Y, Z, W \subseteq F$, if $X \equiv Y$ and $Z \equiv W$, then $X \cup Z \equiv Y \cup W$.*

Corollary 2.3 *For any $X, Y, Z \subseteq F$, if $X \equiv Y$, then $X \cup Z \equiv Y \cup Z$.*

Proposition 2.16 *If $X \subseteq Y \subseteq Z$ and $X \equiv Z$, then $Y \equiv Z$.*

2.5 Level Cuts of a Graded Consequence Relation

Any fuzzy set (and fuzzy relation) can be decomposed (Klir and Yuan 2006) in terms of its level cuts, known as a-cuts for any a belonging to the value set, the range set of the membership function of a fuzzy set. Moreover, study of the level cuts generated from a fuzzy set often provides the meaning of the fuzzy set from a crisp angle. Thus study of the level cuts of a fuzzy set is one of the important aspects of investigation. The notion of graded consequence relation, being a binary fuzzy relation, may thus be studied from the perspective of its level cuts as well. However, a fuzzy set or a fuzzy relation can be represented by its level cuts only when the value set is a complete linear lattice (see Klir and Yuan 2006). So, though level cuts are significant notions fuzzy sets or fuzzy relations do not reduce to the study of their cuts.

Definition 2.13 Given a graded consequence relation $\mid\!\sim$ over a complete residuated lattice $(L, \wedge, \vee, *, \rightarrow_m, 0, 1)$, for each $a \in L$, an operator C_a from $P(F)$ to $P(F)$ is defined as $C_a(X) = \{\beta : gr(X \mid\!\sim \beta) \geq a\}$. C_a is called the a-cut of $\mid\!\sim$.

Proposition 2.17 *For any $a \in L$, and $X \subseteq F$, $X \subseteq C_a(X)$.*

Proof For any $\alpha \in X$, $gr(X \mid\!\sim \alpha) = 1 \geq a$; that is, $\alpha \in C_a(X)$. □

Proposition 2.18 *For $a, b \in L$, if $a \leq b$, then $C_b(X) \subseteq C_a(X)$.*

Proof It is immediate as $\alpha \in C_b(X)$ implies $gr(X \mid\!\sim \alpha) \geq b \geq a$. □

Proposition 2.19 *For any $a \in L$, and $X, Y \subseteq F$, if $X \subseteq Y$, then $C_a(X) \subseteq C_a(Y)$.*

Proof If $\alpha \in C_a(X)$, (GC2) ensures $a \leq gr(X \mid\!\sim \alpha) \leq gr(Y \mid\!\sim \alpha)$. □

Proposition 2.20 *For any $a, b \in L$, and $X \subseteq F$, $C_a C_b(X) \subseteq C_{a *_m b}(X)$.*

Proof Let $\alpha \in C_a C_b(X)$.
Now by (GC3), $gr(X \mid\!\sim \alpha) \geq gr(X \cup C_b(X) \mid\!\sim \alpha) *_m \inf_{\beta \in C_b(X)} gr(X \mid\!\sim \beta)$.
$= gr(C_b(X) \mid\!\sim \alpha) *_m \inf_{\beta \in C_b(X)} gr(X \mid\!\sim \beta)$ (by Proposition 2.17)
$\geq a *_m b$ (by assumption $\alpha \in C_a C_b(X)$ and definition of C_b)
Hence $\alpha \in C_{a *_m b}(X)$. □

Corollary 2.4 $C_a C_a(X) \subseteq C_{a *_m a}(X)$.

Proposition 2.21 *If $X \subseteq C_a(Y)$, and $Y \subseteq C_b(Z)$, then $X \subseteq C_a C_b(Z)$.*

Proof As $Y \subseteq C_b(Z)$ we have $C_a(Y) \subseteq C_a C_b(Z)$. Hence $X \subseteq C_a C_b(Z)$. □

Proposition 2.22 $C_a(X \cup Y) \subseteq C_a(X \cup C_b(Y)) \subseteq C_a C_b(X \cup Y)$.

Proof $Y \subseteq C_b(Y)$ and $X \cup Y \subseteq X \cup C_b(Y)$. Hence, $C_a(X \cup Y) \subseteq C_a(X \cup C_b(Y))$.
Also as $Y \subseteq X \cup Y$, $C_b(Y) \subseteq C_b(X \cup Y)$.
Hence $X \cup C_b(Y) \subseteq X \cup C_b(X \cup Y) = C_b(X \cup Y)$.
So, $C_a(X \cup C_b(Y)) \subseteq C_a C_b(X \cup Y)$. □

Proposition 2.23 $C_a(X \cup Y) \subseteq C_a(C_b(X) \cup C_d(Y)) \subseteq C_a C_b C_d(X \cup Y)$.

Proof $X \cup Y \subseteq C_b(X) \cup C_d(Y)$. So, $C_a(X \cup Y) \subseteq C_a(C_b(X) \cup C_d(Y))$.
Now by Proposition 2.22, $C_a(C_b(X) \cup C_d(Y)) \subseteq C_a C_b(X \cup C_d(Y)) \subseteq C_a C_b C_d (X \cup Y)$. □

As lattice meet $\wedge$ is a special kind of monoidal composition, and is the only one which has idempotent property, below we would explore the special cases of some of the above propositions when $* = \wedge$. That is, in the following sequel, we are going to consider the special kind of complete residuated lattice $(L, \wedge, \vee, \rightarrow_m, 0, 1)$, known as complete Heyting algebra, or complete pseudo-Boolean algebra.

Proposition 2.24 *For $a \leq b$, $C_b C_a(X) = C_a(X)$.*

Proof From Proposition 2.20, we have $C_b C_a(X) \subseteq C_{b \wedge a}(X) = C_a(X)$.
Also, by Proposition 2.17, $C_a(X) \subseteq C_b C_a(X)$. Hence $C_a(X) = C_b C_a(X)$. □

Corollary 2.5 *In particular, $C_a C_a(X) = C_a(X)$.*

Proposition 2.25 *$X \subseteq C_a(Y)$ and $Y \subseteq C_a(Z)$ imply $X \subseteq C_a(Z)$.*

Proof The proof is immediate following Propositions 2.21 and 2.24. □

Proposition 2.26 $C_a(X \cup Y) = C_a(X \cup C_a(Y))$.

Proof The proof directly follows from Propositions 2.22 and 2.24. □

Proposition 2.27 $C_a(X \cup Y) = C_a(C_a(X) \cup C_a(Y))$.

Proof The proof is straightforward from Propositions 2.23 and 2.24. □

So, from Propositions 2.17, 2.19 and 2.24, it comes as a corollary that when we consider the meta-level algebraic structure for a graded consequence relation $\mid\!\sim$ to be a complete pseudo-Boolean algebra, C_a turns out to be a consequence operator in the classical sense. More specifically, in Tarski's axiomatization (Tarski 1956) for the classical notion of consequence, the initial properties, which were imposed on a consequence operator C, are overlap ($X \subseteq C(X)$), idempotence ($C(X) = C(C(X))$) and compactness ($C(X) = \cup_{Y \subseteq X,\ Y finite} C(Y)$). The monotonicity property (if $X \subseteq Y$, then $C(X) \subseteq C(Y)$) follows from the above three properties of a consequence operator in the sense of Tarski (1956). Let us explore the connection of compactness in the context of a graded consequence relation and its level cuts. In Sect. 2.1, the notion of compact fuzzy relation (Definition 2.4) is defined. If a graded consequence relation satisfies the criterion of Definition 2.4, then we call the relation a compact graded consequence relation. Below we define a notion of strong compactness.

Definition 2.14 A compact graded consequence relation $\mid\!\sim$ is said to satisfy the strong compactness if for any $X \subseteq F$ and $\alpha \in F$, $\sup_{X' \subseteq X,\ X' finite} gr(X' \mid\!\sim \alpha)$ attains at some finite $X' \subseteq X$.

It is to be noted here that if a compact graded consequence relation $\mid\!\sim$ satisfies the strong compactness condition, then because of (GC2) we have $gr(X \mid\!\sim \alpha) = gr(X' \mid\!\sim \alpha)$ for some finite $X' (\subseteq X)$.

Theorem 2.5 *Let $\mid\!\sim$ be a graded consequence relation. Then, the following statements are equivalent:*
(i) $C_a(X) = \cup_{X' \subseteq X,\ X' finite} C_a(X')$ for any $a \in L$.
(ii) $\mid\!\sim$ is a strong compact graded consequence relation.

Proof Let us assume (i), and let for some $X \subseteq F$ and $\beta \in F$, $gr(X \mid\!\sim \beta) = a$. Then $\beta \in C_a(X)$. So, by (i) there is some finite $X' \subseteq X$ such that $\beta \in C_a(X')$. That is, $gr(X' \mid\!\sim \beta) \geq a$. So, $a = gr(X \mid\!\sim \beta) \leq gr(X' \mid\!\sim \beta) \leq \sup_{X' \subseteq X,\ X' finite} gr(X' \mid\!\sim \beta)$. Hence $\mid\!\sim$ is a compact graded consequence relation. Moreover, using (GC2), we have $gr(X' \mid\!\sim \beta) \leq gr(X \mid\!\sim \beta) = a$. This inequality holds for any finite $X' (\subseteq X)$. Hence, $\sup_{X' \subseteq X,\ X' finite} gr(X' \mid\!\sim \beta) \leq gr(X \mid\!\sim \beta)$.
Hence $a = gr(X \mid\!\sim \beta) = gr(X' \mid\!\sim \beta) = \sup_{X' \subseteq_f X} gr(X' \mid\!\sim \beta)$ for some $X' \subseteq_f X$. So, $\mid\!\sim$ satisfies strong compactness condition.

Let us assume (ii), and let $\beta \in C_a(X)$ for some $a \in L$, and $X \subseteq F$.

Then $gr(X \mid\!\sim \beta) \geq a$. Now as $\mid\!\sim$ is a strong compact graded consequence relation, there is some finite $X'(\subseteq X)$ such that $gr(X' \mid\!\sim \beta) = gr(X \mid\!\sim \beta) \geq a$.
Hence $\beta \in C_a(X') \subseteq \cup_{X' \subseteq X,\ X' finite} C_a(X')$. Also by Proposition 2.19, $C_a(X') \subseteq C_a(X)$ for any finite $X'(\subseteq X)$. So, we have $\cup_{X' \subseteq X,\ X' finite} C_a(X') \subseteq C_a(X)$.
Thus we obtain (i). □

2.6 Appendix

In this appendix, some prerequisites for residuated lattice discussed in Sect. 2.1 are presented. We assume that the readers are acquainted with the notions of intuitive set theory and relations. We start with a set S and a binary relation R over S.

Definition 2.15 A binary relation R over S is said to be a *partial order relation* if the following conditions hold:

(i) R is *reflexive*, which means for all $x \in S$, $R(x, x)$.
(ii) R is *anti-symmetric*. That is, for any $x, y \in S$, if $R(x, y)$ and $R(y, x)$, then $x = y$.
(iii) R is *transitive*. That is, for any $x, y, z \in S$, if $R(x, y)$ and $R(y, z)$, then $R(x, z)$.

For a partial order relation R over S, the pair (S, R) is said to be a *partially ordered set*, in short pos.

Definition 2.16 Let (S, R) be a pos, and $x, y \in S$.

(i) For a $z \in S$, if $R(z, x)$ and $R(z, y)$, then z is said to be a *lower bound* of x and y.
(ii) For a $z \in S$, if $R(x, z)$ and $R(y, z)$, then z is said to be an *upper bound* of x and y.
(iii) If for a lower bound z of $x, y \in S$, there is no other lower bound z' such that $R(z, z')$, then z is said to be the *greatest lower bound*, in short glb, of x and y. In such case, we write $glb(x, y)$ exists.
(iv) If for an upper bound z of $x, y \in S$, there is no other upper bound z' such that $R(z', z)$, then z is said to be the *least upper bound*, in short lub, of x and y. In such case, we write $lub(x, y)$ exists.
(v) Given $X \subseteq S$, an element $z \in S$ is said to be a lower bound of X if for any $x \in X$, $R(z, x)$. For a lower bound z, if there exists no other lower bound z' of X such that $R(z, z')$, then z is said to be the greatest lower bound of X.
(vi) Given $X \subseteq S$, an element $z \in S$ is said to be an upper bound of X if for any $x \in X$, $R(x, z)$. For an upper bound z, if there exists no other upper bound z' of X such that $R(z', z)$, then z is said to be the least upper bound of X.

Definition 2.17 A pos (S, R) is said to be a *lattice* if for any $x, y \in S$, $lub(x, y)$ and $glb(x, y)$ exist.

Note 2.3 1. Given a pos (S, R) usually $glb(x, y)$ is denoted as $x \wedge y$, and the $lub(x, y)$ is denoted as $x \vee y$.

2. The above definition of lattice also has an equivalent formulation in terms of an algebraic structure $(S, \wedge, \vee)$ where $\wedge$ and $\vee$ are two functions from $S \times S \mapsto S$. An algebraic structure $(S, \wedge, \vee)$ turns out to be a lattice if the following four conditions hold:

   (i) $\wedge$ and $\vee$ are commutative.
   (ii) $\wedge$ and $\vee$ are associative.
   (iii) $\wedge$ and $\vee$ have idempotent property, i.e. $x \wedge x = x$ and $x \vee x = x$.
   (iv) $\wedge$ and $\vee$ satisfy absorption property, i.e. $x \wedge (x \vee y) = x$ and $x \vee (x \wedge y) = x$.

Definition 2.18 1. A pos (S, R) is said to be a *complete lattice* if for any $X \subseteq S$, $glb(X)$ and $lub(X)$ exist.

2. A lattice (S, R) is said to be a *bounded lattice* if for any $x \in S$, there exist $l, u \in S$ such that $R(l, x)$ and $R(x, u)$, and they are, respectively, called as the *least element* and *greatest element* of S.
3. A lattice (S, R) is said to be a *linear* if for any $x, y \in S$, x and y are *comparable*, i.e. either $R(x, y)$ or $R(y, x)$.

Definition 2.19 An algebraic structure $(S, *)$, where $*$ is a binary operation on S, is said to be a *monoid* if the following properties hold:

(i) $*$ is commutative.
(ii) $*$ is associative.
(iii) There is an element $e \in S$ such that e is the *identity* element of S, i.e. for any $x \in S, x * e = x$.

Example 2.5 1. $([0, 1], \leq)$ is a complete, bounded lattice where $\leq$ is a linear order relation.

2. $(P(S), \subseteq)$ is a complete bounded lattice, but $\subseteq$ is not linear.

Definition 2.20 A t-norm is a binary function from $i : [0, 1] \times [0, 1] \mapsto [0, 1]$ satisfying the following properties:

(i) $i(x, y) = i(y, x)$
(ii) $i(x, i(y, z)) = i(i(x, y), z))$
(iii) For $x \leq y, i(x, z) \leq i(y, z)$
(iv) $i(x, 1) = x$.

Note 2.4 1. The notion of t-norm generalizes the notion of classical conjunction, the set-theoretical intersection operation, and more generally speaking, the notion of greatest lower bound. Similarly, there is a notion, called t-conorm, which generalizes the notion of least upper bound, and in particular the notion of classical disjunction and set-theoretical union operation.

2. Lattice meet $(\wedge)$ and join $(\vee)$ operations are, respectively, t-norm and t-conorm.
3. $([0, 1], i)$ forms a monoid.

Definition 2.21 A t-conorm is a binary function from $u : [0, 1] \times [0, 1] \mapsto [0, 1]$ satisfying the following properties:

(i) $u(x, y) = u(y, x)$,
(ii) $u(x, i(y, z)) = u(i(x, y), z))$,
(iii) For $x \leq y$, $u(x, z) \leq u(y, z)$ and
(iv) $u(x, 0) = x$.

In a similar fashion, classical implication on $\{0, 1\}$ is generalized over $[0, 1]$ in fuzzy context. Among different kinds of fuzzy implication, the following one which is defined based on a t-norm is of special interest.

Definition 2.22 Given the monoid $([0, 1], i)$ over the complete lattice $[0, 1]$, a function $R : [0, 1] \times [0, 1] \mapsto [0, 1]$ is said to be an R-implication if for any $x, y \in [0, 1]$, $R(x, y) = \sup\{z : i(x, z) \leq y\}$.

The above-defined R-implication is of special interest because it has a special property $i(x, z) \leq y$ iff $x \leq R(z, y)$, known as *residuation property*. The monoid $([0, 1]i)$ along with the respective R-implication defined by i generates a special kind of lattices, called residuated lattice. Details regarding residuated lattices are presented in this chapter.

References

Castro, J.L., Trillas, E., Cubillo, S.: On consequence in approximate reasoning. J. Appl. Non Class. Log. **4**(1), 91–103 (1994)
Chakraborty, M.K.: Use of fuzzy set theory in introducing graded consequence in multiple valued logic. In: Gupta, M.M., Yamakawa, T. (eds.) Fuzzy Logic in Knowledge-Based Systems, Decision and Control, pp. 247–257. Elsevier Science Publishers, (B.V) North Holland (1988)
Chakraborty, M.K.: Graded consequence: further studies. J. Appl. Non Class. Log. **5**, 227–237 (1995)
Chakraborty, M.K., Basu, S.: Graded consequence and some metalogical notions generalized. Fundam. Inf. **32**, 299–311 (1997)
Chakraborty, M.K., Basu, S.: Introducing grade to some metalogical notions. In: Dubois, D., Klement, E.P., Prade, H. (eds.) Fuzzy Sets, Logics and Reasoning About Knowledge, pp. 85–99. Kluwer Academic Publisher, Netherlands (1999)
Dillworth, R.P., Ward, M.: Residuated lattices. Trans. Amer. Math. Soc. **45**, 335–354 (1939)
Galatos, N., Jipsen, P., Kowalski, T., Ono, H.: Residuated lattices: an algebraic glimpse at substructural logics. In: Studies in Logic and the Foundations of Mathematics, vol. 151. Elsevier (2007)
Goguen, J.A.: L-fuzzy sets. J. Math. Anal. Appl. **18**(1), 145174 (1967)
Hájek, P.: Metamathematics of Fuzzy Logic. Kluwer Academic Publishers, Dordrecht (1998)
Klir, G.J., Yuan, B.: Fuzzy Sets and Fuzzy Logic: Theory and Applications. Prentice Hall of India Private Limited, New Delhi (2006)
Minsky, M: A framework for representation knowledge, MIT-AI Laboratory Memo 306, June, 1974. In: Winston, P. (ed.) The Psychology of Computer Vision. McGraw-Hill (1975) (Reprinted). In: Haugeland, J. (ed.) Mind Design. MIT Press (1981) (Shorter versions). In: Collins, A., Smith, E.E. (eds.) Cognitive Science. Morgan-Kaufmann (1992). ISBN 55860-013-2

Ono, H.: Substructural logics and residuated lattices - an introduction. In: Hendricks, V.F., Malinowski, J. (eds.) Trends in Logic, vol. 20, pp. 177–212. Kluwer Academic Publishers, Netherlands (2003)

Rodríguez, R.O., Esteva, F., Garcia, P., Godo, L.: On implicative closure operators in approximate reasoning. Int. J. Approx. Reason. **33**, 159–184 (2003)

Shoesmith, D.J., Smiley, T.J.: Multiple Conclusion Logic. Cambridge University Press, Cambridge (1978)

Tarski, A.: Methodology of Deductive Sciences. Logic, Semantics, Metamathematics, pp. 60–109 (1956)

Zadeh, L.A.: Fuzzy sets. Inf. Control **8**, 338–353 (1965)

Chapter 3
Introducing Negation (¬) in the Object Language of the Theory of Graded Consequence

Abstract In this chapter, a logical connective negation (¬) is introduced in the object language, and the set of axioms for characterizing a graded consequence relation is extended in the presence of ¬. An axiomatic approach to the notion of graded inconsistency is also introduced, and an equivalence between the notions of graded consequence and graded inconsistency is established. In continuation to the study of level cuts of a graded consequence relation, presented in Chap. 2, a few results considering ¬ in the language are also presented. Reflection of the newly added axioms in the meta-level algebraic structure is explored in an extended algebraic structure of a complete residuated lattice. The properties of this structure, called GC(¬)-algebra, are studied as well.

3.1 Extending the Notion Graded Consequence Relation in the Presence of Negation (¬)

The notions of classical consequence and consistency/inconsistency are equivalent to each other in the sense that assuming one as a primitive concept the other can be obtained. In Surma (1981), the complete axiomatization of the classical notions of consequence and consistency, the object language being endowed with two primitive connectives ¬ (negation) and ⊃ (implication), is proposed. Usually, the notion of inconsistency, and hence consistency, is understood in terms of the notion of consequence and its interrelation with object language negation. So, our present concern is only with ¬ in the object language. A set X of wffs is classically inconsistent if and only if there is a formula α such that α and $\neg\alpha$ are both consequences of X. A set is consistent if and only if it is not inconsistent. In this regard, below we first present a modified set of axioms, both for classical consequence and inconsistency (Dutta 2011), in terms of ¬ only.

Definition 3.1 (*Classical consequence axioms enhanced*) An operator $C : P(F) \mapsto P(F)$ is said to be a consequence operator if and only if it satisfies conditions (C1)–(C5).

M. K. Chakraborty and S. Dutta, *Theory of Graded Consequence*, Logic in Asia: Studia Logica Library, https://doi.org/10.1007/978-981-13-8896-5_3

(C1) $X \subseteq C(X)$.
(C2) If $X \subseteq Y$, then $C(X) \subseteq C(Y)$.
(C3) $C(C(X)) = C(X)$.
(C4) $C(\{\alpha, \neg\alpha\}) = F$.
(C5) $C(X \cup \{\alpha\}) \cap C(X \cup \{\neg\alpha\}) = C(X)$.

Note that (C1), (C2) and (C3) have been presented in Chap. 1; (C4), (C5) are added when negation is present in the object language. Condition (C4) is called explosiveness. It says that every wff follows from the inconsistent set $\{\alpha, \neg\alpha\}$. In fact, it can be shown that every formula follows from any inconsistent set. Condition (C5) is called reasoning by cases. It says that if a wff follows from both the sets $X \cup \{\alpha\}$ and $X \cup \{\neg\alpha\}$ then the formula follows from X itself.

In this chapter, by a consequence we shall mean consequence enhanced with ¬. It is to be noted that at this stage no semantic condition is imposed on ¬.

Definition 3.2 (*Classical inconsistency axioms*) *Incons* is a unary relation defined on the power set $P(F)$ of F (i.e. a subset of $P(F)$), satisfying the following conditions:

(Incons1) If $X \subseteq Y$ and $X \in Incons$ then $Y \in Incons$.
(Incons2) If $X \cup Y \in Incons$ and for each $\alpha \in Y$, $X \cup \{\neg\alpha\} \in Incons$ then $X \in Incons$.
(Incons3) If $\alpha \in F$ then $\{\alpha, \neg\alpha\} \in Incons$.

$X \in Incons$ stands for the meta-linguistic sentence 'X is inconsistent'.

Theorem 3.1 *1. Given C, let Incons be defined by Incons* $= \{X \subseteq F : C(X) = F\}$. *Then Incons satisfies* (*Incons*1) *to* (*Incons*3) *axioms.*
2. Given Incons, let C be defined by, $C(X) = \{\alpha \in F : X \cup \{\neg\alpha\} \in Incons\}$. *Then C satisfies* $(C1)$–$(C5)$ *axioms.*

In order to prove this theorem, we first present one lemma.

Lemma 3.1 *In the presence of (C1) and (C2), the following two conditions are equivalent:*

(i) $Z \subseteq C(X)$ *implies* $C(X \cup Z) \subseteq C(X)$.
(ii) $C(C(X)) = C(X)$.

Proof Let us first prove (i) implies (ii).
$C(X) \subseteq C(C(X))$ [By (C1) and (C2)] …(a)
Now as $X \subseteq C(X)$ [By (C1)] (i) implies $C(X \cup C(X)) \subseteq C(X)$.
That is, $C(C(X)) \subseteq C(X)$ [Since by (C1) $X \subseteq C(X)$] …(b)
Combining (a) and (b), we get $C(C(X)) = C(X)$.
Now to prove (ii) implies (i), let us assume (ii).
Let $\beta \in C(X)$ for all $\beta \in Z$, i.e. $Z \subseteq C(X)$.
Also $X \subseteq C(X)$. [By (C1)]
Then $X \cup Z \subseteq C(X)$.
Therefore, by (C2), $C(X \cup Z) \subseteq C(C(X)) = C(X)$. [By (ii)] □

Proof (1) Incons1: Let $X \subseteq Y$ and $X \in Incons$. Hence $C(X) = F$.
Now by (C2), $C(X) \subseteq C(Y)$ implies $C(Y) = F$, i.e. $Y \in Incons$.

Incons2: Let $X \cup Y \in Incons$ and for each $\alpha \in Y, X \cup \{\neg\alpha\} \in Incons$.
Therefore, $C(X \cup Y) = F$ and for each $\alpha \in Y$, $C(X \cup \{\neg\alpha\}) = F$.
Then for all $\alpha \in Y$, $\alpha \in C(X \cup \{\neg\alpha\})$ as well as $\alpha \in C(X \cup \{\alpha\})$.
So, by (C5), $\alpha \in C(X)$.
Hence for each $\alpha \in Y$, $\alpha \in C(X)$, i.e. $Y \subseteq C(X)$.
Therefore, by Lemma 3.1, we have $C(X \cup Y) \subseteq C(X)$.
This implies $F \subseteq C(X)$. Hence $X \in Incons$.

Incons3: As by (C4), for any α, $C(\{\alpha, \neg\alpha\}) = F$ by definition of *Incons* we have $\{\alpha, \neg\alpha\} \in Incons$.
(2)(C1): Let $\alpha \in X$ then $\{\alpha, \neg\alpha\} \subseteq X \cup \{\neg\alpha\}$.
By Incons3, we have $\{\alpha, \neg\alpha\} \in Incons$. Then by Incons1, it can be stated that $X \cup \{\neg\alpha\} \in Incons$. Hence, by definition we can write $\alpha \in C(X)$.
(C2): Let $X \subseteq Y$ and $\alpha \in C(X)$. Then by definition $X \cup \{\neg\alpha\} \in Incons$.
As $X \cup \{\neg\alpha\} \subseteq Y \cup \{\neg\alpha\}$, by Incons1 $Y \cup \{\neg\alpha\} \in Incons$.
Hence, $\alpha \in C(Y)$ and this implies $C(X) \subseteq C(Y)$.
(C3): Let $Y \subseteq C(X)$. Then for all $\alpha \in Y$, $\alpha \in C(X)$, i.e. $X \cup \{\neg\alpha\} \in Incons$.
Using Incons1, we have for all $\alpha \in Y, X \cup \{\neg\beta\} \cup \{\neg\alpha\} \in Incons$ for any β.
Let $\gamma \in C(X \cup Y)$. Then $X \cup Y \cup \{\neg\gamma\} \in Incons$.
That is for all $\alpha \in Y, X \cup \{\neg\gamma\} \cup \{\neg\alpha\} \in Incons$ and $X \cup \{\neg\gamma\} \cup Y \in Incons$.
Hence by Incons2, $X \cup \{\neg\gamma\} \in Incons$. This implies $\gamma \in C(X)$, and hence we have $C(X \cup Y) \subseteq C(X)$.
As in Lemma 3.1 we have already proved that $C(C(X)) = C(X)$ is equivalent to $Y \subseteq C(X)$ implies $C(X \cup Y) \subseteq C(X)$. So, instead of proving $C(C(X)) = C(X)$ it is sufficient to prove $Y \subseteq C(X)$ implies $C(X \cup Y) \subseteq C(X)$.
(C4): By Incons3, we have that $\{\alpha, \neg\alpha\} \in Incons$. So, using Incons1, we can have $\{\alpha, \neg\alpha\} \cup \{\neg\beta\} \in Incons$ for any β.
Hence for any β, $\beta \in C(\{\alpha, \neg\alpha\})$, i.e. $C(\{\alpha, \neg\alpha\}) = F$.
(C5): Let $\beta \in C(X \cup \{\alpha\})$ and $\beta \in C(X \cup \{\neg\alpha\})$.
Then $X \cup \{\alpha\} \cup \{\neg\beta\} \in Incons$ and $X \cup \{\neg\alpha\} \cup \{\neg\beta\} \in Incons$.
Let us consider $Y = \{\alpha\}$.
Then by Incons2, for $\alpha \in Y, X \cup \{\neg\beta\} \cup \{\neg\alpha\} \in Incons$ and $X \cup \{\neg\beta\} \cup Y \in Incons$ imply $X \cup \{\neg\beta\} \in Incons$. Hence $\beta \in C(X)$. □

Inconsistency is usually looked upon as a curse in classical deductive systems. If one arrives at an inconsistency, everything is lost in the sense that everything is inferable, thus making the whole enterprise trivial.

For philosophers like Parmenides, Plato, Bradley and others, consistency is not just an epistemological desideratum, a regulative principle of system building, but an intrinsic feature of reality so much so that whatever is laden with inconsistency is relegated to the level of appearance only. Following this tradition, logicians have so far taken great care to keep inconsistency out of their theories.

The negative attitude towards inconsistency notwithstanding, there are thinkers who do not think that inconsistency dooms the prospect of an intelligible dialogue or of building a non-trivial theory altogether. One of the dominant themes of Wittgenstein's writings on foundation and philosophy of mathematics is to dispel the misconception that contradiction destroys everything that it makes no sense. He holds that if contradiction were found in arithmetic this would not show that it was never a genuine arithmetic, it would only persuade us to change our conception of certainty. He has the foresight to say,
"*Indeed even at this stage, I predict a time when there will be mathematical investigation of calculi containing contradictions, and people will actually be proud of having emancipated themselves from consistency.*" [Wittgenstein (1975), [147]]

In Buddhist tradition great Master Nagarjuna adopted an argumentation method called 'Catuskoti' (Katsura 2008; Chattopadhyay 2015) or 'four corners' in which four possibilities were considered regarding applicability of a property P to an object a, viz. a is P, a is non-P, a is neither P nor non-P, and a is both P and non-P. The fourth one is a clear acceptance of the possibility of actual contradiction (recall Sect. 1.3 of Introduction). In the present times, various approaches of 'inconsistency-tolerant' systems have been proposed. Ours being degree-theoretic approach, in this chapter we have proposed the notion of graded inconsistency (and thereby graded consistency). Relationship of this notion with graded consequence has been worked out in detail. In this way, inconsistency has been accommodated in many-valued context.

As the notion of graded consequence is introduced as a generalized notion of classical consequence, our aim in this chapter is to extend the theory in such a way that the classical result of equivalence between consequence and inconsistency can be retained in the graded context as well. In this regard, we first consider an extension of the notion of graded consequence relation in the presence of a negation $\neg$ in the object language.

In Chap. 2, $\mid\!\sim$ has been introduced as a fuzzy relation between $P(F)$ and F with grades in a complete residuated lattice $(L, \wedge, \vee, *_m, \rightarrow_m, 0, 1)$. By interpreting $gr(X \mid\!\sim \alpha)$, an element of the value set L, as 'the grade to which α is a consequence of X', the postulates (GC1), (GC2) and (GC3) are presented as graded counterparts of overlap, monotonicity and cut condition of the classical notion of consequence. Assuming the presence of negation ($\neg$) in the object language $\mid\!\sim$ is postulated by the following axioms.

Definition 3.3 *(Extended notion of graded consequence)* A fuzzy relation $\mid\!\sim$, which assumes values from a complete residuated lattice $(L, \wedge, \vee, *_m, \rightarrow_m, 0, 1)$, and is defined over $P(F) \times F$ for a set F of wffs, is said to be an extended graded consequence relation if the following axioms are satisfied:

(GC1) If $\alpha \in X$ then $gr(X \mid\!\sim \alpha) = 1$.
(GC2) If $X \subseteq Y$ then $gr(X \mid\!\sim \alpha) \leq gr(Y \mid\!\sim \alpha)$.
(GC3) $inf_{\beta \in Y}\, gr(X \mid\!\sim \beta) *_m gr(X \cup Y \mid\!\sim \alpha) \leq gr(X \mid\!\sim \alpha)$.
(GC4) There is some $k \in L$ and $k > 0$ such that $inf_{\alpha,\beta} gr(\{\alpha, \neg\alpha\} \mid\!\sim \beta) = k$.

(GC5) $gr(X \cup \{\alpha\} \mid\!\sim \beta) *_m gr(X \cup \{\neg\alpha\} \mid\!\sim \beta) *_m c \leq gr(X \mid\!\sim \beta)$, for some $c \in L$ and $c > 0$.

(GC4) and (GC5), respectively, are the graded counterparts of explosiveness condition and reasoning by cases. Through (GC4), it is admitted that from a set of the form $\{\alpha, \neg\alpha\}$ any β can follow to a non-zero degree, but that is not necessarily of the full extent. The above form of (GC5) can equivalently be written as follows. For some $c > 0$, $gr(X \cup \{\alpha\} \mid\!\sim \beta) *_m gr(X \cup \{\neg\alpha\} \mid\!\sim \beta) \leq c \rightarrow_m gr(X \mid\!\sim \beta)$. So, (GC5) generalizes the classical reading that if β follows from $X \cup \{\alpha\}$ and β follows from $X \cup \{\neg\alpha\}$ then β follows from X at least to a degree $c > 0$, where c may not be the topmost element.

In Chap. 2 we have defined $INCONS_{abs}$, the notion of graded inconsistency of a set X in terms of semantics. Here, in Definition 3.4, the notion of inconsistency is presented axiomatically. The two notions are equivalent in classical logic, but not so in the general case. The link between the two notions is shown in Theorems 3.2 and 3.4.

On the other hand, the notion of inconsistency, in the graded context, has been assumed to be a fuzzy subset *INCONS* of $P(F)$. Given any set of formulae X, $INCONS(X)$ is the inconsistency degree of X. With this understanding, as a generalization of the classical notion of inconsistency, *INCONS* is postulated (Chakraborty and Basu 1997; Chakraborty and Dutta 2010; Dutta 2014) by the following axioms.

Definition 3.4 (*Notion of graded inconsistency*) Given a complete residuated lattice $(L, \wedge, \vee, *_m, \rightarrow_m, 0, 1)$, *INCONS* is a fuzzy subset of $P(F)$ satisfying the following axioms:

(INC1) If $X \subseteq Y$ then $INCONS(X) \leq INCONS(Y)$.
(INC2) $\inf_{\alpha \in Y} INCONS(X \cup \{\neg\alpha\}) *_m INCONS(X \cup Y) *_m c \leq INCONS(X)$, for some $c > 0$.
(INC3) There is some $k > 0$ such that $inf_\alpha INCONS(\{\alpha, \neg\alpha\}) = k$.

INCONS is said to be a graded notion of inconsistency, and $INCONS(X) \in L$, for any $X \subseteq F$, denotes the *degree of inconsistency* of X.

It may be observed that (INC1)–(INC3) are, respectively, the many-valued generalizations of (Incons1)–(Incons3). In graded context, (INC1) ensures that the degree of inconsistency of Y would be at least as much as the degree of inconsistency of X for any X, Y such that $X \subseteq Y$. (INC2) generalizes *Incons*2: if $X \cup Y$ is inconsistent and $X \cup \{\neg\alpha\}$ is inconsistent for any $\alpha \in Y$, then X is inconsistent at least to a positive grade c. (INC3) ascertains that for any α, the set $\{\alpha, \neg\alpha\}$ is inconsistent to some non-zero extent.

Definition 3.5 1. Let $\mid\!\sim$ be a fuzzy relation between $P(F)$ and F satisfying (GC1)–(GC5). We define *INCONS*, a fuzzy subset over $P(F)$, as follows:
$INCONS(X) = inf_\alpha gr(X \mid\!\sim \alpha)$.
2. Let *INCONS* be a notion of graded inconsistency following Definition 3.4. Using *INCONS*, a fuzzy relation $\mid\!\sim$ between $P(F)$ and F is defined as follows:
$gr(X \mid\!\sim \alpha) = 1$ if $\alpha \in X$; otherwise

$$= INCONS(X \cup \{\neg\alpha\}), \text{ if } \alpha \text{ does not start with } \neg, \text{ and}$$
$$= INCONS(X \cup \{\neg\alpha\}) = INCONS(X \cup \{\beta\}) \text{ if } \alpha = \neg\beta.$$

The following theorems establish the connection between the notions of graded consequence and graded inconsistency.

Theorem 3.2 *Let $\mid\!\sim$ be a graded consequence relation as given in Definition 3.3, and $INCONS(X)$ be defined following (1) of Definition 3.5. Then INCONS satisfies (INC1)–(INC3).*

Proof Let $\mid\!\sim$ be a graded consequence relation following Definition 3.3.
(INC1): Let $X \subseteq Y$. Then by (GC2), $gr(X \mid\!\sim \alpha) \leq gr(Y \mid\!\sim \alpha)$.
Hence, $\inf_\alpha gr(X \mid\!\sim \alpha) \leq \inf_\alpha gr(Y \mid\!\sim \alpha)$, i.e. $INCONS(X) \leq INCONS(Y)$.

(INC2): From (GC3) we have the following:
$\inf_\alpha gr(X \cup Y \mid\!\sim \alpha) *_m \inf_{\beta\in Y} gr(X \mid\!\sim \beta) \leq \inf_\alpha gr(X \mid\!\sim \alpha)$. ...(i)
Also, from (GC5) for each $\beta \in Y$, for some $c > 0$ we have
$gr(X \cup \{\beta\} \mid\!\sim \beta) *_m gr(X \cup \{\neg\beta\} \mid\!\sim \beta) *_m c \leq gr(X \mid\!\sim \beta)$.
That is, by (GC1), $gr(X \cup \{\neg\beta\} \mid\!\sim \beta) *_m c \leq gr(X \mid\!\sim \beta)$ for each $\beta \in Y$.
So, $\inf_\alpha gr(X \cup \{\neg\beta\} \mid\!\sim \alpha) *_m c \leq gr(X \cup \{\neg\beta\} \mid\!\sim \beta) *_m c \leq gr(X \mid\!\sim \beta)$ for any $\beta \in Y$.
That is, $\inf_{\beta\in Y} \inf_\alpha gr(X \cup \{\neg\beta\} \mid\!\sim \alpha) *_m c \leq \inf_{\beta\in Y} gr(X \mid\!\sim \beta)$ ...(ii)
Hence using (i) and (ii), $\inf_\alpha gr(X \cup Y \mid\!\sim \alpha) *_m \inf_{\beta\in Y} \inf_\alpha gr(X \cup \{\neg\beta\} \mid\!\sim \alpha) *_m c$
$\leq \inf_\alpha gr(X \cup Y \mid\!\sim \alpha) *_m \inf_{\beta\in Y} gr(X \mid\!\sim \beta) \leq \inf_\alpha gr(X \mid\!\sim \alpha)$.
By definition of *INCONS*, we have the following:
$INCONS(X \cup Y) *_m \inf_{\beta\in Y} INCONS(X \cup \{\neg\beta\}) *_m c \leq INCONS(X)$.
(INC3): By (GC4) there exists $k > 0$ such that $inf_{\alpha,\beta} gr(\{\alpha, \neg\alpha\} \mid\!\sim \beta) = k$.
That is, $inf_\alpha INCONS(\{\alpha, \neg\alpha\}) = k$. □

Note 3.1 (i) The definition of *INCONS* may be marked. We have considered $\alpha = \neg\beta$ and α as a formula not starting with $\neg$ separately. This is because, in general, β and $\neg\neg\beta$ may not behave alike.
(ii) It can be shown that the fuzzy relation $\mid\!\sim$ defined in Definition 3.5 satisfies (GC1) and (GC2), but does not satisfy (GC3). So, though from a graded consequence relation $\mid\!\sim$ we can derive *INCONS*, a graded notion of inconsistency (cf. Theorem 3.2), the converse is not true. The equivalence between these notions can be obtained in some special cases (cf. Theorems 3.3 and 3.4).

Theorem 3.3 *Let $c \in L$ be such that for all $x \leq c$, $x *_m c = x$. Let INCONS be a fuzzy set over $P(F)$ such that INCONS satisfies all the axioms INC1 to INC3, and for any $X \subseteq F$ either $INCONS(X) = 1$ or $INCONS(X) \leq c$. Then defining $\mid\!\sim$ according to (2) of Definition 3.3, the obtained $\mid\!\sim$ satisfies all the axioms of a graded consequence relation.*

Proof With the definition of $\mid\!\sim$, given in terms of *INCONS*, we see (GC1) follows immediately from the definition.
(GC2): Let $X \subseteq Y$. If $gr(Y \mid\!\sim \alpha) = 1$ we are done. Let $gr(Y \mid\!\sim \alpha) \neq 1$.
Then $\alpha \notin Y$, and hence $\alpha \notin X$. So, irrespective of the structural form of α, we can write $gr(X \mid\!\sim \alpha) = INCONS(X \cup \{\neg\alpha\})$ and $gr(Y \mid\!\sim \alpha) = INCONS(Y \cup \{\neg\alpha\})$.
By INC1, we know $INCONS(X \cup \{\neg\alpha\}) \leq INCONS(Y \cup \{\neg\alpha\})$.
So, $gr(X \mid\!\sim \alpha) \leq gr(Y \mid\!\sim \alpha)$.

(GC3): We want to show $inf_{\beta \in Y}\, gr(X \mid\!\sim \beta) *_m gr(X \cup Y \mid\!\sim \alpha) \leq gr(X \mid\!\sim \alpha)$. If $gr(X \mid\!\sim \alpha) = 1$, we are done. So, let $gr(X \mid\!\sim \alpha) \neq 1$. That is, $\alpha \notin X$.
Now two cases arise. (i) $\alpha \in Y$ (ii) $\alpha \notin Y$.
(i) Let $\alpha \in Y$. Then $gr(X \cup Y \mid\!\sim \alpha) = 1$.
Hence, $inf_{\beta \in Y}\, gr(X \mid\!\sim \beta) *_m gr(X \cup Y \mid\!\sim \alpha) = inf_{\beta \in Y}\, gr(X \mid\!\sim \beta)$.
$\leq gr(X \mid\!\sim \alpha)$ [since $\alpha \in Y$]

(ii) Let $\alpha \notin Y$. So, $gr(X \cup Y \mid\!\sim \alpha) = INCONS(X \cup Y \cup \{\neg\alpha\})$ and as $\alpha \notin X$, $gr(X \mid\!\sim \alpha) = INCONS(X \cup \{\neg\alpha\})$. Now consider any $\beta \in Y$.
If for all $\beta \in Y, \beta \in X$, then $inf_{\beta \in Y} gr(X \mid\!\sim \beta) = 1$ and $gr(X \cup Y \mid\!\sim \alpha) = gr(X \mid\!\sim \alpha)$ as $X \cup Y = X$. Hence, the required inequality for (GC3) holds.
Let for some β, $\beta \in Y$ and $\beta \notin X$, and hence $gr(X \mid\!\sim \beta) = INCONS(X \cup \{\neg\beta\})$.
So, $\inf_{\beta \in Y} gr(X \mid\!\sim \beta) = \inf_{\beta \in Y, \beta \notin X} gr(X \mid\!\sim \beta)$ as $gr(X \mid\!\sim \beta) = 1$ for $\beta \in X$.
Hence $\inf_{\beta \in Y} gr(X \mid\!\sim \beta) *_m gr(X \cup Y \mid\!\sim \alpha)$
$= \inf_{\beta \in Y} gr(X \mid\!\sim \beta) *_m gr(X \cup Y \mid\!\sim \alpha)$ for $\beta \notin X$.
$= \inf_{\beta \in Y} INCONS(X \cup \{\neg\beta\}) *_m INCONS(X \cup Y \cup \{\neg\alpha\})$
$\leq \inf_{\beta \in Y} INCONS(X \cup \{\neg\alpha\} \cup \{\neg\beta\}) *_m INCONS(X \cup Y \cup \{\neg\alpha\})$ (by INC1)
$= \inf_{\beta \in Y} INCONS(X \cup \{\neg\alpha\} \cup \{\neg\beta\}) *_m INCONS(X \cup Y \cup \{\neg\alpha\}) *_m c$
$\leq INCONS(X \cup \{\neg\alpha\})$. (by INC2)
$= gr(X \mid\!\sim \alpha)$ [since $\alpha \notin X$].
So, combining the cases, $\inf_{\beta \in Y} gr(X \mid\!\sim \beta) *_m gr(X \cup Y \mid\!\sim \alpha) \leq gr(X \mid\!\sim \alpha)$.
(GC4): For any $\beta \neq \alpha, \neg\alpha$, $gr(\{\alpha, \neg\alpha\} \mid\!\sim \beta) = INCONS(\{\alpha, \neg\alpha\} \cup \{\neg\beta\})$
$\geq INCONS(\{\alpha, \neg\alpha\})$ (by INC1).
Hence, $\inf_{\beta \neq \alpha, \neg\alpha} gr(\{\alpha, \neg\alpha\} \mid\!\sim \beta) \geq INCONS(\{\alpha, \neg\alpha\})$.
As for $\beta = \alpha$ or $\beta = \neg\alpha$, we have $gr(\{\alpha, \neg\alpha\} \mid\!\sim \beta) = 1$, combining all possibilities we can claim for any α, $inf_\beta\, gr(\{\alpha, \neg\alpha\} \mid\!\sim \beta) \geq INCONS(\{\alpha, \neg\alpha\})$.
Hence, $inf_{\alpha,\beta}\, gr(\{\alpha, \neg\alpha\} \mid\!\sim \beta) \geq \inf_\alpha INCONS(\{\alpha, \neg\alpha\}) = k > 0$ (by INC3).
So, there is some $k' > 0$, such that $inf_{\alpha,\beta}\, gr(\{\alpha, \neg\alpha\} \mid\!\sim \beta) = k'$.

(GC5): If $\beta \notin X \cup \{\alpha\}$ as well as $\beta \notin X \cup \{\neg\alpha\}$, then
$gr(X \cup \{\alpha\} \mid\!\sim \beta) *_m gr(X \cup \{\neg\alpha\} \mid\!\sim \beta)$
$= INCONS(X \cup \{\alpha\} \cup \{\neg\beta\}) *_m INCONS(X \cup \{\neg\alpha\} \cup \{\neg\beta\})$ (by definition)
$\leq INCONS(X \cup \{\neg\beta\})$ (by INC2 where $Y = \{\alpha\}$)
$= gr(X \mid\!\sim \beta)$ [since $\beta \notin X \cup \{\alpha\}$ implies $\beta \notin X$].
We have to now consider three cases: (i) $\beta \in X \cup \{\alpha\}$ (ii) $\beta \in X \cup \{\neg\alpha\}$.
(iii) $\beta \in X \cup \{\alpha\}, X \cup \{\neg\alpha\}$.

(i) Let $\beta \in X \cup \{\alpha\}$. Then either $\beta \in X$ or $\beta = \alpha$. If $\beta \in X$ then the required inequality follows immediately as $gr(X \mid\sim \beta) = 1$. So let $\beta \notin X$ and $\beta = \alpha$.
$gr(X \cup \{\alpha\} \mid\sim \beta) *_m gr(X \cup \{\neg\alpha\} \mid\sim \beta) = 1 *_m gr(X \cup \{\neg\beta\} \mid\sim \beta)$
$= INCONS(X \cup \{\neg\beta\})$. [since $\beta \notin X \cup \{\neg\beta\}$]
$= gr(X \mid\sim \beta)$ [since $\beta \notin X$]
For case (ii), i.e. $\beta \notin X$ and $\beta = \neg\alpha$, we now proceed as follows.
$gr(X \cup \{\alpha\} \mid\sim \beta) *_m gr(X \cup \{\neg\alpha\} \mid\sim \beta) = gr(X \cup \{\alpha\} \mid\sim \neg\alpha) *_m 1$
$= INCONS(X \cup \{\alpha\})$ [since $\neg\alpha \notin X \cup \{\alpha\}$]
$= gr(X \mid\sim \neg\alpha)$ [since $\neg\alpha \notin X$]
$= gr(X \mid\sim \beta)$.
(iii) $\beta \in X \cup \{\alpha\}, X \cup \{\neg\alpha\}$ imply $\beta \in X$. Hence $gr(X \mid\sim \beta) = 1$, and the inequality holds trivially. □

Theorem 3.4 *1. Let a graded consequence relation $\mid\sim$ following Definition 3.3 with $c = 1$ in (GC5) be given. Let $INCONS(X) = inf_\alpha gr(X \mid\sim \alpha)$. Then INCONS satisfies (INC1)–(INC3) with the restriction of $c = 1$ in (INC2).*
2. Given INCONS with $c = 1$ in (INC2), let us define $\mid\sim$ following Definition 3.5. Then $\mid\sim$ is a graded consequence relation following Definition 3.3 with the restriction of $c = 1$ in (GC5).

Proof 1. We only need to check INC2 considering $c = 1$, as the rest are the same as Theorem 3.2. For INC2 also it is enough to check that putting $c = 1$, in the previous proof of INC2 of Theorem 3.2 we can obtain the required proof.
2. Let *INCONS* satisfy INC1 to INC3 with the restriction of $c = 1$ in INC2. Then with the definition of $\mid\sim$, given in terms of *INCONS*, proofs of GC1, GC2, GC4 and GC5 follow the same steps as given in the proof of Theorem 3.3. For GC3 also, putting $c = 1$, we can obtain the proof directly from the proof of Theorem 3.3.

Before entering into the discussion on what do these properties semantically entail on the meta-level algebraic structure, here we shall present a few results (Chakraborty and Dutta 2010) in continuation of the study of level cuts made in Sect. 2.5 of Chap. 2. Recall that for any $a \in L$, the a-cut C_a of a graded consequence relation $\mid\sim$ is defined by $C_a(X) = \{\alpha : gr(X \mid\sim \alpha) \geq a\}$.

Theorem 3.5 *Let $\mid\sim$ be a graded consequence relation for which there is a $k \neq 0$ such that $inf_{\alpha,\beta}\, gr(\{\alpha, \neg\alpha\} \mid\sim \beta) = k$. Then for each $a \in L$, if $a \leq k$ then $C_a(\{\alpha, \neg\alpha\}) = F$ for any α and if $a > k$ or a is non-comparable with k then there is some α, for which $C_a(\{\alpha, \neg\alpha\}) \neq F$.*

Proof Let $inf_{\alpha,\beta}\, gr(\{\alpha, \neg\alpha\} \mid\sim \beta) = k > 0$.
For any $a \in L$, either $a \leq k$ or $a > k$ or a is non-comparable with k.
(I) Let $a \leq k$. Then $inf_{\alpha,\beta}\, gr(\{\alpha, \neg\alpha\} \mid\sim \beta) = k \geq a$.
That is, $inf_\beta\, gr(\{\alpha, \neg\alpha\} \mid\sim \beta) \geq a$ for any α.
That is, for any α, $gr(\{\alpha, \neg\alpha\} \mid\sim \beta) \geq a$ holds for any β.
Hence $C_a(\{\alpha, \neg\alpha\}) = F$ for any α.

(II) Let $a > k$. As $inf_\beta\ gr(\{\alpha, \neg\alpha\} \mid\!\sim \beta) \geq k$ for any α, the claim is that there must exist some α_1 such that either (A) $a > inf_\beta\ gr(\{\alpha_1, \neg\alpha_1\} \mid\!\sim \beta) \geq k$ or
(B) $inf_\beta\ gr(\{\alpha_1, \neg\alpha_1\} \mid\!\sim \beta)$ is non-comparable with a.
Because if not, then for all α, $inf_\beta gr(\{\alpha, \neg\alpha\} \mid\!\sim \beta) \geq a > k$ that is
$inf_{\alpha,\beta}\ gr(\{\alpha, \neg\alpha\} \mid\!\sim \beta) \geq a > k$ and this contradicts the assumption.
(A) Let $a > inf_\beta\ gr(\{\alpha_1, \neg\alpha_1\} \mid\!\sim \beta) = b$ (say) $\geq k$.
Then $C_b(\{\alpha_1, \neg\alpha_1\}) = F$ and $C_a(\{\alpha_1, \neg\alpha_1\}) \subseteq C_b(\{\alpha_1, \neg\alpha_1\})$ by Proposition 2.18.
We claim that $C_a(\{\alpha_1, \neg\alpha_1\}) \subset C_b(\{\alpha_1, \neg\alpha_1\})$ where $\subset$ denotes the proper subset relation. If not, then $C_a(\{\alpha_1, \neg\alpha_1\}) = F$ and $gr(\{\alpha_1, \neg\alpha_1\} \mid\!\sim \beta) \geq a$ for all β.
That is, $inf_\beta gr(\{\alpha_1, \neg\alpha_1\} \mid\!\sim \beta) \geq a > b$. Hence contradiction arises.
$C_a(\{\alpha_1, \neg\alpha_1\}) \neq F$.
(B) Let $inf_\beta gr(\{\alpha_1, \neg\alpha_1\} \mid\!\sim \beta) = b$ be non-comparable with a and $l = sup\{a, b\}$.
That is, $a < l$ and $b < l$.
As $gr(\{\alpha_1, \neg\alpha_1\} \mid\!\sim \beta) \geq b$ for all β, there is no γ such that $gr(\{\alpha_1, \neg\alpha_1\} \mid\!\sim \gamma) = a$
That is, for any β, $gr(\{\alpha_1, \neg\alpha_1\} \mid\!\sim \beta) \geq a$ implies $gr(\{\alpha_1, \neg\alpha_1\} \mid\!\sim \beta) > a$.
Now, the claim is that $C_a(\{\alpha_1, \neg\alpha_1\}) = C_l(\{\alpha_1, \neg\alpha_1\})$.
As $a < l$, $C_l(\{\alpha_1, \neg\alpha_1\}) \subseteq C_a(\{\alpha_1, \neg\alpha_1\})$.
To show that $C_a(\{\alpha_1, \neg\alpha_1\}) = C_l(\{\alpha_1, \neg\alpha_1\})$, we have to show there exists no β such that $a < gr(\{\alpha_1, \neg\alpha_1\} \mid\!\sim \beta) < l$ or $a < gr(\{\alpha_1, \neg\alpha_1\} \mid\!\sim \beta)$ but $gr(\{\alpha_1, \neg\alpha_1\} \mid\!\sim \beta)$ is non-comparable with l.
If possible let $a < gr(\{\alpha_1, \neg\alpha_1\} \mid\!\sim \beta) = a' < l$, for some β.
Then $a' \geq b$ and $a' > a$. These imply $a' > b$ and $a' > a$, that is, $a' \geq sup\{a, b\} = l$.
This contradicts the assumption that $a' < l$.
So there is no β such that $a < gr(\{\alpha_1, \neg\alpha_1\} \mid\!\sim \beta) < l$.
If possible, let $a < gr(\{\alpha_1, \neg\alpha_1\} \mid\!\sim \beta) = a'$ and a' be non-comparable to l.
As $gr(\{\alpha_1, \neg\alpha_1\} \mid\!\sim \beta) = a' \geq b$ and $a' > a$, $sup\{a, b\} = l \leq a'$.
This again contradicts the assumption that a' is non-comparable to l.
So, there is no β such that $a < gr(\{\alpha_1, \neg\alpha_1\} \mid\!\sim \beta)$ and $gr(\{\alpha_1, \neg\alpha_1\} \mid\!\sim \beta)$ is non-comparable with l.
So, we come to the conclusion that $C_a(\{\alpha_1, \neg\alpha_1\}) = C_l(\{\alpha_1, \neg\alpha_1\})$.
Now $C_l(\{\alpha_1, \neg\alpha_1\}) \subset C_b(\{\alpha_1, \neg\alpha_1\}) = F$ [Since $b < l$].
Hence $C_l(\{\alpha_1, \neg\alpha_1\}) = C_a(\{\alpha_1, \neg\alpha_1\}) \neq F$.
(III) Let a be non-comparable with k and $b = sup\{a, k\}$. That is, $a < b$, $k < b$.
As $a < b$, $C_b(\{\alpha, \neg\alpha\}) \subseteq C_a(\{\alpha, \neg\alpha\})$ for any α.
As $inf_\beta gr(\{\alpha, \neg\alpha\} \mid\!\sim \beta) \geq k$ for any α, $gr(\{\alpha, \neg\alpha\} \mid\!\sim \beta) \geq k$ holds for any α, β.
That is, there is no α such that for some β, $gr(\{\alpha, \neg\alpha\} \mid\!\sim \beta) = a$. That is, for any α, $gr(\{\alpha, \neg\alpha\} \mid\!\sim \beta) \geq a$ implies $gr(\{\alpha, \neg\alpha\} \mid\!\sim \beta) > a$.
The claim is that, for any α, $C_a(\{\alpha, \neg\alpha\}) = C_b(\{\alpha, \neg\alpha\})$.
If not, then there is some α_1 such that for some β, $a < gr(\{\alpha_1, \neg\alpha_1\} \mid\!\sim \beta) < b$ or $a < gr(\{\alpha_1, \neg\alpha_1\} \mid\!\sim \beta)$ and $gr(\{\alpha_1, \neg\alpha_1\} \mid\!\sim \beta)$ is non-comparable with b.
Subcase (a): Let $a < gr(\{\alpha_1, \neg\alpha_1\} \mid\!\sim \beta) = l < b$.
Then $l \geq k$ and $l > a$ and hence $l \geq sup\{a, k\} = b$.
This contradicts the assumption $l < b$.
Hence, there is no α_1 such that $a < gr(\{\alpha_1, \neg\alpha_1\} \mid\!\sim \beta) < b$.

Subcase (b): Let $a < gr(\{\alpha_1, \neg\alpha_1\} \mid\sim \beta) = l$ and $gr(\{\alpha_1, \neg\alpha_1\} \mid\sim \beta) = l$ be non-comparable with b. Now $l \geq k$ and $l > a$ and hence $l \geq sup\{a, k\} = b$.
This contradicts the assumption that b is non-comparable to l.
That is, there is no α_1 such that $a < gr(\{\alpha_1, \neg\alpha_1\} \mid\sim \beta)$ and $gr(\{\alpha_1, \neg\alpha_1\} \mid\sim \beta)$ is non-comparable with b.
Hence, from two subcases we have $C_a(\{\alpha, \neg\alpha\}) = C_b(\{\alpha, \neg\alpha\})$ for any α.
But as $b > k$ for some α_2, $C_b(\{\alpha_2, \neg\alpha_2\}) \subset C_k(\{\alpha_2, \neg\alpha_2\}) = F$ [By (I) and (II)]
That is, $C_b(\{\alpha_2, \neg\alpha_2\}) = C_a(\{\alpha_2, \neg\alpha_2\}) \neq F$. Hence for some α, $C_a(\{\alpha, \neg\alpha\}) \neq F$. □

Note 3.2 The argument made in the above theorem is at metameta-level or level-2. At this level the theory of graded consequence uses two-valued sentences. It may be argued that the above proof could be shortened by proving '$C_a(\{\alpha, \neg\alpha\}) = F$ if and only if $a \leq k$' which is immediate. But in order to consider the above statement and the statement of Theorem 3.5 to be equivalent, one needs to consider the equivalence between '$\exists x A(x)$' and 'not $\forall x$ (not $A(x)$)', as well as the presence of the rule 'not not α implies α'. So far we had no need to specify the logic of metameta-level. However, we would like to keep the logic employed at level-2 more inclined towards constructive logic. The above-mentioned principles are not accepted in any form of constructivism. As $GC4$ suggests that for each α, $inf_\beta\ gr(\{\alpha, \neg\alpha\} \mid\sim \beta) \geq k$, one can see that for each α, a value say, $b(\in L) \geq k$ is associated, and hence with respect to the condition of (GC4), F can be mapped onto a subset of L. From the proof of the theorem, it is clear that it does not simply ascertain the existence in a non-constructive manner; rather, it determines the range of values, i.e. a's which are associated to the formulas α_1 for which $C_a(\{\alpha_1, \neg\alpha_1\}) \neq F$.

Proposition 3.1 *Let $\mid\sim$ be a graded consequence relation with a threshold $c > 0$ for GC5. Then, for each $b \in L$, and α, $C_b(X \cup \{\alpha\}) \cap C_b(X \cup \{\neg\alpha\}) \subseteq C_{b *_m b *_m c}(X)$.*

Proof Let $\beta \in C_b(X \cup \{\alpha\})$ and $\beta \in C_b(X \cup \{\neg\alpha\})$.
Hence $b *_m b \leq gr(X \cup \{\alpha\} \mid\sim \beta) *_m gr(X \cup \{\neg\alpha\} \mid\sim \beta)$.
$b *_m b *_m c \leq gr(X \cup \{\alpha\} \mid\sim \beta) *_m gr(X \cup \{\neg\alpha\} \mid\sim \beta) *_m c \leq gr(X \mid\sim \beta)$.
So, $\beta \in C_{b *_m b *_m c}(X)$. Thus, $C_b(X \cup \{\alpha\}) \cap C_b(X \cup \{\neg\alpha\}) \subseteq C_{b *_m b *_m c}(X)$. □

Corollary 3.1 *For a graded consequence relation $\mid\sim$ satisfying GC5 with $c = 1$, and any $b \in L$, $C_b(X \cup \{\alpha\}) \cap C_b(X \cup \{\neg\alpha\}) \subseteq C_{b *_m b}(X)$.*

Corollary 3.2 *When the lattice meet $\wedge = *_m$, and a graded consequence relation $\mid\sim$ satisfies GC5 with $c = 1$, $C_b(X \cup \{\alpha\}) \cap C_b(X \cup \{\neg\alpha\}) = C_b(X)$, for any α.*

Proof For $*_m = \wedge$, $C_b(X \cup \{\alpha\}) \cap C_b(X \cup \{\neg\alpha\}) \subseteq C_b(X)$, following Corollary 3.1. And the converse direction holds by Proposition 2.19 of Chap. 2. That is, combining both the directions, we get $C_b(X \cup \{\alpha\}) \cap C_b(X \cup \{\neg\alpha\}) = C_b(X)$, for any α. □

From the properties of level cuts obtained in Sect. 2.5 of Chap. 2 and the above theorems, it can be noticed that considering $*_m = \wedge$ and $c = 1$, C_a's for $a \leq k$ behave like classical explosive consequence operation. On the other hand, for $a > k$ or a

non-comparable with k, C_a's are of non-explosive nature. Pictorial representation of the properties of level cuts of a graded consequence relation can be presented as follows.

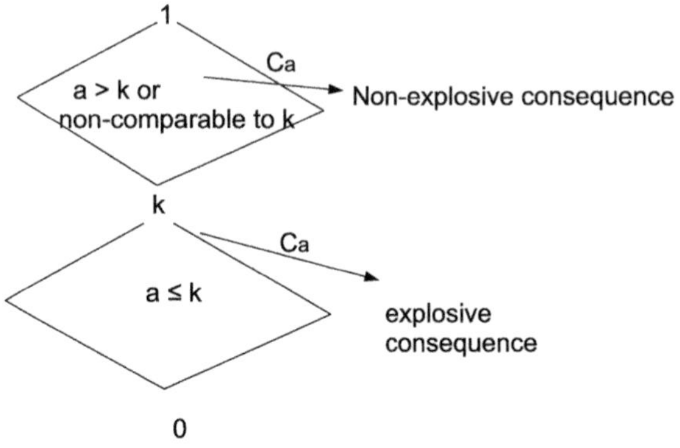

Note 3.3 Nowadays, 'inconsistency-tolerant system' is a prevalent terminology in the logic community. Inconsistency-tolerant systems allow the background logic to remain reasonably sound while drawing conclusions from an inconsistent set of premises. In literature, there is a well-known non-classical approach generating a bunch of paraconsistent logics (Arruda 1989; Priest and Routley 1989) which emerge to negate the classical law of explosive consequence, which is the principle that for any formula α, the set $\{\alpha, \neg\alpha\}$ yields any formula β. In this respect, the name of J. Łukasiewicz (cf. Priest and Routley 1989) comes to the forefront. He was of the opinion that the law of contradiction, i.e. $\neg(\alpha$ and $\neg\alpha)$, is not valid in general. It is natural for logics in which wffs can assume values other than 0 and 1 that $(\alpha \wedge \neg\alpha)$ is not 0 always. In the theory of graded consequence, one of the postulates to characterize the behaviour of $\neg$ with respect to the notion of consequence has been stated in such a manner that in this context, for every formula α, from $\{\alpha, \neg\alpha\}$ any formula β can follow to a non-zero grade k which is not necessarily the greatest value 1. In that sense, the theory of graded consequence allows inconsistency to be tolerated. The above diagram, followed from Theorem 3.5, reflects the same. It is to be marked also that the logics with consequence operator C_a, where $a > k$ or non-comparable with k, are all paraconsistent two-valued logics. This gives an interesting interpretation of the value k appearing in the axioms (GC4) and (Inc3).

3.2 Semantic Import of (GC4) and (GC5) on the Meta-structure

GC4 and GC5 are introduced as generalizations of the rules 'explosiveness condition' (or 'ex falso qudolibet') and 'reasoning by cases', respectively. Accordingly, the notion of graded inconsistency has been formulated with axioms INC1 to INC3. Characterization of any postulate leads us towards its representation in terms of semantics. Reflection of the same fact is observed in Chap. 2 where after characterizing graded notion of consequence, its semantical viability has been established through the representation theorem (Theorem 2.1). Axioms GC4 and GC5 also need to be presented in terms of their semantic import. Besides, finding natural examples is obligatory as well from the point of view of theory building and potential applications.

In Chap. 2, we have talked about the meta-theory of the proposed notion of graded consequence. We have not entered into any specific object language. To look into the feasibility of GC4 and GC5 from the perspective of semantics, we need to check the algebraic properties of the object language as well as the meta-language.

In order to achieve the above objective, in this section we shall investigate the conditions on the truth value set and/or in the collection $\{T_i\}_{i\in I}$ of valuations that necessarily follow when GC4 or GC5 or both are incorporated in the set of axioms of graded consequence. In this sense, these are the necessary implications of the axioms. Conversely, we shall also look for conditions, which when satisfied by the truth value set and/or the collection $\{T_i\}_{i\in I}$ of valuations, the axioms will hold non-trivially. It is to be noted that in the classical two-valued logic GC4 and GC5 always hold and the values of both k and c are 1. By 'non-triviality', we mean the situations when k and c are not necessarily 1. In fact, really interesting situations are those when $k \neq 1$ or $c \neq 1$. A number of examples are presented to show that there are quite a few such cases and these examples may be utilized by the users. So far we have assumed the meta-level structure to be any complete residuated lattice. It will be observed that the necessary and the sufficient conditions, obtained in this chapter, impose some further conditions on our choice of the value sets depicting some interrelations between the two algebraic structures (for object and meta-level languages), if the axioms are to be satisfied.

From the representation theorems (Theorem 2.1) presented in Chap. 2, we know that for any graded consequence relation $|\!\sim$ there corresponds a collection $\{T_i\}_{i\in I}$ of fuzzy subsets (not necessarily unique) so that $|\!\sim\, = \,\approx\!\!|_{\{T_i\}_{i\in I}}$, and given any collection of fuzzy subsets $\{T_i\}_{i\in I}$, $\approx\!\!|_{\{T_i\}_{i\in I}}$, the fuzzy relation generated by the collection turns out to be a graded consequence relation. It should be noted that, while dealing with graded counterparts of overlap, dilution and cut, i.e. (GC1), (GC2) and (GC3), we need not specify the nature of fuzzy subsets $\{T_i\}_{i\in I}$ generating the consequence relation. But when the object-level connectives come into the scenario, we would stick to those fuzzy subsets which satisfy the truth functionality condition; that is, truth values of the compound statements are determined from truth values of its constituent formulae and truth functions corresponding to the involved connectives. Moreover, to retain classical case as a special case of the logical rules from graded context, we choose those truth functional valuations, i.e. fuzzy subsets $\{T_i\}_{i\in I}$ which behave classically on $\{0, 1\}$-valued structure.

A collection of fuzzy sets, which are truth functional as well as classical value-preserving, will be said to have property (TC).

Let $\neg_o$ be the operator for object-level negation, and $(L, \wedge, \vee, *_m, \rightarrow_m, 0, 1)$, a complete residuated lattice, be the structure for the meta-level. Then for any graded consequence relation we get the following theorems (Chakraborty and Dutta 2010).

Theorem 3.6 *(The necessary condition for GC4) Let* $\{T_i\}_{i\in I}$ *be a collection of fuzzy subsets over the set of all formulae with property (TC). Let* $\approx\!\!\!|_{\{T_i\}_{i\in I}}$ *satisfy GC4. Then either no formula receives the value* 0 *under any* T_i *or for any non-zero* $T_i(\alpha) \wedge \neg_o T_i(\alpha)$ *there is a non-zero* $z \in L$ *such that* $(T_i(\alpha) \wedge \neg_o T_i(\alpha)) *_m z = 0$.

Proof Let $inf_{\alpha,\beta} gr(\{\alpha, \neg\alpha\} \approx\!\!\!|_{\{T_i\}_{i\in I}} \beta) > 0$.
That is, for all α, β, $gr(\{\alpha, \neg\alpha\} \approx\!\!\!|_{\{T_i\}_{i\in I}} \beta) > 0$.
So, $\inf_i\{(T_i(\alpha) \wedge T_i(\neg\alpha)) \rightarrow_m T_i(\beta)\} > 0$, i.e. assuming $\neg_o$ to be the operator used to compute $\neg$, we have $inf_i\{(T_i(\alpha) \wedge \neg_o T_i(\alpha)) \rightarrow_m T_i(\beta)\} > 0$.
So, for all i and for all α, β, $\{(T_i(\alpha) \wedge \neg_o T_i(\alpha)) \rightarrow_m T_i(\beta)\} > 0$. ...(1)
As L is a residuated lattice $\{(T_i(\alpha) \wedge \neg_o T_i(\alpha)) \rightarrow_m T_i(\beta)\} \geq T_i(\beta)$ holds. ...(2)
Now there are two cases: *(I)*: $T_i(\beta) > 0$ for all β or *(II)*: $T_i(\beta) = 0$ for some β.
(I): For all β, $T_i(\beta) > 0$ immediately gives the condition that no formula receives the value 0 under any T_i.
(II): In this case, we have $\{(T_i(\alpha) \wedge \neg_o T_i(\alpha)) \rightarrow_m T_i(\beta)\} > 0$ when $T_i(\beta) = 0$.
Again two subcases arise. (A) $(T_i(\alpha) \wedge \neg_o T_i(\alpha)) = 0$ or (B) $(T_i(\alpha) \wedge \neg_o T_i(\alpha)) \neq 0$.
(A) In this subcase $\{(T_i(\alpha) \wedge \neg_o T_i(\alpha)) \rightarrow_m T_i(\beta)\} = 1 > 0$.
(B) Let $(T_i(\alpha) \wedge \neg_o T_i(\alpha)) \neq 0$.
Since (1) implies $\{(T_i(\alpha) \wedge \neg_o T_i(\alpha)) \rightarrow_m T_i(\beta)\} > 0$ we must have $Max\{z/(T_i(\alpha) \wedge \neg_o T_i(\alpha)) *_m z \leq 0\} > 0$.

Hence, there is some $z \neq 0$, such that for $(T_i(\alpha) \wedge \neg_o T_i(\alpha)) \neq 0$, $(T_i(\alpha) \wedge \neg_o T_i(\alpha)) *_m z = 0$. □

Note 3.4 We should note that the classical value-preserving property of T_i's plays a role to obtain the above necessary condition. $(T_i(\alpha) \wedge \neg_o T_i(\alpha))$ can be 1 only when both $T_i(\alpha)$ and $\neg_o T_i(\alpha)$ are 1; but that cannot be the case as we assume T_i's to preserve the nature of classical valuation at 0, 1. If we do not put the said constraint, then $T_i(\alpha) \wedge \neg_o T_i(\alpha)$ could be 1, and that in turn gives $(T_i(\alpha) \wedge \neg_o T_i(\alpha)) *_m z \leq 0$ implies $z = 0$.

From now onwards whenever we state any collection of $\{T_i\}_{i\in I}$ or a graded consequence relation $\approx\!\!\!|_{\{T_i\}_{i\in I}}$ generated by the collection $\{T_i\}_{i\in I}$, we would mean the collection with property (TC).

Corollary 3.3 *Given a complete residuated lattice* $(L, \wedge, \vee, *_m, \rightarrow_m, 0, 1)$ *for the meta-structure and a structure* $(L, \neg_o)$ *for the object language, if GC4 holds for any* $\approx\!\!\!|_{\{T_i\}_{i\in I}}$, *then for any non-zero* $b \wedge \neg_o b$ *of* L, *there exists some non-zero* z *in* L *such that* $(b \wedge \neg_o b) *_m z = 0$.

Proof As for any $\approx\!\!\!|_{\{T_i\}_{i\in I}}$ GC4 holds, in particular, for a singleton collection containing an arbitrary T_i also GC4 holds. Now, if α is a propositional variable, $T_i(\alpha)$ can assume any value b of L. Hence, by the previous theorem for any non-zero $(b \wedge \neg_o b)$ there exists a non-zero z such that $(b \wedge \neg_o b) *_m z = 0$. □

Note 3.5 It is to be noted that the necessary condition, obtained in Corollary 3.3, is not a property available in any complete residuated lattice. If we consider Gödel algebra, then for no non-zero value of $b \wedge \neg_o b \in [0, 1]$ there will be non-zero $z \in [0, 1]$ such that $(b \wedge \neg_o b) \wedge z = 0$ (cf. Example 3.5).

Theorem 3.7 *(A sufficient condition for GC4) Let* $(L, \wedge, \vee, *_m, \rightarrow_m, 0, 1)$ *be a complete residuated lattice, and* $\neg_o$ *be the operator for object language negation satisfying the following conditions:*

(i) there is some $k'(\neq 1)$*, such that* $b \wedge \neg_o b \leq k'$ *for any b in L and*
(ii) if the above $k' \neq 0$ *then there exists* $z(\neq 0)$ *such that* $k' *_m z = 0$.

Then for any graded consequence relation $\approx_{\{T_i\}_{i\in I}}$ *with the above structure for the meta-language, GC4 holds.*

Proof Let us assume that L is a complete residuated lattice and (i) and (ii) hold.
Let us consider the collection $\{T_i\}_{i\in I}$ generating $\approx_{\{T_i\}_{i\in I}}$.
We prove the theorem in two stages.
Stage 1: Claim: For any α, β, i, $\{(T_i(\alpha) \wedge \neg_o T_i(\alpha)) \rightarrow_m T_i(\beta)\} > 0$.
Let us consider two cases. Case-I: $T_i(\beta) \neq 0$ Case-II: $T_i(\beta) = 0$.
Case-I: If $T_i(\beta) \neq 0$ then as $\{(T_i(\alpha) \wedge \neg_o T_i(\alpha)) \rightarrow_m T_i(\beta)\} \geq T_i(\beta)$,
$\{(T_i(\alpha) \wedge \neg_o T_i(\alpha)) \rightarrow_m T_i(\beta)\} > 0$.
Case-II: Let $T_i(\beta) = 0$.
Now two subcases arise. (A) $T_i(\alpha) \wedge \neg_o T_i(\alpha) = 0$ (B) $T_i(\alpha) \wedge \neg_o T_i(\alpha) \neq 0$.
Subcase (A): $T_i(\alpha) \wedge \neg_o T_i(\alpha) = 0$ immediately implies
$\{(T_i(\alpha) \wedge \neg_o T_i(\alpha)) \rightarrow_m T_i(\beta)\} = 1 > 0$.
Subcase (B): Let $T_i(\alpha) \wedge \neg_o T_i(\alpha) \neq 0$ and $T_i(\alpha) = b$.
According to the condition (i), $b \wedge \neg_o b \leq k'$ for all $b \in L$.
Hence $(b \wedge \neg_o b) \rightarrow_m 0 \geq k' \rightarrow_m 0$. Now as k' is non-zero by condition (ii), we have
$(b \wedge \neg_o b) \rightarrow_m 0 \geq k' \rightarrow_m 0 = \max\{z/k' *_m z \leq 0\} > 0$.
Combining (I) and (II), we have for any α, β, i, $(T_i(\alpha) \wedge \neg_o T_i(\alpha)) \rightarrow_m T_i(\beta) > 0$.

Stage 2: By condition (i), we have $T_i(\alpha) \wedge \neg_o T_i(\alpha) \leq k'$ for any i, where $k' \neq 1$.
Then $(T_i(\alpha) \wedge \neg_o T_i(\alpha)) \rightarrow_m T_i(\beta) \geq k' \rightarrow_m T_i(\beta) > 0$.
[If $k' = 0$ then $k' \rightarrow_m T_i(\beta)$ is 1 and if $k' \neq 0$ then also repeating the same argument as in case-II of stage 1 we can show that $k' \rightarrow_m T_i(\beta) > 0$.]
I.e. for any i, $(T_i(\alpha) \wedge \neg_o T_i(\alpha)) \rightarrow_m T_i(\beta) \geq k' \rightarrow_m T_i(\beta) \geq k' \rightarrow_m inf_i T_i(\beta) > 0$.
[Reason for $k' \rightarrow_m inf_i T_i(\beta) > 0$ is the same as the case for $k' \rightarrow_m T_i(\beta)$.]
I.e. $inf_i\{(T_i(\alpha) \wedge \neg_o T_i(\alpha)) \rightarrow_m T_i(\beta)\} \geq k' \rightarrow_m inf_i T_i(\beta) > 0$.
I.e. $gr(\{\alpha, \neg\alpha\} \mid\!\sim \beta) \geq k' \rightarrow_m inf_i T_i(\beta) > 0$ for any β.
I.e. $gr(\{\alpha, \neg\alpha\} \mid\!\sim \beta) \geq k' \rightarrow_m inf_\beta\{inf_i T_i(\beta)\} > 0$ for any β.
[Applying the same reason as for $k' \rightarrow_m T_i(\beta) > 0$ and $k' \rightarrow_m inf_i T_i(\beta) > 0$.]
I.e. $inf_\beta\, gr(\{\alpha, \neg\alpha\} \mid\!\sim \beta) \geq k' \rightarrow_m inf_\beta\{inf_i T_i(\beta)\} > 0$ holds for all α.
I.e. $inf_{\alpha,\beta}\, gr(\{\alpha, \neg\alpha\} \mid\!\sim \beta) \geq k' \rightarrow_m inf_\beta\{inf_i T_i(\beta)\} > 0$.
That is, $inf_{\alpha,\beta}\, gr(\{\alpha, \neg\alpha\} \mid\!\sim \beta) = k > 0$ □

Note 3.6 A non-zero element a is said to be a zero divisor or divisor of zero if there exists a non-zero element b such that $a *_m b = b *_m a = 0$. In Theorem 3.6, the necessary condition happens to be that each non-zero $T_i(\alpha) \wedge \neg_o T_i(\alpha)$ has a zero divisor, whereas in Theorem 3.7 the specific element k' is assumed to have a zero divisor.

Moreover, from the proof of Theorem 3.7 it can be noticed that the value of k of GC4, i.e. the value of the threshold for $inf_{\alpha,\beta}\, gr(\{\alpha, \neg\alpha\} \mid\!\sim \beta)$ depends upon the value of $inf_\beta\{inf_i T_i(\beta)\}$, which varies from one collection ($\{T_i\}_{i\in I}$) to another. But irrespective of collection we can say that $k \geq k' \rightarrow_m 0$ as for any collection $\{T_i\}_{i\in I}$, $k' \rightarrow_m inf_\beta\{inf_i T_i(\beta)\} \geq k' \rightarrow_m 0$.

Example 3.1 Let $L = \{0, \frac{1}{5}, \frac{2}{5}, \frac{3}{5}, \frac{4}{5}, 1\}$, and the operator for object-level negation be $\neg_o b = 1 - b$.
Meta-level operators $*_m$ and $\rightarrow_m$ are of Łukasiewicz, i.e. $a *_m b = \max(0, a + b - 1)$ and $a \rightarrow_m b = \min(1, 1 - a + b)$.

b	$\neg_o b$	$b \wedge \neg_o b$
0	1	0
$\frac{1}{5}$	$\frac{4}{5}$	$\frac{1}{5}$
$\frac{2}{5}$	$\frac{3}{5}$	$\frac{2}{5}$
$\frac{3}{5}$	$\frac{2}{5}$	$\frac{2}{5}$
$\frac{4}{5}$	$\frac{1}{5}$	$\frac{1}{5}$
1	0	0

$*_m$	0	$\frac{1}{5}$	$\frac{2}{5}$	$\frac{3}{5}$	$\frac{4}{5}$	1
0	0	0	0	0	0	0
$\frac{1}{5}$	0	0	0	0	0	$\frac{1}{5}$
$\frac{2}{5}$	0	0	0	0	$\frac{1}{5}$	$\frac{2}{5}$
$\frac{3}{5}$	0	0	0	$\frac{1}{5}$	$\frac{2}{5}$	$\frac{3}{5}$
$\frac{4}{5}$	0	0	$\frac{1}{5}$	$\frac{2}{5}$	$\frac{3}{5}$	$\frac{4}{5}$
1	0	$\frac{1}{5}$	$\frac{2}{5}$	$\frac{3}{5}$	$\frac{4}{5}$	1

Possible values for $b \wedge \neg_o b$ are $0, \frac{1}{5}, \frac{2}{5}$. Now to check the necessary condition for GC4, we need to check for each non-zero $b \wedge \neg_o b$ whether there is a $z (\neq 0)$ such that $(b \wedge \neg_o b) *_m z = 0$. For $b \wedge \neg_o b = \frac{1}{5}$ possible values of z are $\frac{1}{5}, \frac{2}{5}, \frac{3}{5}, \frac{4}{5}$ and for $b \wedge \neg_o b = \frac{2}{5}$ possible values of z are $\frac{1}{5}, \frac{2}{5}, \frac{3}{5}$. Again to check the sufficient condition for GC4, we see from the table of $\neg_o$, $b \wedge \neg_o b \leq \frac{2}{5}$ and from that of $*_m$ the possible non-zero, non-unit values of z are $\frac{1}{5}, \frac{2}{5}, \frac{3}{5}$. Hence, as Theorem 3.7 suggests, $k \geq \frac{2}{5} \rightarrow_m 0 = \frac{3}{5}$. The exact value of k depends upon the collection of $\{T_i\}_{i\in I}$, taken.

Example 3.2 Consider the same value set L as in Example 3.1. Let the object level be computed by $\neg_o b = b \rightarrow_o 0$, where $\rightarrow_o$ is the largest R-implication (Klir et al. 2006) defined by
$a \rightarrow_o b = b$ when $a = 1$
$\quad = 1$, otherwise.
That is, the operator $\neg_o$ assigns 0 to 1 and 1 to all other elements. Meta-level $*_m$ and $\rightarrow_m$ are of Łukasiewicz. Then as for each $a \in L$, $a \wedge \neg_o a \leq \frac{4}{5}$, $k \geq \frac{4}{5} \rightarrow_m 0 = \frac{1}{5}$.

Example 3.3 In [0, 1] taking the operator for the object-level ¬ as $\neg_o a = 1 - a$ and $(*_m, \rightarrow_m)$ of Łukasiewicz (Hájek 1998), it can be shown that GC4 holds with a non-zero value k. In this case, for any $a \in [0, 1]$, $a \wedge \neg_o a \leq \frac{1}{2}$ and hence $k \geq \frac{1}{2} \rightarrow_m 0 = \frac{1}{2}$.

Example 3.4 In [0, 1] computing the operator for the object level ¬ by Gödel negation (Hájek 1998), viz. $\neg_o a = 1$, when $a = 0$
$= 0$, otherwise,
and meta-linguistic conjunction and implication by any adjoint pair $(*_m, \rightarrow_m)$, it can be shown that GC4 holds with $k = 1$, as for any $a \in [0, 1]$, $a \wedge \neg_o a = 0$ implies $k \geq 0 \rightarrow_m 0 = 1$.

Example 3.5 Let the meta-level structure be $\langle [0, 1], \wedge, \rightarrow_m, 0, 1 \rangle$, i.e. Gödel algebra where $(\wedge, \rightarrow_m)$ forms the adjoint pair, that is, $\rightarrow_m$ is defined by
$a \rightarrow_m b = 1$ if $a \leq b$
$= b$, otherwise.
Let the operator for object-level negation $\neg_o$ be $\neg_o x = 1 - x$. Here, $a \wedge \neg_o a$ can have values other than 0, and $k' = \frac{1}{2}$. But no non-zero element in [0, 1] has a divisor of zero. So, following Corollary 3.3 GC4 does not hold in this context.

The following theorem presents the necessary and sufficient condition for the notion graded consequence to be explosive to the full extent.

Proposition 3.2 *Let $\{T_i\}_{i \in I}$ be a given collection and $\mid\!\sim$ be the graded consequence relation defined with respect to the collection. Let $\neg_o$ be the operator used to compute the object-level negation. Then for any α, $inf_\beta gr(\{\alpha, \neg\alpha\} \mid\!\sim \beta) = 1$ if and only if for any i, $T_i(\alpha) \wedge \neg_o T_i(\alpha) \leq inf_\beta T_i(\beta)$.*

Proof For any α, $inf_\beta gr(\{\alpha, \neg\alpha\} \mid\!\sim \beta) = 1$ if and only if
for any α, $gr(\{\alpha, \neg\alpha\} \mid\!\sim \beta) = 1$ for any β, if and only if
for any α, $inf_i\{(T_i(\alpha) \wedge \neg_o T_i(\alpha)) \rightarrow_m T_i(\beta)\} = 1$ for any β, if and only if
for any α, $(T_i(\alpha) \wedge \neg_o T_i(\alpha)) \rightarrow_m T_i(\beta) = 1$ for any β, for any i, if and only if
for any α, $(T_i(\alpha) \wedge \neg_o T_i(\alpha)) \leq T_i(\beta)$ for any β, for any i.
Let us take a fixed α and then vary β. Thus we get $(T_i(\alpha) \wedge \neg_o T_i(\alpha)) \leq inf_\beta T_i(\beta)$. □

Corollary 3.4 *For any $\approx_{\{T_i\}_{i \in I}}$, GC4 holds with $k = 1$ if and only if $a \wedge \neg_o a = 0$ for any $a \in L$.*

Proof Let us consider an $a \in L$. As we are considering all possible T_i's, we choose a singleton collection $\{T_i\}$ such that $T_i(\alpha) = a$ and $T_i(\beta) = 0$ for two atomic formulae α and β. Now, from Proposition 3.2, for any α, $(T_i(\alpha) \wedge \neg_o T_i(\alpha)) \leq \inf_\beta T_i(\beta) = 0$. So, we have $a \wedge \neg_o a = 0$. As this is true for any arbitrary $a \in L$, we obtain the result. □

Theorem 3.8 *(A sufficient condition for GC5) Let $(L, \wedge, \vee, *_m, \rightarrow_m, 0, 1)$, a complete residuated lattice, be the meta-level structure, and $\neg_o$ be the operator corresponding to the object-level negation ¬ satisfying the following conditions:*

(i) there is an element $c(\neq 0)$ *in L such that* $x \leq c$ *or* $x > c$ *for all x in L,*
(ii) $x \vee \neg_o x \geq c$ *for all* $x \in L$, *and*
(iii) $c *_m z = z$ *for all* $z \leq c$.

Then for any graded consequence relation $\approx_{\{T_i\}_{i\in I}}$, *with the above structure for the meta-language GC5 holds.*

Proof In order to prove this theorem, we first need to prove the following lemma.

Lemma 3.2 *For the element c, mentioned in Theorem 3.8,* $c \to_m x = x$ *for all* $x < c$.

Proof of Lemma 3.2: Let $x < c$. Now $c \to_m x = max\{z/c *_m z \leq x\}$.
Now three cases arise. (I) $z \leq x$ (II) z is non-comparable to x (III) $z > x$.
(I) If $z \leq x$ then $c *_m z \leq c *_m x = x$. Hence for all $z \leq x$, $z \in \{\omega/c *_m \omega \leq x\}$.
(II) Let z be non-comparable to x. Then either $z < c$ or $z \geq c$.
But $z \geq c > x$ contradicts the assumption that z is non-comparable to x.
Now for $z < c$, if possible let $c *_m z \leq x$. Then $z \leq x$. [Since $z < c$ implies $c *_m z = z$]

This again contradicts the assumption that z is non-comparable to x.
Hence for no z non-comparable to x, $z \in \{\omega/c *_m \omega \leq x\}$.
(III) Let $z > x$. Then two subcases arise. (A) $x < z \leq c$ (B) $x < c < z$.
(A) As $x < c$ as well as $z \leq c$, $c *_m x = x$ and $c *_m z = z$.
Then $c *_m x < c *_m z$. That is $c *_m z > x$. Therefore, the z's satisfying $x < z \leq c$ do not belong to the set $\{\omega/c *_m \omega \leq x\}$.
(B) Let $x < c < z$. Therefore, $c *_m z \geq c *_m c = c > x$. That is, z's such that $x < c < z$, do not belong to the set $\{\omega/c *_m \omega \leq x\}$.
Hence combining (I), (II) and (III), we conclude $c \to_m x = x$ for all $x < c$.

Main proof: $gr(X \cup \{\alpha\} \approx_{\{T_i\}_{i\in I}} \beta) *_m gr(X \cup \{\neg\alpha\} \approx_{\{T_i\}_{i\in I}} \beta) *_m c$
$= inf_i\{(inf_{\gamma\in X} T_i(\gamma) \wedge T_i(\alpha)) \to_m T_i(\beta)\} *_m inf_i\{(inf_{\gamma\in X} T_i(\gamma) \wedge T_i(\neg\alpha)) \to_m T_i(\beta)\} *_m c$.
$\leq inf_i[\{(inf_{\gamma\in X} T_i(\gamma) \wedge T_i(\alpha)) \to_m T_i(\beta)\} *_m \{(inf_{\gamma\in X} T_i(\gamma) \wedge T_i(\neg\alpha)) \to_m T_i(\beta)\}] *_m c$.
$\leq inf_i[\{(inf_{\gamma\in X} T_i(\gamma) *_m T_i(\alpha)) \to_m T_i(\beta)\} *_m \{(inf_{\gamma\in X} T_i(\gamma) *_m T_i(\neg\alpha)) \to_m T_i(\beta)\}] *_m c$. [Since $x *_m y \leq x \wedge y$ and $\to_m$ is antitone in the first variable.]
$= inf_i[\{(inf_{\gamma\in X} T_i(\gamma) *_m T_i(\alpha)) \to_m T_i(\beta)\} *_m \{(inf_{\gamma\in X} T_i(\gamma) *_m \neg_o T_i(\alpha)) \to_m T_i(\beta)\}] *_m c$.
$= inf_i[\{T_i(\alpha) \to_m (inf_{\gamma\in X} T_i(\gamma) \to_m T_i(\beta))\} *_m \{\neg_o T_i(\alpha) \to_m (inf_{\gamma\in X} T_i(\gamma) \to_m T_i(\beta))\}] *_m c$.
[Since in a residuated lattice $(x *_m y) \to_m z = x \to_m (y \to_m z) = y \to_m (x \to_m z)$.]
$\leq inf_i\{(T_i(\alpha) \vee \neg_o T_i(\alpha)) \to_m (inf_{\gamma\in X} T_i(\gamma) \to_m T_i(\beta))\} *_m c$.
[Since in a residuated lattice $(x \to_m z) *_m (y \to_m z) \leq (x \vee y) \to_m z$ holds.]
$\leq inf_i\{c \to_m (inf_{\gamma\in X} T_i(\gamma) \to_m T_i(\beta))\} *_m c$. [Using (ii)].
$= inf_i[\{c \to_m (inf_{\gamma\in X} T_i(\gamma) \to_m T_i(\beta)) : c > inf_{\gamma\in X} T_i(\gamma) \to_m T_i(\beta)\} \cup \{c \to_m (inf_{\gamma\in X} T_i(\gamma) \to_m T_i(\beta)) : c \leq inf_{\gamma\in X} T_i(\gamma) \to_m T_i(\beta)\}] *_m c$. [Using (i)].

$= inf_i\{c \rightarrow_m (inf_{\gamma \in X} T_i(\gamma) \rightarrow_m T_i(\beta)) : c > inf_{\gamma \in X} T_i(\gamma) \rightarrow_m T_i(\beta)\} \wedge$
$inf_i\{c \rightarrow_m (inf_{\gamma \in X} T_i(\gamma) \rightarrow_m T_i(\beta)) : c \leq inf_{\gamma \in X} T_i(\gamma) \rightarrow_m T_i(\beta)\} *_m c.$
[Since $inf(\{a_i\} \cup \{b_j\}) = inf\{a_i\} \wedge inf\{b_j\}$.]
$= inf_i\{c \rightarrow_m (inf_{\gamma \in X} T_i(\gamma) \rightarrow_m T_i(\beta)) : c > inf_{\gamma \in X} T_i(\gamma) \rightarrow_m T_i(\beta)\} *_m c.$
$\leq inf_i\{c \rightarrow_m (inf_{\gamma \in X} T_i(\gamma) \rightarrow_m T_i(\beta)) : c > inf_{\gamma \in X} T_i(\gamma) \rightarrow_m T_i(\beta)\} \wedge c.$
$= inf_i\{inf_{\gamma \in X} T_i(\gamma) \rightarrow_m T_i(\beta) : c > inf_{\gamma \in X} T_i(\gamma) \rightarrow_m T_i(\beta)\} \wedge c$ [By Lemma 3.2]
$= inf_i\{inf_{\gamma \in X} T_i(\gamma) \rightarrow_m T_i(\beta) : c > inf_{\gamma \in X} T_i(\gamma) \rightarrow_m T_i(\beta)\}$
$\wedge\, inf_i\{inf_{\gamma \in X} T_i(\gamma) \rightarrow_m T_i(\beta) : c \leq inf_{\gamma \in X} T_i(\gamma) \rightarrow_m T_i(\beta)\} \wedge c$
$= gr(X \approx_{\{T_i\}_{i\in I}} \beta) \wedge c$
$\leq gr(X \approx_{\{T_i\}_{i\in I}} \beta).$ □

Example 3.6 Taking the meta-level structure as $\langle [0, 1], \wedge, \rightarrow_m, 0, 1\rangle$, i.e. Gödel algebra where $(\wedge, \rightarrow_m)$ forms the adjoint pair that is $\rightarrow_m$ is defined by
$a \rightarrow_m b = 1$ if $a \leq b$
$= b$, otherwise
and operator for object-level negation $\neg_o$ as $\neg_o x = 1 - x$ one can verify the aforesaid sufficient condition for GC5. In this case c would be $\frac{1}{2}$.

Theorem 3.9 *(Necessary condition for GC5) Let* $(L, \wedge, \vee, *_m, \rightarrow_m, 0, 1)$*, a complete residuated lattice, be the meta-level structure, and* $(L, \neg_o)$ *be the object-level structure. Let for any* $\approx_{\{T_i\}_{i\in I}}$ *GC5 hold. Then for all* b, d *such that* $b \nleq d$ *or* $\neg_o b \nleq d$, $(b \rightarrow_m d) *_m (\neg_o b \rightarrow_m d) \leq c \rightarrow_m d$.

Proof Let for any $\approx_{\{T_i\}_{i\in I}}$,
$gr(X \cup \{\alpha\} \approx_{\{T_i\}_{i\in I}} \beta) *_m gr(X \cup \{\neg\alpha\} \approx_{\{T_i\}_{i\in I}} \beta) *_m c \leq gr(X \approx_{\{T_i\}_{i\in I}} \beta)$ hold, i.e.
$inf_i\{(inf_{\gamma \in X} T_i(\gamma) \wedge T_i(\alpha)) \rightarrow_m T_i(\beta)\} *_m inf_i\{(inf_{\gamma \in X} T_i(\gamma) \wedge T_i(\neg\alpha)) \rightarrow_m T_i(\beta)\}$
$*_m c \leq inf_i\{inf_{\gamma \in X} T_i(\gamma) \rightarrow_m T_i(\beta)\}$.
I.e. $inf_i\{(a_i \wedge b_i) \rightarrow_m d_i\} *_m inf_i\{(a_i \wedge \neg_o b_i) \rightarrow_m d_i\} *_m c \leq inf_i\{a_i \rightarrow_m d_i\}$.
[Assuming $inf_{\gamma \in X} T_i(\gamma) = a_i$, $T_i(\alpha) = b_i$ and $T_i(\beta) = d_i$.]
Hence $inf_i(b_i \rightarrow_m d_i) *_m inf_i(\neg_o b_i \rightarrow_m d_i) *_m c$
$\leq inf_i\{(a_i \wedge b_i) \rightarrow_m d_i\} *_m inf_i\{(a_i \wedge \neg_o b_i) \rightarrow_m d_i\} *_m c \leq inf_i\{a_i \rightarrow_m d_i\}$.
Hence $inf_i\{b_i \rightarrow_m d_i : b_i \nleq d_i\} *_m inf_i\{\neg_o b_i \rightarrow_m d_i : \neg_o b_i \nleq d_i\} *_m c \leq \{a_i \rightarrow_m d_i\}$ for each i.
The expression in the left-hand side of the inequality is independent of the a_i's occurred in the right-hand side of the expression. That is, the inequality holds for any $a_i \in L$.
Then in particular for $a_i = 1$, for each i we have
$inf_i\{b_i \rightarrow_m d_i : b_i \nleq d_i\} *_m inf_i\{\neg_o b_i \rightarrow_m d_i : \neg_o b_i \nleq d_i\} *_m c \leq \{1 \rightarrow_m d_i\} = d_i$.
That is, $inf_i\{b_i \rightarrow_m d_i : b_i \nleq d_i\} *_m inf_i\{\neg_o b_i \rightarrow_m d_i : \neg_o b_i \nleq d_i\} \leq c \rightarrow_m d_i$, for each i.
From the above inequality, it is clear that the value of the left-hand side expression depends on those i's such that
either $b_i \nleq d_i$ and $\neg_o b_i \leq d_i$
or $b_i \leq d_i$ and $\neg_o b_i \nleq d_i$
or $b_i \nleq d_i$ and $\neg_o b_i \nleq d_i$.

In particular, for a singleton collection $\{T_i\}_{i\in I}$ the above case reduces to
$(b \rightarrow_m d) *_m (\neg_o b \rightarrow_m d) \leq c \rightarrow_m d$ if $b \nleq d$ or $\neg_o b \nleq d$, for each $b, d \in L$. □

Note 3.7 The necessary condition for GC5 is that, for $b \nleq d$ or $\neg_o b \nleq d$, Here it should be noted that$(b \rightarrow_m d) *_m (\neg_o b \rightarrow_m d) \leq c \rightarrow_m d$. So, putting $c = 1$, we get if $b \nleq d$ or $\neg_o b \nleq d$, then $(b \rightarrow_m d) *_m (\neg_o b \rightarrow_m d) \leq d$. This is the necessary condition for GC5 in the special case when $c = 1$, which we may call GC5^C (GC5 classical). Also we know $d \leq c \rightarrow_m d$. So, we can conclude that the necessary condition for GC5^C implies the necessary condition for GC5 in general case. There are several t-norms satisfying the above-mentioned necessary conditions for GC5^C, and hence GC5 too. That is, in the following cases, we cannot deny the existence of GC5^C and GC5. But these conditions are not sufficient to ensure GC5^C.

Example 3.7 Let the meta-structure be $\langle [0, 1], *_m, \rightarrow_m, 0, 1\rangle$ where
$a *_m b = \frac{ab}{max\{a,b,\frac{1}{2}\}}$ [That is $*_m$ is by Dubois–Prade (Klir et al. 2006)] and $\neg_o a = 1 - a$.
That is, $a *_m b = min\{a, b\}$, if $max\{a, b, \frac{1}{2}\} \neq \frac{1}{2}$
$= \frac{ab}{\frac{1}{2}}$, if $max\{a, b, \frac{1}{2}\} = \frac{1}{2}$
So, three cases arise. (I) $a > \frac{1}{2}$, (II) $a < \frac{1}{2}$ and (III) $a = \frac{1}{2}$.
(I) If $a > \frac{1}{2}$ then $\neg_o a < \frac{1}{2}$.
Also as $a > \frac{1}{2}$ for any z, $max\{a, z, \frac{1}{2}\} \neq \frac{1}{2}$, i.e. $a *_m z = min\{a, z\}$.
So for $a > b$, $a *_m b = min\{a, b\} = b$.
Now, for $z \leq b$, $a *_m z \leq a *_m b = b$. Hence for all z such that $z \leq b$, $z \in \{x : a *_m x \leq b\}$.
Let $z > b$, then $a *_m z > a *_m b = b$. [Since $a *_m z = min\{a, z\} > b$.]
I.e. $max\{x : a *_m x \leq b\} = b$, as there is no $z(> b)$ belongs to the set $\{x : a *_m x \leq b\}$.
Hence for $a > b$, $a \rightarrow_m b = b$, i.e. $(a \rightarrow_m b) *_m (\neg_o a \rightarrow_m b) \leq b$ where $a > b$.
Here it should be noted that for $a > \frac{1}{2}$, the case for $\neg_o a > b$ needs not be checked separately.
(II) If $a < \frac{1}{2}$ then $\neg_o a > \frac{1}{2}$.
In this case, arguing as case (I) we can show that for $\neg_o a > b$, $\neg_o a \rightarrow_m b = b$, i.e. $(a \rightarrow_m b) *_m (\neg_o a \rightarrow_m b) \leq b$ where $\neg_o a > b$.
(III) If $a = \frac{1}{2}$, $\neg_o a = \frac{1}{2}$.
Here we need to check the value of $(a \rightarrow_m b) *_m (\neg_o a \rightarrow_m b)$ for $b < \frac{1}{2}$.
Now $max\{\neg_o a, b, \frac{1}{2}\} = \frac{1}{2} = \neg_o a$. So, $\neg_o a *_m b = \frac{\neg_o a.b}{\neg_o a} = b$.
Also, for $z \leq b$, $\neg_o a *_m z \leq \neg_o a *_m b = b$. Hence for all $z \leq b$, $z \in \{x : \neg_o a *_m x \leq b\}$.
Let $z > b$, then two subcases arise - (a) $b < z \leq \frac{1}{2}$ (b) $b < \frac{1}{2} < z$.
(a) If $b < z \leq \frac{1}{2}$ then $max\{\neg_o a, z, \frac{1}{2}\} = \frac{1}{2} = \neg_o a$ and hence $\neg_o a *_m z = z > b$.
(b) If $b < \frac{1}{2} < z$, $max\{\neg_o a, z, \frac{1}{2}\} = z \neq \frac{1}{2}$ and hence $\neg_o a *_m z = min\{\neg_o a, z\} = \neg_o a > b$.
Combining these two subcases, we have for $z > b$, $\neg_o a *_m z > b$.
Hence $\neg_o a \rightarrow_m b = max\{x : \neg_o a *_m x \leq b\} = b$.

Similarly, it can be shown that $a \to_m b = b$ and hence $(a \to_m b) *_m (\neg_o a \to_m b) \le b$ for $a > b$ and $\neg_o a > b$.

Example 3.8 Let the meta-structure be $\langle [0, 1], *_m, \to_m, 0, 1\rangle$,
where $a *_m b = 0$, if $a + b \le 1$
$= min\{a, b\}$, otherwise [Fodor's t-norm] (Klir et al. 2006)
and the operator $\neg_o$ for object-level negation $\neg$ be defined by $\neg_o a = 1 - a$.
Now three cases arise—(I) $a < \frac{1}{2}$, (II) $a = \frac{1}{2}$ and (III) $a > \frac{1}{2}$.
(I) If $a < \frac{1}{2}$ then $\neg_o a > \frac{1}{2}$. Let $\neg_o a > b$.
Hence two subcases arise. (A) $a > b$ (B) $a \le b < \neg_o a$.
(I) (A) Let $a > b$, i.e. $b < a, \neg_o a$.
So, $\neg_o a + b < \neg_o a + a = 1$ hence $\neg_o a *_m b = 0$.
Then for all z, $z \le b$, $\neg_o a *_m z = 0$, i.e. $z \in \{x : \neg_o a *_m x \le b\}$ for all $z \le b$.
Let $z > b$. Then either $z > a > b$ or $a \ge z > b$.
For the first case, i.e. $z > a > b$, $\neg_o a + z > \neg_o a + a = 1$.
So, $\neg_o a *_m z = min\{\neg_o a, z\} > b$ and hence any z such that $z > a > b$ does not belong to the set $\{x : \neg_o a *_m x \le b\}$.
Now for the second case, i.e. $a \ge z > b$, $1 = \neg_o a + a \ge \neg_o a + z$, i.e. $\neg_o a *_m z = 0$.
Hence $z \in \{x : \neg_o a *_m x \le b\}$ where $a \ge z > b$.
I.e. z such that $z \le b < a$ and $a \ge z > b$, belongs to $\{x : \neg_o a *_m x \le b\}$ where $a < \frac{1}{2}$.
So, $\neg_o a \to_m b = \{x : \neg_o a *_m x \le b\} \le a$ for $a < \frac{1}{2}$ and $a > b$. …(i)
On the other hand, as $b < a < \frac{1}{2}$, $a + b < 1$, i.e. $a *_m b = 0$.
Hence, for all z such that $z \le b < a < \neg_o a$, $a *_m z = 0$, i.e. $z \in \{x : a *_m x \le b\}$.
If $z > b$, then again either $z > \neg_o a > b$ or $\neg_o a \ge z > b$.
For the first option $a + z > a + \neg_o a = 1$, i.e. $a *_m z = min\{a, z\} > b$.
So, any z such that $z > \neg_o a > b$ does not belong to the set $\{x : a *_m x \le b\}$.
For the second option, $1 = \neg_o a + a \ge z + a$, i.e. $a *_m z = 0$.
Hence $z \in \{x : a *_m x \le b\}$ for all z such that $\neg_o a \ge z > b$.
That is, all z such that $z \le b < \neg_o a$ or $\neg_o a \ge z > b$, z belongs to $\{x : a *_m x \le b\}$ where $a < \frac{1}{2}$ and $a > b$.
Hence $a \to_m b = max\{x : a *_m x \le b\} \le \neg_o a$ for $a < \frac{1}{2}$ and $a > b$. …(ii)
Hence, from (i) and (ii), $(a \to_m b) + (\neg_o a \to_m b) \le \neg_o a + a = 1$, i.e.
$(a \to_m b) *_m (\neg_o a \to_m b) = 0 \le b$.
(B) Now let $a \le b < \neg_o a$. Then $1 = \neg_o a + a \le \neg_o a + b$.
That is, either $\neg_o a + b = 1$ or $\neg_o a + b > 1$.
If $\neg_o a + b = 1$ then $\neg_o a * b = 0$ and for all z such that $z \le b$, $\neg_o a *_m z = 0$.
If $z > b$ then $z > b \ge a$, i.e. $\neg_o a + z > \neg_o a + b \ge \neg_o a + a = 1$.
So, $\neg_o a *_m z = min\{\neg_o a, z\} > b$, i.e. any z such that $z > b \ge a$ does not belong to the set $\{x : \neg_o a *_m x \le b\}$.
On the other hand, if $\neg_o a + b > 1$ then $\neg_o a *_m b = min\{\neg_o a, b\} = b$ and for all z such that $z \le b$, $\neg_o a *_m z \le \neg_o a * b = b$.
But if $z > b$, i.e. $z > b \ge a$, i.e. $\neg_o a + z > \neg_o a + b \ge \neg_o a + a = 1$.
So, $\neg_o a *_m z = min\{\neg_o a, z\} > b$, i.e. any z such that $z > b \ge a$ does not belong to the set $\{x : \neg_o a *_m x \le b\}$.

Thus from the above cases we have $\neg_o a \rightarrow_m b = b$, i.e. $(a \rightarrow_m b) *_m (\neg_o a \rightarrow_m b) \leq b$.
(II) Let $a, \neg_o a = \frac{1}{2}$. Then we need to check $(a \rightarrow_m b) *_m (\neg_o a \rightarrow_m b)$ for $b < \frac{1}{2}$. As $b < \frac{1}{2}, a + b < a + \frac{1}{2} = 1$, i.e. $a *_m b = 0$, and for $z \leq b$, $a *_m z = 0$. ...(iii)
If $z > b$, then two subcases arise-(C) $z > \frac{1}{2} > b$ (D) $\frac{1}{2} \geq z > b$.
(C) If $z > \frac{1}{2} > b$ then $\frac{1}{2} + z > \frac{1}{2} + \frac{1}{2} = 1$, i.e. $a + z > 1$.
So, $a *_m z = min\{a, z\} = a = \frac{1}{2} > b$. Hence z does not belong to the set $\{x : a *_m x \leq b\}$.
(D) If $\frac{1}{2} \geq z > b$ then $1 = \frac{1}{2} + \frac{1}{2} \geq \frac{1}{2} + z$, i.e. $a + z \leq 1$ and hence $a *_m z = 0$.
That is, all z such that $\frac{1}{2} \geq z > b$ belong to the set $\{x : a *_m x \leq b\}$. ...(iv)
So, combining (iii) and (iv) we have $a \rightarrow_m b = max\{x : a *_m x \leq b\} \leq \frac{1}{2}$ for $a = \frac{1}{2}$.
As $\neg_o a = \frac{1}{2}$ with similar argument it can be shown that $\neg_o a \rightarrow_m b \leq \frac{1}{2}$.
Hence $(a \rightarrow_m b) *_m (\neg_o a \rightarrow_m b) \leq \frac{1}{2} + \frac{1}{2} = 1$ implies $(a \rightarrow_m b) *_m (\neg_o a \rightarrow_m b) = 0$.
(III) This case is the same as the case for (I). Here $\neg_o a < \frac{1}{2}$ and $a > \frac{1}{2}$. Hence, assuming $a > b$, we can check all possible relations between $\neg_o a$ and b. Argument will be exactly the same as case (I).
Hence, it is proved that $(a \rightarrow_m b) *_m (\neg_o a \rightarrow_m b) \leq b$ if $a > b$ or $\neg_o a > b$.

Note 3.8 So the structures given in Examples 3.7 and 3.8 ensure the necessary condition for GC5 as well as $GC5^C$. But this condition is not sufficient. In some cases, imposition of some conditions on the collection $\{T_i\}_{i \in I}$ may help to ensure the availability of $GC5^C$. The following example shows that Fodor's structure is not sufficient to ensure $GC5^C$.

Example 3.9 Let us consider a collection of fuzzy subsets, viz. $\{T_1, T_2, T_3\}$ whose value assignment to a set of formulae, viz. $\{\alpha, \beta, \alpha_1, \alpha_2, \alpha_3\}$ is given by the following table:

	α	β	α_1	α_2	α_3
T_1	.7	.5	.9	.8	.9
T_2	.6	.4	.8	.8	.9
T_3	.8	.7	.5	.7	.9

Let the operator $\neg_o$ for object-level negation be defined by $\neg_o a = 1 - a$ and the meta-structure be $\langle [0, 1], *_m, \rightarrow_m, 0, 1 \rangle$, where $*_m$ is the Fodor's t-norm. Let $X = \{\alpha_1, \alpha_2, \alpha_3\}$.
Now, we have to calculate the residua of $*_m$, i.e. $\rightarrow_m$. The definition of $*_m$ is given by, $a *_m b = 0$ if $a + b \leq 1$.
$= a \wedge b$ otherwise.
Now for any $a, b \in L$, two cases arise. (I) $a > b$ and (II) $a \leq b$.
(I) Let $a > b$. Then again two subcases arise. (A) $a + b \leq 1$ and (B) $a + b > 1$.
(A) If $a + b \leq 1$ then $a *_m b = 0$. Now for any $z \in [0, 1]$ either $z \leq b$ or $z > b$.
For $z \leq b$, $a *_m z \leq a *_m b = 0$. Hence these z's belong to the set $\{x : a *_m x \leq b\}$.
For $z > b$, $a + z > a + b$. Now as $a + b \leq 1$, two subcases arise.
(i) $a + z > 1 \geq a + b$ and (ii) $1 \geq a + z \geq a + b$.

(i) If $a + z > 1 \geq a + b$ then $a *_m z = a \wedge z > b$. [Since $z > b$ and $a > b$.]
Hence these z's do not belong to the set $\{x : a *_m x \leq b\}$.
(ii) If $1 \geq a+z \geq a+b$ then $a *_m z = 0$. So, these z's belong to the set $\{x : a *_m x \leq b\}$.
Hence, for case (I) only those z's who satisfy $z \leq b$ or $z > b$ and $1 \geq a + z \geq a + b$, belong to $\{x : a *_m x \leq b\}$. $1 \geq a + z$ implies $z \leq 1 - a$.
That is, for $a > b$ and $a + b \leq 1$, z such that $z \leq b$ or $b < z \leq 1 - a$ belongs to the set $\{x : a *_m x \leq b\}$.
Hence $a \rightarrow_m b = max\ \{x : a *_m x \leq b\} = 1 - a$, if $a > b$ and $a + b \leq 1$.
(B) Let $a + b > 1$. Then $a *_m b = a \wedge b = b$. [Since $a > b$.]
Let $z \leq b$. Then $a *_m z \leq a *_m b = b$.
And for $z > b$, $a + z > a + b > 1$. So, $a *_m z = a \wedge z > b$. [Since $a > b$, $z > b$.]
Hence for $a > b$ and $a + b > 1$ only those z's for which $z \leq b$ holds, belong to $\{x : a *_m x \leq b\}$.
Hence $a \rightarrow_m b = max\ \{x : a *_m x \leq b\} = b$ for $a > b$ and $a + b > 1$.
(II) Let $a \leq b$. Then as $a *_m 1 = a$ and $a \leq b$, $1 \in \{x : a *_m x \leq b\}$.
Hence $a \rightarrow_m b = 1$ for $a \leq b$.
So, $\rightarrow_m$ is defined by $a \rightarrow_m b = 1$ if $a \leq b$
$= b$ if $a > b$ and $a + b > 1$
$= 1 - a$ if $a > b$ and $a + b \leq 1$.
Then $gr(X \cup \{\beta\} \mid\!\sim \alpha) = inf\{.5 \rightarrow_m .7, .4 \rightarrow_m .6, .5 \rightarrow_m .8\} = 1$.
$gr(X \cup \{\neg\beta\} \mid\!\sim \alpha) = inf\{.5 \rightarrow_m .7, .6 \rightarrow_m .6, .3 \rightarrow_m .8\} = 1$.
$gr(X \mid\!\sim \alpha) = inf\{.8 \rightarrow_m .7, .8 \rightarrow_m .6, .5 \rightarrow_m .8\} = .6$.
Hence $gr(X \cup \{\beta\} \mid\!\sim \alpha) *_m gr(X \cup \{\neg\beta\} \mid\!\sim \alpha) > gr(X \mid\!\sim \alpha)$.

Thus, Examples 3.8 and 3.9 show that Fodor's structure satisfies the necessary condition for $GC5^C$, but it is not sufficient for obtaining the same.

Let us now present what suffices to have $GC5^C$.

Theorem 3.10 *(A sufficient condition for* $GC5^C$*) Let* $(L, *_m, \rightarrow_m, 0, 1)$ *be a complete residuated lattice satisfying* $x \vee \neg_o x = 1$ *for all* $x \in L$. *Then for any graded consequence relation* $\approx_{\{T_i\}_{i\in I}}$ *with the above structure for the meta-language GC5 holds.*

Proof $gr(X \cup \{\alpha\} \approx_{\{T_i\}_{i\in I}} \beta) *_m gr(X \cup \{\neg\alpha\} \approx_{\{T_i\}_{i\in I}} \beta)$
$= inf_i\{(inf_{\gamma\in X} T_i(\gamma) \wedge T_i(\alpha)) \rightarrow_m T_i(\beta)\} *_m inf_i\{(inf_{\gamma\in X} T_i(\gamma) \wedge T_i(\neg\alpha)) \rightarrow_m T_i(\beta)\}$.
$\leq inf_i[\{(inf_{\gamma\in X} T_i(\gamma) \wedge T_i(\alpha)) \rightarrow_m T_i(\beta)\} *_m \{(inf_{\gamma\in X} T_i(\gamma) \wedge T_i(\neg\alpha)) \rightarrow_m T_i(\beta)\}]$.
[Since $inf_i\{a_i\} *_m inf_i\{b_i\} \leq inf_i\{a_i *_m b_i\}$ (see (r11) in Chap. 2).]
$\leq inf_i[\{(inf_{\gamma\in X} T_i(\gamma) *_m T_i(\alpha)) \rightarrow_m T_i(\beta)\} *_m \{(inf_{\gamma\in X} T_i(\gamma) *_m T_i(\neg\alpha)) \rightarrow_m T_i(\beta)\}]$.
[Since $a *_m b \leq a \wedge b$ and $\rightarrow_m$ is antitone in the first variable.]
$= inf_i[\{(inf_{\gamma\in X} T_i(\gamma) *_m T_i(\alpha)) \rightarrow_m T_i(\beta)\} *_m \{(inf_{\gamma\in X} T_i(\gamma) *_m \neg_o T_i(\alpha)) \rightarrow_m T_i(\beta)\}]$.
$= inf_i[\{T_i(\alpha) \rightarrow_m (inf_{\gamma\in X} T_i(\gamma) \rightarrow_m T_i(\beta))\} *_m \{\neg_o T_i(\alpha) \rightarrow_m (inf_{\gamma\in X} T_i(\gamma) \rightarrow_m T_i(\beta))\}]$.
[Since in a residuated lattice $(x *_m y) \rightarrow_m z = x \rightarrow_m (y \rightarrow_m z) = y \rightarrow_m (x \rightarrow_m z)$.]
$\leq inf_i\{(T_i(\alpha) \vee \neg_o T_i(\alpha)) \rightarrow_m (inf_{\gamma\in X} T_i(\gamma) \rightarrow_m T_i(\beta))\}$.
[Since in a residuated lattice $(x \rightarrow_m z) *_m (y \rightarrow_m z) \leq (x \vee y) \rightarrow_m z$ holds].
$= inf_i\{1 \rightarrow_m (inf_{\gamma\in X} T_i(\gamma) \rightarrow_m T_i(\beta))\}$. [Since $x \vee \neg_o x = 1$] $=$
$gr(X \approx_{\{T_i\}_{i\in I}} \beta)$. □

Below, we present some examples where $GC5^C$ hold.

Example 3.10 Let us consider [0,1] with respect to any adjoint pair $(*_m, \rightarrow_m)$ as the meta-structure for a graded consequence relation. Let $\neg$ be computed by the operator $\neg_o$, defined by $\neg_o a = 0$ for $a > 0$ and 1 otherwise (i.e. Gödel negation). Then the sufficient condition for GC5 can be easily verified.

Example 3.11 Let us consider a complete residuated lattice $([0, 1], *_m, \rightarrow_m, 0, 1)$, generated by any t-norm $*_m$, with
$\neg_o a = 0$, if $a = 1$
$= 1$, otherwise.
(i.e. negation corresponding to the largest R-implication which is defined by
$a \rightarrow_o b = b$ if $a = 1$
$= 1$, otherwise.)
In this case, $x \vee \neg_o x = 1$ for all $x \in [0, 1]$.
Then $gr(X \cup \{\alpha\} \mid\!\sim \beta) *_m gr(X \cup \{\neg\alpha\} \mid\!\sim \beta) \leq gr(X \mid\!\sim \beta)$.

The motivation for introducing GC5 with some non-zero factor c may be visible from the respective sufficient conditions for GC5 and the same in the special case with $c = 1$. In Theorem 3.8, we can see all values of the form $x \vee \neg_o x$ has to be greater or equal to some non-zero threshold c. On the other hand, in the special case of GC5 with $c = 1$, which actually renders the standard principle of reasoning by cases, Theorem 3.10 demands $x \vee \neg_o x$ to be exactly 1 as a sufficient condition. Reasoning by cases, in its standard form, reflects the classical nature of dichotomy that a formula and its negation have to have. In many-valued context, this does not work. For example, in none of these standard many-valued logics, viz. Łukasiewicz logic, Gödel logic and Product logic, reasoning by cases holds. Theorem 3.8 also shows that in the graded context c behaves like a monoidal identity for all elements lying below it, and all elements of the lattice are comparable to c. Combining all these conditions, which c has to satisfy in order to ensure GC5, it is visible that c, in graded context, is the least element of the *truth zone* (cf. Sect. 3.3). Significance of (GC4) and (GC5) as generalizations of *explosiveness condition* and *reasoning by cases*, respectively, will be more clear in Sect. 3.3.

3.3 GC($\neg$) Algebra and its Properties

Classical logic witnesses the simultaneity of GC4 and GC5 in two-valued context. In this section, we would like to explore some structures for graded consequence where GC4 and GC5 both hold simultaneously, that is, these structures would work as models for all the axioms GC1, GC2, GC3, GC4 and GC5. Later, with the help of these examples, we will arrive at a sufficient condition for the meta-level algebraic structure, putting constraints on the structure of complete residuated lattice, so that GC4 and GC5 hold together.

Example 3.12 Let the meta-level truth structure be a complete pseudo-Boolean algebra (i.e. a complete residuated lattice with lattice meet as the monoidal operation) $(L, \wedge, \rightarrow_m, 0, 1)$ where the set L consists of the elements, viz. $0, a, k', c, b, 1$. The order relation of the lattice is given by the following diagram.

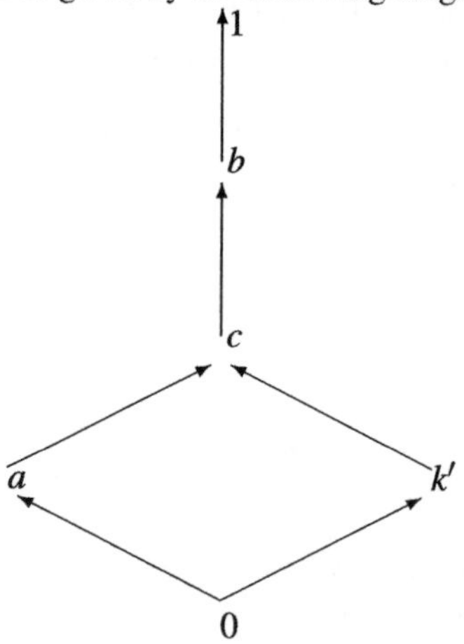

Let $\neg_o$ be the operator for object-level negation defined by the following table:

x	$\neg_o x$	$x \wedge \neg_o x$	$x \vee \neg_o x$
0	1	0	1
a	k'	0	c
k'	b	k'	b
c	k'	k'	c
b	k'	k'	b
1	0	0	1

The negation operator $\neg_o$ satisfies the following conditions:
(1) $x \leq y$ implies $\neg_o y \leq \neg_o x$, (2) $x \leq \neg_o \neg_o x$, (3) $x \wedge \neg_o x \leq k'$ and (4) $x \vee \neg_o x \geq c$.
GC4: Here we see that for any x, $x \wedge \neg_o x \leq k'$ and $k' \wedge a = 0$. So the sufficient condition for GC4 holds. Hence for any collection $\{T_i\}_{i\in I}$, for any α, β and for any T_i of the collection $\{T_i\}_{i\in I}$, $(T_i(\alpha) \wedge \neg_o T_i(\alpha)) \rightarrow_m T_i(\beta) \geq k' \rightarrow_m 0 = a > 0$, where $\rightarrow_m$ is the residua of $\wedge$. That is, for any α, β, $gr(\{\alpha, \neg\alpha\} \mid\!\sim \beta) \geq a$.
Hence for any β, $inf_\alpha gr(\{\alpha, \neg\alpha\} \mid\!\sim \beta) \geq a$. …(i)
Now let $\{T_1, T_2, T_3\}$ be a particular collection of fuzzy subsets over F and for some α, β, the values assigned to α, β are as follows:

	α	$\neg_o\alpha$	β
T_1	1	0	b
T_2	a	k'	0
T_3	k'	b	a

With respect to $\{T_i\}_{i=1,2,3}$, $gr(\{\alpha, \neg\alpha\} \mathrel{|\!\sim} \beta) = inf_i\{(T_i(\alpha) \wedge T_i(\neg\alpha)) \rightarrow_m T_i(\beta)\}$
$= a > 0.$

So, with respect to the collection $\{T_i\}_{i=1,2,3}$, $inf_\beta gr(\{\alpha, \neg\alpha\} \mathrel{|\!\sim} \beta) \leq a$. ...(ii)
Hence, combining (i) and (ii), we can conclude that with respect to $\{T_1, T_2, T_3\}$, $inf_\beta gr(\{\alpha, \neg\alpha\} \mathrel{|\!\sim} \beta) = a > 0$.
GC5: To check GC5, it suffices to check the sufficient condition for GC5. In this example, the possible values for $x \vee \neg_o x$ are $c, b, 1$, i.e. $x \vee \neg_o x \geq c$ and either $x \leq c$ or $x > c$ for all x in L. Also $c \wedge z = z$ for all $z \leq c$. Also $c \rightarrow_m 0 = 0$, $c \rightarrow_m a = a$ and $c \rightarrow_m k' = k'$ verify Lemma 3.2 which states that for all $z \leq c$, $c \rightarrow_m z = z$. So, GC5 is satisfied here.

Example 3.13 Let us consider $(L, \wedge, \vee, *_m, \rightarrow_m, 0, 1)$, a residuated lattice to be the meta-structure for a graded consequence relation where $L = \{0, a, b, k', c, 1\}$. The order relation and the table for $*_m$ and $\rightarrow_m$ are given as follows:

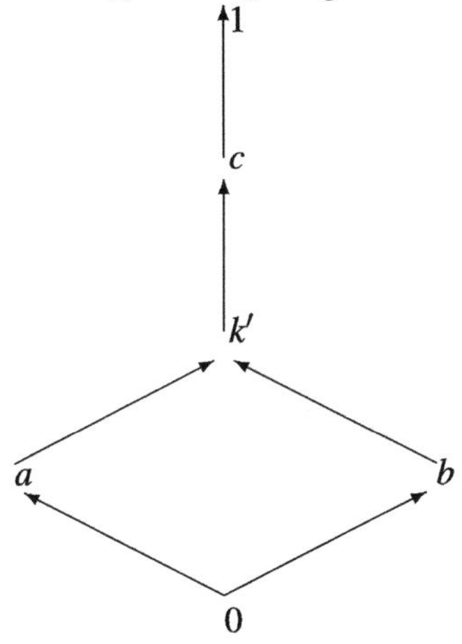

$*_m$	0	a	b	k'	c	1
0	0	0	0	0	0	0
a	0	0	0	0	a	a
b	0	0	0	0	b	b
k'	0	0	0	0	k'	k'
c	0	a	b	k'	c	c
1	0	a	b	k'	c	1

$\rightarrow_m$	0	a	b	k'	c	1
0	1	1	1	1	1	1
a	k'	1	k'	1	1	1
b	k'	k'	1	1	1	1
k'	k'	k'	k'	1	1	1
c	0	a	b	k'	1	1
1	0	1	b	k'	c	1

Let us define $\neg_o$, the operator for the object-level negation by the following table:

$\neg_o$	0	a	b	k'	c	1
	1	c	c	c	k'	0

$x \wedge \neg_o x \leq k'$ and $x \vee \neg_o x \geq c$.

Also $\neg_o$ satisfies $x \leq y$ implies $\neg_o y \leq \neg_o x$ and $x \leq \neg_o \neg_o x$.

Also from the table of $*_m$ it is clear that $c *_m z = z$ for all $z \leq c$ and for all $x \in L$ either $x \leq c$ or $x > c$. Hence, in this example it can be proved that GC5 holds. On the other hand, as $x \wedge \neg_o x \leq k'$ and $k'(\neq 0)$ has a zero divisor in L, GC4 holds for any graded consequence relation with $inf_\beta\ gr(\{\alpha, \neg\alpha\} \mid\!\sim \beta) \geq k' \rightarrow_m 0 = k' > 0$.

Example 3.14 Let us consider $(L, \wedge, \vee, *_m, \rightarrow_m, 0, 1)$, a residuated lattice to be the meta-structure of a graded consequence relation where $L = \{0, a, k', c, d, 1\}$. Along with the order relation of the lattice the tables for $*_m$ and $\neg_o$ also are given below:

$$0 \longrightarrow a \longrightarrow k' \longrightarrow c \longrightarrow d \longrightarrow 1$$

$*_m$	0	a	k'	c	d	1
0	0	0	0	0	0	0
a	0	0	0	a	a	a
k'	0	0	k'	k'	k'	k'
c	0	a	k'	c	c	c
d	0	a	k'	c	c	d
1	0	a	k'	c	d	1

$\neg_o$	0	a	k'	c	d	1
	1	d	c	k'	a	0

Hence, $x \wedge \neg_o x \leq k'$, $x \vee \neg_o x \geq c$, $x \leq y$ implies $\neg_o y \leq \neg_o x$ and $x = \neg_o \neg_o x$. As L is a linear structure, condition (i) of the Theorem 3.8 holds immediately. Also, $x \vee \neg_o x \geq c$ holds and the table for $*_m$ shows that for such a c the condition (iii) of 3.8 holds. So this example ensures the availability of GC5. On the other hand, $x \wedge \neg_o x \leq k'$ satisfies the first condition of Theorem 3.7 and from the table for $*_m$ the condition (ii) of 3.7 is also ensured as k' has a divisor of zero, and that is a. That is, for all α, β and for all i, $(T_i(\alpha) \wedge T_i(\neg\alpha)) \rightarrow_m T_i(\beta) \geq k' \rightarrow_m 0$. Hence, GC4 holds with the threshold $k' \rightarrow_m 0 = a > 0$.

From the above examples, it can be noticed that both the sufficient conditions for (GC4) and (GC5) of Theorems 3.7 and 3.8 are satisfied independently. This leads us towards a sufficient condition for both GC4 and GC5 to hold together.

Theorem 3.11 *(A sufficient condition for both GC4 and GC5) Let the meta-level structure be* $(L, *_m, \rightarrow_m, 0, 1)$*, a complete residuated lattice, and* $\neg_o$ *be the operator for object-level negation satisfying the following conditions:*

(*i*) *there is an element* $c(\neq 0) \in L$ *such that for all* $x \in L$, $x \leq c$ *or* $x > c$,
(*ii*) $c *_m z = z$ *for all* $z \leq c$,
(*iii*) $x \vee \neg_o x \geq c$ *for all* $x \in L$, *and*
(*iv*) *there is an element* k' $(0 \leq k' < c) \in L$ *such that* $x \wedge \neg_o x \leq k'$ *for all* $x \in L$ *and if* $k' \neq 0$ *then* $k' *_m p = 0$ *for some* $p(\neq 0)$.

Then for any $\approx\!\!\!|_{\{T_i\}_{i \in I}}$ *defined with respect to the structure* $(L, *_m, \rightarrow_m, 0, 1)$, *GC4 and GC5 hold simultaneously.*

Proof GC4 is ascertained from the condition (iv). (See Theorem 3.7)
Also, conditions (i), (ii) and (iii) assure the existence of GC5. (See Theorem 3.8)
The condition $k' < c$ actually ascertains the consistency between the conditions $c *_m z = z$ for all $z \leq c$ and $k' *_m p = 0$ for some $p \neq 0$.
According to the condition (i), we can have either $k' < c$ or $k' = c$ or $k' > c$.
Let us explore the consequences if instead of $k' < c$, $k' = c$ or $k' > c$ were assumed.
(I) If possible let $k' = c$. Then $k' *_m z = z$ for all $z \leq k'$.
As for k' there exists $p \neq 0$ such that $k' *_m p = 0$, $k' *_m p \neq p$ and hence $p > k'$.
Since, if $p \leq k' = c$, $k' *_m p = p$. Also as $k' *_m p = 0$, $p = 0$. So, contradiction arises.
Since $k' *_m z = z$ for all $z \leq k'$, $k' *_m k' = k'$. Hence we have $k' *_m p \geq k' *_m k' = k'$.
i.e. $0 \geq k'$, i.e. $k' = c = 0$. This contradicts (i). So $k' \neq c$.
(II) If possible let $k' > c$. We take p as above, i.e. for $p(\neq 0)$, $k' *_m p = 0$.
Now two subcases arise. (a) $p \leq c$ and (b) $p > c$.
(a) Let $p \leq c$. Now as $k' > c$, $k' *_m p \geq c *_m p = p$. [Since $p \leq c$]
I.e. $0 \geq p$. Again contradiction arises.
(b) Let $p > c$. So, as $k' > c$, $k' *_m p \geq c *_m c = c$. [By (ii)]
That is $0 \geq c$. This contradicts the condition (i).
Hence from (I) and (II) we can conclude that $k' < c$ ensures the consistency among the conditions, which suffice to have GC4 and GC5 simultaneously. □

From the algebraic semantics, as obtained in Theorem 3.11, it can be noted that in the meta-level structure $(L, \wedge, \vee, *_m, \rightarrow_m, 0, 1)$ apart from the top (1) and the least (0) two more elements need to be designated, and designations of these two elements are directly related to the algebraic semantics of the object language negation. Let us list the properties of the algebraic structure, required to be a model for a graded consequence relation satisfying GC1, GC2, GC3, GC4 and GC5, in the definition (Dutta 2014) below.

Definition 3.6 An algebraic structure $(L, \wedge, \vee, *_m, \rightarrow_m, k', c, 0, 1)$, along with an operator $\neg_o$, is said to be a $GC(\neg)$ algebra if

(i) $(L, \wedge, \vee, *_m, \rightarrow_m, 0, 1)$ is a complete residuated lattice,
(ii) $k', c \in L$ are such that $0 \leq k' < c \leq 1$,
(iii) $L = Z_T \cup Z_T^c$ such that $Z_T = \{x \in L : c \leq x \leq 1\}$ and $Z_T^c = \{x \in L : 0 \leq x < c\}$,
(iv) $(Z_T^c \cup \{c\}, *_m, c)$ is a commutative monoid,
(v) $x \vee \neg_0 x \geq c$ for all $x \in L$,
(vi) $x \wedge \neg_o x \leq k'$ for all $x \in L$ and

(vii) if $k' \neq 0$, then there is a $z(\neq 0) \in L$ such that $k' *_m z = 0$.

Usually by an algebraic structure, we mean a non-empty set and a finite number of operations defined over it. In the context of GC(¬) algebra, two different algebraic structures $(L, \wedge, \vee, \neg_o, 0, 1)$ and $(L, \wedge, \vee, *_m, \rightarrow_m, 0, 1)$ are involved; that is, two sets of operations, one for the object level and the other for the meta-level, are defined over the same set. The former one is for the object-level language and the latter one is for the meta-level language. So, while studying the meta-linguistic property of GCT, in the presence of ¬ in the object language, we prefer not to explicitly mention the object-level operator $\neg_o$ within the body of the algebraic structure of GC(¬) algebra (cf. Definition 3.6).

Classically, 0 is considered to be the algebraic counterpart for falsity, the value which is always taken by a formula of the form 'α *and not-*α', representing the law of contradiction, and 1 denotes the algebraic counterpart for truth, the value which is always taken by a formula of the form 'α *or not-*α', representing the law of excluded middle. In the present context, we have two more values, apart from the least and the top, to be distinguished; k' can be considered as the threshold for the law of contradiction, and c can be considered as the threshold for the law of excluded middle. In order to explain the reason behind considering such names for k' and c, let us assume that the object language has conjunction and disjunction the respective algebraic operators being $\otimes$ and $\oplus$. We know that for any standard monoidal conjunction $\otimes$ (e.g. t-norm), and corresponding disjunction $\oplus$ (e.g. co-norm), $x \otimes \neg_o x \leq x \wedge \neg_o x \leq k'$ and $c \leq x \vee \neg_o x \leq x \oplus \neg_o x$ hold. Thus, the names for k' and c are used to address the above-mentioned intuition.

We write $Z_T^c = Z_F \cup Z_M$, where $Z_F = \{x \in Z_T^c : x \leq k'\}$ and $Z_M = \{x \in Z_T^c : x > k'$ or x is non-comparable to $k'\}$. We may call Z_T, Z_F and Z_M as the *truth zone*, *falsity zone* and *middle zone*, respectively.

The idea of truth zone and falsity zone is similar to the algebraic model of classical linear logic, viz., CL-algebra (Sen 2000; Troelstra 1992). In this algebra four constants $T, 1, 0, \perp$ are indicated. The characteristics of the subsets $\{x : 1 \leq x \leq T\}$ and $x : \perp \leq x \leq 0\}$ are akin to (but not exactly the same) the two zones presented here.

A diagram representing GC(¬)-algebra may look like as follows.

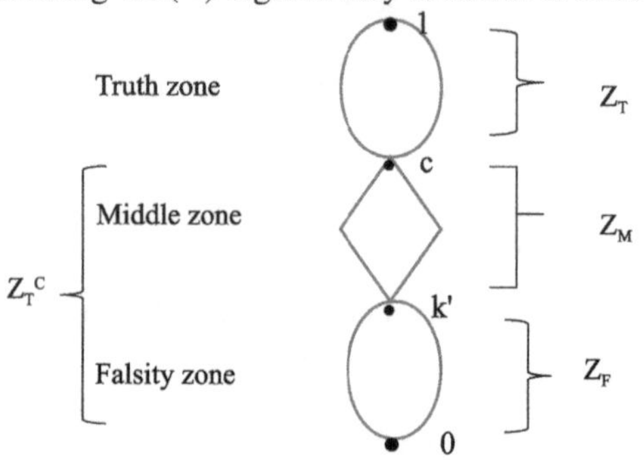

Below, we present a series of properties (Dutta 2014) of this newly obtained algebraic structure.

Proposition 3.3 *For a* $GC(\neg)$*-algebra* L*, if* $a \in Z_T^c \cup \{c\}$*,* $b \in Z_T$*, then* $a *_m b^n = a$ *where* b^n *represents* n *times composition of* b *with itself with respect to the operation* $*_m$ *and* n *is any positive integer.*

Proof We prove the theorem by applying induction on n, the power of b.
For $n = 1$, we know $a *_m b \leq a$. Now as $b \in Z_T$, $c \leq b$, and hence $a *_m c \leq a *_m b$, i.e. $a \leq a *_m b$ as $a \in Z_T^c \cup \{c\}$ and c is the monoidal identity for $Z_T^c \cup \{c\}$.
So, $a *_m b = a$. Let $a *_m b^{n-1} = a$ hold for some positive integer $n - 1$.
So, it is immediate that $a *_m b^n = (a *_m b^{n-1}) *_m b = a *_m b = a$. □

Proposition 3.4 $(Z_T, *_m, 1)$ *is a commutative monoid.*

Proposition 3.5 $(Z_T^c \cup \{c\}, \wedge, \vee, 0, c)$ *is a complete lattice.*

Proof Let $\{a_i\}_{i \in I}$ be a collection of elements from $Z_T^c \cup \{c\} \subseteq L$. Then $a_i \leq c$ for each $i \in I$. As L is a complete lattice, $\vee_{i \in I} a_i$ exists. Now $a_i \leq c$ for each $i \in I$ implies $\vee_{i \in I} a_i \leq c$, i.e. $\vee_{i \in I} a_i \in Z_T^c \cup \{c\}$. Similarly, for L being a complete lattice $\wedge_{i \in I} a_i$ exists, and $\wedge_{i \in I} a_i \leq \vee_{i \in I} a_i \leq c$. Hence $\wedge_{i \in I} a_i \in Z_T^c \cup \{c\}$. So, $(Z_T^c \cup \{c\}, \wedge, \vee, 0, c)$ is a complete lattice. □

Proposition 3.6 $(Z_T, \wedge, \vee, c, 1)$ *is a complete lattice.*

Below we explore whether these substructures, such as Z_T, $Z_T^c \cup \{c\}$, of L also form residuated lattices with respect to the restrictions of $\rightarrow_m$, say $\rightarrow_{Z_T}$ and $\rightarrow_{Z_T^c \cup \{c\}}$ on the respective sets.

Theorem 3.12 $(Z_T^c \cup \{c\}, \wedge, \vee, *_m, \rightarrow_{Z_T^c \cup \{c\}}, 0, c)$ *forms a complete residuated lattice where for* $x, y \in Z_T^c \cup \{c\}$*,* $x \rightarrow_{Z_T^c \cup \{c\}} y = \{z \in Z_T^c \cup \{c\} : x *_m z \leq y\}$*.*

Proof As by condition (iv) of Definition 3.6 $(Z_T^c \cup \{c\}, *_m, c)$ is a monoid, in order to prove the theorem we need to show that $(*_m, \rightarrow_{Z_T^c \cup \{c\}})$ forms an adjoint pair on $Z_T^c \cup \{c\}$, i.e. for any $x, y, z \in Z_T^c \cup \{c\}$, $x *_m z \leq y$ iff $x \leq z \rightarrow_{Z_T^c \cup \{c\}} y$. That is, we have to show for any $x, y \in Z_T^c \cup \{c\}$, $x \rightarrow_{Z_T^c \cup \{c\}} y \in Z_T^c \cup \{c\}$. In this regard, we consider two cases—(i) $x \not\leq y$ and (ii) $x \leq y$.
(i) Let $x \not\leq y$. Now as $x *_m 0 = 0 \leq y$, $0 \in \{z \in Z_T^c \cup \{c\} : x *_m z \leq y\}$, i.e. $\{z \in Z_T^c \cup \{c\} : x *_m z \leq y\}$ is non-empty, and by Proposition 3.5 supremum of the set exists in $Z_T^c \cup \{c\}$.
Let $z' \in \{z \in Z_T^c \cup \{c\} : x *_m z \leq y\}$. So, $x *_m z' \leq y$ implies $z' \leq x \rightarrow_m y$ by condition (i) of Definition 3.6.
That is, $x \rightarrow_m y$ is an upper bound of all z's such that $\{z \in Z_T^c \cup \{c\} : x *_m z \leq y\}$.
We want to show $x \rightarrow_m y \leq c$. If not, then by condition (iii) of Definition 3.6, $c < x \rightarrow_m y$.
So, $x = x *_m c \leq x *_m (x \rightarrow_m y) = y$ as L is a residuated lattice. This contradicts the assumption that $x \not\leq y$. Hence $x \rightarrow_m y \leq c$, and $x \rightarrow_m y \in Z_T^c \cup \{c\}$.
Hence $x \rightarrow_m y = \sup\{z \in Z_T^c \cup \{c\} : x *_m z \leq y\} = x \rightarrow_{Z_T^c \cup \{c\}} y$ for $x \not\leq y$.
(ii) Let $x \leq y$. Therefore, $x *_m c = x \leq y$ and $c \in \{z \in Z_T^c \cup \{c\} : x *_m z \leq y\}$. Also, for all $z \in Z_T^c \cup \{c\}$, $z \leq c$, and hence $x *_m z \leq x *_m c = x \leq y$.
So, $\sup\{z \in Z_T^c \cup \{c\} : x *_m z \leq y\} = c$.

Hence, combining (i) and (ii) we have for any $x, y \in Z_T^c \cup \{c\}$,
$x \rightarrow_{Z_T^c \cup \{c\}} y = x \rightarrow_m y \in Z_T^c \cup \{c\}$ if $x \nleq y$, and
$= c \in Z_T^c \cup \{c\}$ if $x \leq y$. □

Theorem 3.13 *$(Z_T, \wedge, \vee, *_m, \rightarrow_{Z_T}, c, 1)$ forms a complete residuated lattice where for $x, y \in Z_T$, $x \rightarrow_{Z_T} y = \{z \in Z_T : x *_m z \leq y\}$.*

Proof To prove this theorem, we have to show for any $x, y \in Z_T$, $x \rightarrow_{Z_T} y$ exists. Let us consider the following two cases—(i) $x \nleq y$ and (ii) $x \leq y$.
(i) Let $x \nleq y$. As $x *_m c \leq c \leq y$, we have $c \in \{z \in Z_T : x *_m z \leq y\}$. Hence, the set is non-empty, and by Proposition 3.6 the supremum exists.
Now for any $z \in \{z \in Z_T : x *_m z \leq y\}$, $z \leq x \rightarrow_m y$ by condition (i) of Definition 3.6. We want to show that $c \leq x \rightarrow_m y$, i.e. $x \rightarrow_m y \in Z_T$. If not, then $x \rightarrow_m y < c$ by condition (iii) of Definition 3.6. Now L being a complete residuated lattice, $y \leq x \rightarrow_m y < c$. This contradicts the assumption that $y \in Z_T$.
That is, $x \rightarrow_m y \geq c$, or in other words, $x \rightarrow_m y \in Z_T$.
Hence, $\sup\{z \in Z_T : x *_m z \leq y\} = x \rightarrow_m y = x \rightarrow_{Z_T} y$ for $x \nleq y$.
(ii) Let $x \leq y$. That is, $x *_m 1 = x \leq y$, and $1 \in \{z \in Z_T : x *_m z \leq y\}$.
That is, $x \rightarrow_{Z_T} y = \sup\{z \in Z_T : x *_m z \leq y\} = 1$ for $x \leq y$.
Hence, combining (i) and (ii), we have for any $x, y \in Z_T$,
$x \rightarrow_{Z_T} y = x \rightarrow_m y \in Z_T$ if $x \nleq y$, and
$= 1 \in Z_T$ if $x \leq y$. □

Theorem 3.14 *If $x, z \neq 0$ and $x *_m z = 0$, then $x, z < c$.*

Proof Let us arbitrarily fix $x \neq 0$. Now for any $z \in L$, two cases arise—(i) $z \geq c$ and (ii) $z < c$.
(i) If $z \geq c$, then $0 = x *_m z \geq x *_m c$, i.e. $x *_m c = 0$. Now three subcases arise.
(a) $x \in Z_T^c$, (b) $x \in Z_T \setminus \{c\}$ and (c) $x = c$.
(a) For $x \in Z_T^c$, $x *_m c = x = 0$. This contradicts the assumption that $x \neq 0$.
(b) For $x \in Z_T \setminus \{c\}$, $x *_m c \in Z_T$, as Z_T is closed under $*_m$ by Proposition 3.4. So, $x *_m c \neq 0$. This also contradicts the assumption that $x *_m c = 0$.
(c) For $x = c$, $0 = x *_m c = c *_m c = c$. This contradicts the condition (ii) of Definition 3.6. So, from all the subcases of (i), we can conclude that $z < c$.
Similar is the argument for $x < c$. □

Proposition 3.7 *If $k' *_m z = 0$ for some $z\ (\neq 0) \leq k'$, then for any non-zero $x, y \leq z$, $x *_m y = 0$.*

Proof The proof follows directly from the monotonicity property of $*_m$. □

Proposition 3.8 *If for some $z \neq 0$, $k' *_m z = 0$, then $x *_m z = 0$ for all $x \leq k'$.*

Proof The proof follows directly from the monotonicity property of $*_m$. □

Theorem 3.15 *Let for some $a\ (< c) \in Z_T^c$, (i) there is no z such that $a < z < c$, and (ii) for every $x < a$, there is a y non-comparable to a such that $a *_m y = x$. Then a is an idempotent element of L.*

Proof Let $a(< c) \in Z_T^c$ be such that conditions (i) and (ii) of the stated theorem hold. We know $a *_m a \leq a$. If possible let $a *_m a = a_1 < a$. Then by (ii), there is an element $b_1 \in L$ such that b_1 is non-comparable to a and $a *_m b_1 = a_1$.
Then both $a, b_1 \in \{z : a *_m z \leq a_1\}$. Let us consider an element z such that $a, b_1 < z$. Then by (i) $z \geq c$. That is, $z \in Z_T$. So, by Proposition 3.3, $a *_m z = a\ (> a_1)$. Therefore, there is no element $z > a, b_1$ such that $z \in \{x : a *_m x \leq a_1\}$. Hence $a \rightarrow_m a_1 = \max\{x : a *_m x \leq a_1\}$ does not exist. This contradicts the fact that L is a residuated lattice. Hence $a *_m a = a$. □

Proposition 3.9 *If* $x \in Z_T$, *then* $\neg_o x \leq k'$.

Proof As $x \in Z_T$, $x \geq c$. If $\neg_o x \nleq k'$ then either (i) $\neg_o x > k'$ or (ii) $\neg_o x$ is non-comparable to k'. For (i) two subcases arise. (a) $k' < \neg_o x \leq c$ and (b) $k' < c < \neg_o x$.
(a) As $\neg_o x \leq c \leq x$, $x \wedge \neg_o x = \neg_o x > k'$. This contradicts the condition (vi) of Definition 3.6.
(b) For $x \geq c$ and $\neg_o x > c$, $x \wedge \neg_o x \geq c > k'$. This also contradicts the condition (vi) of Definition 3.6.
Now for (ii) either $\neg_o x < c$ or $\neg_o x \geq c$. If $\neg_o x < c \leq x$, $x \wedge \neg_o x = \neg_o x$, that is, $x \wedge \neg_o x$ is non-comparable to k'. This contradicts the condition (vi) of Definition 3.6.
If $\neg_o x \geq c$ and $x \geq c$, $x \wedge \neg_o x \geq c > k'$. This also contradicts the condition (vi) of Definition 3.6.
Hence, combining all the cases we have $\neg_o x \leq k'$. □

Theorem 3.16 *For* $x \leq k'$, *either* $\neg_o x \in Z_T$ *or* $\neg_o x$ *is non-comparable to* x *and* $x \vee \neg_o x = c$.

Proof As $x \leq k'$, $\neg_o x \nleq k'$ [since $\neg_o x \leq k'$ implies $x \vee \neg_o x \leq k' < c$].
Therefore, either (i) $\neg_o x > k'$ or (ii) $\neg_o x$ is non-comparable to k'.
(i) For $\neg_o x > k'$ two subcases arise. (a) $c > \neg_o x > k' \geq x$ and (b) $\neg_o x \geq c > k' \geq x$.
For (a), $x \vee \neg_o x = \neg_o x < c$. This contradicts the condition (v) of Definition 3.6.
For (b), $x \wedge \neg_o x = x \leq k'$, and $x \vee \neg_o x = \neg_o x \geq c$. So, $\neg_o x \in Z_T$.
(ii) Let $\neg_o x$ be non-comparable to k'. Now as $x \leq k'$ and $k' < c$, $x < c$. So either $x \leq \neg_o x$ or x is non-comparable to $\neg_o x$ as $\neg_o x < x \leq k'$ cannot be the case.
Now if $x \leq \neg_o x$, by the condition (v) of Definition 3.6 $x \vee \neg_o x = \neg_o x \geq c > k'$. This contradicts the assumption that $\neg_o x$ is non-comparable to k'.
Hence x and $\neg_o x$ are non-comparable. Now as $x < c$ and $x, \neg_o x$ are non-comparable by condition (iii) of Definition 3.6, $\neg_o x \notin Z_T$. Hence $\neg_o x < c$, which implies $x \vee \neg_o x \leq c$. Now combining this with the condition (v) of Definition 3.6, we have $x \vee \neg_o x = c$. □

Theorem 3.17 *If* $k' < x < c$, *then* x *and* $\neg_o x$ *are non-comparable, and* $x \vee \neg_o x = c$.

Proof First we want to prove x and $\neg_o x$ are non-comparable. If not, then either $x \leq \neg_o x$ or $\neg_o x < x$. For $x \leq \neg_o x$, $x \wedge \neg_o x = x > k'$. This contradicts the condition (vi)

of Definition 3.6. On the other hand for $\neg_o x < x$, $x \vee \neg_o x = x < c$. This contradicts the condition (v) of Definition 3.6. Hence, x and $\neg_o x$ are non-comparable.
Now as $k' < x < c$ and $x, \neg_o x$ are non-comparable neither $\neg_o x \leq k'$ nor $c \leq \neg_o x$. Then either (1) $k' < \neg_o x < c$ or (2) k' and $\neg_o x$ are non-comparable and $\neg_o x < c$.
(1) For $k' < x < c$ and $k' < \neg_o x < c$ we have $x \vee \neg_o x \leq c$. Then by using condition (v) of Definition 3.6 we have $x \vee \neg_o x = c$.
(2) In this case also $x < c$, $\neg_o x < c$, so using similar argument as (1) we can prove $x \vee \neg_o x = c$. □

Proposition 3.10 *For a linearly ordered complete residuated lattice L, if* $z(\neq 0) \in L$ *has a divisor of zero then* $\neg_o z \geq c$.

Proof As $z(\neq 0)$ has a divisor of zero by Theorem 3.14, $z < c$.
Now either $k < z$ or $z \leq k$. If $k < z < c$, then by Theorem 3.17 z and $\neg_o z$ are non-comparable. But as L is linear this cannot be possible. So $z \leq k$.
Now for $z \leq k$, as L is linearly ordered by Theorem 3.16, $\neg_o z \geq c$. □

Proposition 3.11 *If for some* $x \in L$, $k' < x < c$, *then* Z_M *is not a linearly ordered structure and* Z_M *is not a singleton set.*

Proof Let $x \in Z_M$ and $k' < x < c$. Then by Theorem 3.17, x and $\neg_o x$ are non-comparable, and $x \vee \neg_o x = c$. As $x < c$ and $x, \neg_o x$ are non-comparable, $\neg_o x \notin U_1$.
If $\neg_o x \in Z_F$, $\neg_o x \leq k' < x$. This contradicts the fact that x and $\neg_o x$ are non-comparable. So, $\neg_o x \in Z_M$. So, x and $\neg_o x$ are non-comparable to each other and they belong to Z_M. This proves the proposition. □

From the above properties of GC(¬)-algebra, a comparison between classical scenario and that of graded scenario can be made.

1. In classical logic, $k' = 0$ and $c = 1$. c is the identity element with respect to the operator conjoining values, as well as the topmost element of the value set. In the graded context, either $k' = 0$ or $k' \neq 0$. In the second case, k' has a divisor of zero (i.e. there is some $z(\neq 0)$ such that $k' *_m z = 0$). k' can never be 1. Moreover either $c = 1$ or $c \neq 1$. It behaves as an identity for all the elements lying below it. Besides, for any element x of the value set either $x \leq c$ or $x > c$, and c can never be 0.
2. Classically, the value 'true (1)' behaves as an identity element when conjoined to the value 'false (0)'. In graded context, every element of the truth zone is behaving like an identity for the rest of the elements.
3. Classical counterpart of 'negation of true is false' and 'negation of false is true'. This is captured in the graded context as (i) negation of any element of the truth zone lies in the falsity zone, and (ii) negation of an element of the falsity zone either is a member of the truth zone or it belongs to the middle zone. Moreover, in a linear structure, negation of an element belonging to the falsity zone belongs to the truth zone.
4. In the classical context, there are only two values, and they are linearly ordered. In graded context, it turns out that for a linear structure there cannot be any element lying in the middle zone.

3.4 Possible Three-Valued and Four-Valued GC(¬)-Algebras

From the properties discussed in Sect. 3.3 it can be noticed that both, $\neg_o$, the operator for object language negation, and $(*_m, \rightarrow_m)$, the adjoint pair corresponding to the meta-linguistic conjunction and implication, are affecting each other. In this section, we shall observe how the above properties work as the guideline for constructing GC(¬)-algebras for finite value sets; in particular, we shall discuss all possible cases where the value set is of cardinality three and four.

Let us first consider L to be a three-valued set $\{0, \frac{1}{2}, 1\}$ with natural order. The structure is represented by the following diagram.

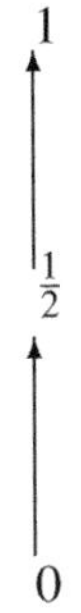

So, maintaining the classical properties of negation we can have the following three choices.

(1)

	$\neg_o$
0	1
$\frac{1}{2}$	$\frac{1}{2}$
1	0

(2)

	$\neg_o$
0	1
$\frac{1}{2}$	0
1	0

(3)

	$\neg_o$
0	1
$\frac{1}{2}$	1
1	0

(1) The first case yields $k' = c = \frac{1}{2}$ as $x \wedge \neg_o x \leq \frac{1}{2}$ as well as $x \vee \neg_o x \geq \frac{1}{2}$. So, following the condition (ii) of Definition 3.6 this choice of $\neg_o$ has to be eliminated. (2) In the second case, for all $x \in L$, $x \wedge \neg_o x = 0$ and $x \vee \neg_o x \geq \frac{1}{2}$. Hence $k' = 0$ and $c = \frac{1}{2}$. So, from the properties of $*_m$, discussed in the previous section, $x *_m 0 = 0$ and $x *_m 1 = x$ are obvious. As $\frac{1}{2}$ plays the role of c, by condition (iv) of Definition 3.6, $\frac{1}{2} *_m \frac{1}{2} = \frac{1}{2}$. So, we have the following definition of $*_m$.

$*_m$	0	$\frac{1}{2}$	1
0	0	0	0
$\frac{1}{2}$	0	$\frac{1}{2}$	$\frac{1}{2}$
1	0	$\frac{1}{2}$	1

This eventually gives us the Gödel algebra for the meta-level algebraic structure, and Gödel negation for the object language negation. That is, GC(¬)-algebra here is a Gödel algebra with Gödel negation for $\neg_o$.

(3) For the third case, as $x \wedge \neg_o x \leq \frac{1}{2}$ and $x \vee \neg_o x = 1$, $k' = \frac{1}{2}$ and $c = 1$. So, from the condition (vii) of Definition 3.6, there must be a non-zero divisor z such that $\frac{1}{2} *_m z = 0$, and the only possibility for such an z is $\frac{1}{2}$ itself. Thus we have the following table for $*_m$:

$*_m$	0	$\frac{1}{2}$	1
0	0	0	0
$\frac{1}{2}$	0	0	$\frac{1}{2}$
1	0	$\frac{1}{2}$	1

This table for $*_m$ is nothing but Łukasiewicz t-norm, and in the literature (Klir et al. 2006) $\neg_o$ is known as drastic negation. So, in this case we obtain the GC(¬)-algebra as the Łukasiewicz algebra with the drastic negation for $\neg_o$.

Let us now consider L to be a four-valued set $\{0, a, b, 1\}$. So, there are only two possible lattice structures.

(a) Non-linear

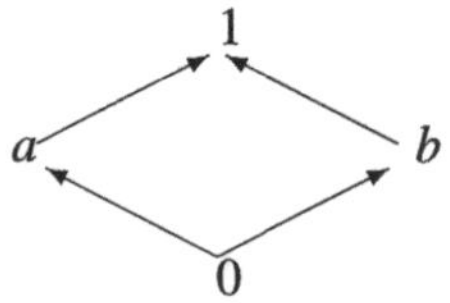

(b) Linear

(a) For the non-linear lattice structure, keeping the classical nature of negation intact, there could be sixteen different choices for $\neg_o$. As a and b are non-comparable, neither of them can be c of Definition 3.6. So, c has to be 1. So, according to condition (v) of Definition 3.6, either $\neg_o a = \neg_o b = 1$ or $\neg_o a = b$ and $\neg_o b = a$. In the first case, the possible values for $x \wedge \neg_o x$ would be 0, a, b, which yields k' to be 0. In the other case for $\neg_o$ we also have $k' = 0$ and $c = 1$. But by Theorem 3.17, $\neg_o a$ has to be b and $\neg_o b$ has to be a. So, we have only one choice for negation.

(1)

$\neg_o$	
0	1
b	a
a	b
1	0

Hence, the constraints of GC(¬)-algebra yield the following table for $*_m$. Rows corresponding to 0 and 1 are immediate. As a and b are non-comparable, $a *_m b$ has to be 0. It can be also noticed that a and b are such that there is no element z such that $a < z < 1(= c)$ and $b < z < 1(= c)$. Moreover, for 0, the only element lying below a and b, $a *_m b = 0$. That means a and b satisfy all the condition of Theorem 3.15. Hence, a and b are idempotent elements.

$*_m$	0	b	a	1
0	0	0	0	0
b	0	b	0	b
a	0	0	a	a
1	0	b	a	1

The table for $*_m$ represents the lattice meet operation. Hence, the GC(¬)-algebra in this context is a Heyting algebra with flip-flop negation.

(b) Following the same line of argument, an extensive study gives the following nine different cases of GC(¬)-algebras with a linear four-valued set. Let us first consider the following four possible operations for object language negation.

(1)

	$\neg_o$
0	1
b	a
a	b
1	0

(2)

	$\neg_o$
0	1
b	1
a	0
1	0

(3)

	$\neg_o$
0	1
b	a
a	0
1	0

(4)

	$\neg_o$
0	1
b	1
a	b
1	0

All these above four cases give $k' = b$ and $c = a$. The induced meta-level $*_m$ is as follows:

$*_m$	0	b	a	1
0	0	0	0	0
b	0	0	b	b
a	0	b	a	a
1	0	b	a	1

In the above table, we need to explain the rows for a and b. As by the condition (iv) of Definition 3.6, a plays the role of monoidal identity for all $x \leq a$, $a *_m b = b$ and $a *_m a = a$. On the other hand, as $k' = b \neq 0$, by condition (vii) of Definition 3.6, b must have a divisor of zero, and for that only possibility is b itself. So, $b *_m b = 0$. The above $*_m$ for meta-linguistic conjunction along with any of the four possible operations for object language negation would form a GC(¬)-algebra.

Another possible operation for object language negation is the following:

	$\neg_0$
0	1
b	1
a	1
1	0

Here $k' = a$ and $c = 1$. As $k' = a \neq 0$, by condition (vii) of Definition 3.6, either $a *_m b = 0$ or $a *_m a = 0$. If $a *_m b = 0$, then by monotonicity $b *_m b \leq a *_m b = 0$. In this case, $a *_m a$ can assume either a or b or 0. Thus, we have Tables (5), (6) and (7), respectively. In case (7), by monotonicity $a *_m a = 0$ yields $a *_m b = b *_m b = 0$.

(5)

$*_m$	0	b	a	1
0	0	0	0	0
b	0	0	0	b
a	0	0	a	a
1	0	b	a	1

(6)

$*_m$	0	b	a	1
0	0	0	0	0
b	0	0	0	b
a	0	0	b	a
1	0	b	a	1

(7)

$*_m$	0	b	a	1
0	0	0	0	0
b	0	0	0	b
a	0	0	0	a
1	0	b	a	1

Assuming $\{0, b, a, 1\}$ as $\{0, \frac{1}{3}, \frac{2}{3}, 1\}$ it can be noticed that (6) is the table for the restriction of Łukasiewicz t-norm on four-valued chain, and $\neg_o$ is the operator for drastic negation.

Below we have one more possible operation for object language negation:

	$\neg_0$
0	1
b	0
a	0
1	0

Here $k' = 0$ and $c = b$. As b is the monoidal identity for $x \leq b$, $b *_m b = b$. Then by monotonicity $b = b *_m b \leq a *_m b$. As $a *_m b \leq b$, we have $a *_m b = b$. Again as by monotonicity $a *_m a \geq a *_m b = b$, either $a *_m a = a$ or $a *_m a = b$. This yields the following two tables for $*_m$:

(8)

$*_m$	0	b	a	1
0	0	0	0	0
b	0	b	b	b
a	0	b	a	a
1	0	b	a	1

(9)

$*_m$	0	b	a	1
0	0	0	0	0
b	0	b	b	b
a	0	b	b	a
1	0	b	a	1

From the above two possible $*_m$'s for GC(¬)-algebra, we can notice that Table (8) is the same as the table for lattice meet. So, considering $\{0, b, a, 1\}$ as $\{0, \frac{1}{3}, \frac{2}{3}, 1\}$, (8) turns out to be a GC(¬)-algebra where the meta-level algebraic structure is a Gödel algebra, and the object language negation is the Gödel negation.

References

Arruda, A.I.: Aspects of the historical development of paraconsistent logic. In: Priest, G., Routley, R., Norman, J. (eds.) Paraconsistent Logic: Essays on the Inconsistent, pp. 99–129. Philosophia Verlag, München, Handen, Wien (1989)

Chakraborty, M.K., Basu, S.: Graded consequence and some metalogical notions generalized. Fundam. Inform. **32**, 299–311 (1997)

Chakraborty, M.K., Dutta, S.: Graded consequence revisited. Fuzzy Sets Syst. **161**, 1885–1905 (2010)

Chattopadhyay, M.: Nagarjuner Darshan Parikrama (Philosophy of Nagarjuna). Jadavpur University, Kolkata (2015)

Dutta, S.: Algebra of negation fragment of a logic with graded notion of consequence. In: S. Bhattacharya Halder, S. Bhowmik (eds.), Rough Sets, Fuzzy Sets and Soft Computing, pp. 203–212. Narosa Publishing House, New Delhi (2014)

Dutta, S.: Graded Consequence as the Foundation of Fuzzy Logic. Doctoral Thesis, University of Calcutta (2011)

Hájek, P.: Metamathematics of Fuzzy Logic. Kluwer Academic Publishers, Dordrecht (1998)

Katsura, S.: Nagarjuna and the Tetralemma, in Buddhist studies. In: Silk, J.A. (ed.) The legacy of Godjin M. Nagas. Motilal Banarsidass Pvt. Ltd., Delhi (2008)

Klir, George J., Yuan, Bo.: Fuzzy Sets And Fuzzy Logic: Theory and Applications. Prentice Hall of India Private Limited, New Delhi (2006)

Priest, G., Routley, R.: A preliminary history of paraconsistent and dialethic approaches. In: Priest, G., Routley, R., Norman, J. (eds.) Paraconsistent Logic: Essays on the Inconsistent, pp 3–75. Philosophia Verlag, München, Hamden, Wien (1989)

Sen, J.: Doctoral thesis on 'Some embeddings in linear logic and related issues'. University of Calcutta (2000)

Surma, S.J.: The growth of logic out of the foundational research in mathematics. In: Agazzi, E. (ed.) Modern Logic–A Survey, pp. 15–33. D. Reidel Publishing Co., Dordrecht (1981)
Troelstra, A.S.: Lectures on Linear Logic, vol. 29. CSLI, Standford (1992)
Wittgenstein, L.: Philosophical Remarks. Blackwell, London (1975)

Chapter 4
Proof Theoretic Rules in Graded Consequence: From Semantic Perspective

Abstract The proof theory of a logic deals with the rules of inference corresponding to the logical connectives in the object language. As the theory of graded consequence is not a particular logic, rather a general scheme for logics dealing with multilayered many-valuedness; in this chapter, we shall present the necessary and sufficient conditions for getting specific rules corresponding to logical connectives of the object language. That is, we shall add connectives, say #, in the language and explore the necessary and sufficient conditions to get hold of particular rules concerning #. This study will help us to present a general scheme for generating different logics based on GCT. We shall show that the interrelation between the object-level algebraic structure and the meta-level algebraic structure determines the proof theory of a logic. That the many-valued logics can be obtained as a special case of this scheme will be presented at the end.

4.1 General Scheme for Proof Theory in GCT

From the angle of some philosophy of logic, sequent calculi and natural deduction systems provide meaning of the notion of logical deduction and logical connectives through the inferential rules presented in them. That is, semantics evolves through syntax. The present study analyses this issue from the perspective of graded consequence.

Some forms of proof theory for classical logic and intuitionistic logic, viz. natural deduction systems (NJ for intuitionistic logic and NK for classical logic) and sequent calculi (LJ for intuitionistic logic and LK for classical logic) were introduced by Gentzen in the year 1934 (Gentzen 1969). Besides the natural deduction systems and sequent calculi, proof theory also includes Hilbert-type proof construction from axioms and rules of inference. The three kinds of proof theories are equivalent in the case of intuitionistic and classical logics in the sense that from a set of wffs (premises) the derivability of the same set of wffs may be established in all the approaches. Also, no semantics is involved in the procedures adopted in either of them. But it should be mentioned that each approach has its own merits and demerits. For example, sequent calculus can handle the decidability issue of logic without entering into the

M. K. Chakraborty and S. Dutta, *Theory of Graded Consequence*, Logic in Asia: Studia Logica Library, https://doi.org/10.1007/978-981-13-8896-5_4

semantics. This is possible because of the proof pattern of the sequent calculus which yields an algorithm for finding out a proof of a formula. Natural deduction systems can deal with premises of arbitrary cardinality, and is closer to mathematical methods of argumentation. On the other hand, axiomatic method reveals before the eye the basic wffs which can be taken as the building blocks of the theory and the basic units of inference.

Calculus of natural deduction was introduced by Gentzen to have a formal system which comes, in his own words, 'as close as possible to actual reasoning'. In the same paper (Gentzen 1969), Gentzen introduced the sequent calculi both for intuitionistic logic and classical logic. As presented in Gentzen (1969), a sequent is an expression of the form $A_1, A_2, \ldots, A_m \rightarrow B_1, B_2, \ldots, B_n$, where $A_1, A_2, \ldots, A_m, B_1, B_2, \ldots, B_n$ are formulae and $\rightarrow$, like comma, is an auxiliary symbol, i.e. a non-logical symbol. After defining sequent in this manner, at the very next paragraph Gentzen remarked that $A_1, A_2, \ldots, A_m \rightarrow B_1, B_2, \ldots, B_n$ has exactly the same informal meaning as the formula $A_1 \wedge A_2 \wedge \ldots \wedge A_m \supset B_1 \vee B_2 \vee \ldots \vee B_n$. In this remark, the distinction between meta-level and object-level implications turns out to be opaque.

For the investigation of syntactic rules in the context of graded consequence, distinguishing the object-level and meta-level languages is essential. Since the notion of consequence is graded here it is also necessary to consider the algebraic structures required to give the semantics at the two levels. So, our position in this respect is to read the $\rightarrow$ of a sequent clearly as the meta-linguistic relation $\vdash$. Hence, to avoid confusion, in the rest of the presentation instead of $\rightarrow$ we will use $\vdash$.

Gentzen's original version of the theories (Gentzen 1969) has undergone several modifications subsequently (Katsura 2008; Lemmon 1965; Dubois and Prade 2004; Takeuti 1987). From time to time, the pattern of sequent calculus presentation has varied keeping the basic idea intact. Initially, as mentioned, of a sequent $\Gamma \vdash \Delta$, Γ, Δ were assumed to be finite sequences of formulae, but later on, in some presentations, Γ, Δ are considered as multisets (Troelstra 1992), even as sets (Barwise 1982; Lemmon 1965) also. It is obvious that if instead of finite sequence of formulae, a multiset is considered, the role of exchange rule (Troelstra 1992) becomes redundant, and if a set is considered in place of a sequence then exchange and contraction rules (Barwise 1982; Lemmon 1965) become redundant. So, keeping convenience of use in mind, various presentations of sequent calculi have come up with respective modifications.

We will consider a classical sequent as an expression of the form $X \vdash \alpha$, where X is a set of formulae and α is a single formula. By $X \vdash Y$ we shall mean $X \vdash \alpha$ for all $\alpha \in Y$ and $X, \alpha \vdash \beta$ means $X \cup \{\alpha\} \vdash \beta$. We first present in this form the proof theory of intuitionistic logic and then switch to the presentation of classical logic. Subsequently, we recast this into a form that can be generalized to our case that means in the context of graded consequence.

Table 4.1 represents the structural rules for both intuitionistic and classical propositional logics. These are, in fact, represented in Tarskian axioms (see Chap. 1). Table 4.2 represents the non-structural or logical rules for intuitionistic logic. This table will make the scheme somewhat clearer. The first and second columns contain, respectively, the rules of intuitionistic logic as natural deduction (that is in the

Table 4.1 Structural rules

$\alpha \vdash \alpha$	(Reflexivity or overlap)
$\dfrac{X \vdash \alpha}{X \cup Y \vdash \alpha}$	(Monotonicity or dilution)
$\dfrac{X \vdash Y \quad X \cup Y \vdash \alpha}{X \vdash \alpha}$	(Transitivity or cut)

form of introduction–elimination of connectives) and as sequent calculus (that is the approach of left–right rules of connectives). The third column contains a modified version of the calculi. It is this version that will be generalized in the context of GCT.

All the three rules presented under the heading 'structural rules' are not called so in the literature of existing proof theories. The first one is termed as 'axiom'. The third one, which is a version of the rule 'cut', usually in the literature is not mentioned as the structural rule. However, these may be termed as 'structural rules' since they do not involve logical connectives; in this sense, these are basic to all deductive systems. However, recent development of substructural logics and non-monotonic logics question, and even do not accept one or more of the rules, viz. exchange, weakening and contraction (Ono 2003). But, in this work, we shall stick to these rules, and generalize them in many-valued context.

Regarding the non-structural rules, none of the columns presented in the table below is exactly the same as Gentzen's original version for natural deduction and sequent calculus for intuitionistic logic. First of all, definition of a sequent is quite different in our case. Second, though the first column looks similar to Gentzen's natural deduction presentation, rules for negation are taken differently. The first column is the natural deduction system of Gentzen with a modification in the rules for $\neg$, the second column is a modified version of Gentzen's sequent calculus and the third column depicts the form of natural deduction system which carries the same idea as that of column 1 but differs in presentation.

Note 4.1 1. In each of the classical presentations (Gentzen 1969; Dubois and Prade 2004), explicitly or implicitly, the structural rules, in our sense, are also assumed.

2. The rules for negation, i.e. ($\neg$-I) and ($\neg$-E) are not the same as those originally taken by Gentzen Gentzen (1969). In Gentzen (1969), the 'Introduction' and 'Elimination' rules for negation ($\neg$) are given as follows:

   ($\neg$-I_G): $\dfrac{\begin{matrix}[\alpha]\\ \vdots \\ \Lambda\end{matrix}}{\neg\alpha}$ ($\neg$-E_G): $\dfrac{\alpha, \neg\alpha}{\Lambda}$

   where Λ is the propositional constant, viz. falsum.

   Also the rule for Λ was given by (Λ_G): $\dfrac{\Lambda}{\beta \quad \text{(any)}}$.

   Now, in a language where Λ, a formula representing the contradiction, is not present, replacing every occurrence of Λ by $\gamma \& \neg\gamma$ for some formula γ it can be shown that ($\neg$-I) is equivalent to ($\neg$-I_G) and ($\neg$-E) is equivalent to ($\neg$-E_G).

3. However, the advantage of taking the propositional constant falsum (Λ) lies in achieving decidability of the calculus (or logic). But we are more concerned with the 'nature' of the object-level logical connectives rather than decidability. The wff Λ is not present in our language. The current formulation has a direct counterpart in the logic of graded consequence. Besides, we are not dealing with the proof theory of any specific logic developed from the angle of graded consequence. Rather we are presenting some 'principles' on which a concrete proof theory for a specific logic may be developed. While doing that it might be helpful in some cases to incorporate falsum (Λ) within the language, and accordingly some other rules for negation introduction might be suitable.
4. In the original presentation of sequent calculus, Gentzen used the following rules as ($\neg$-right(G)) and ($\neg$-left(G)), respectively.

   $\dfrac{\alpha, \Gamma \vdash \Delta}{\Gamma \vdash \Delta, \neg\alpha}$ $\quad$ $\dfrac{\Gamma \vdash \Delta, \alpha}{\neg\alpha, \Gamma \vdash \Delta.}$ $\quad$ [Γ, Δ are finite sequences of formulae]

   But since we stick to the form of placing exactly one formula in the right-hand side of $\vdash$, the rules are modified as presented in column 2.

Theorem 4.2, below, establishes that the presentation in column 2 is equivalent to the presentation in column 3. Furthermore, columns 1 and 2 can be proved to be equivalent in the sense that every derivation in one calculus can be transformed into an equivalent derivation of the other calculus; proof of Theorem 4.1 demonstrates that. Thus, the three forms of proof systems in columns 1, 2 and 3 are mutually equivalent. The current formulation, specifically of column 2 and 3, has a direct counterpart in the logic of graded consequence.

Though all these three columns differ in their ways of presenting a rule, the basic essence of a derivation remains the same keeping analogy with the structure of the rules of the respective systems. In case of Column 1, the notion of derivation is same as it is in a standard natural deduction system. Whereas in case of Column 2, a derivation is a sequence of finitely many sets of sequents, say $S_1, S_2, \ldots, S_n$. Each member of S_i, $1 \le i \le n$, is either appeared as an assumption or following a rule from some of the members of some of $S_1, \ldots S_{i-1}$ in the sequence. In the body of the derivation, two consecutive sets of sequents S_{i-1} and S_i are presented above and below a horizontal line; this indicates that S_i is obtained from S_{i-1} as a one-step derivation. In case of Column 1, a derivation usually ends to a single target formula, and in case of Column 2, a derivation ends to a single sequent. In case of Column 1, we can observe that some of the rules as a premise assume itself the derivation of some formula. For instance, one can notice the rule for $\underline{\vee}$-E, i.e. disjunction elimination. The presence of the formula $\alpha \underline{\vee} \beta$ along with the formula γ, which is assumed to be proved both from α and β, ensures the derivation of γ from $\alpha \underline{\vee} \beta$. Thus, a lower level derivation is involved in the derivation of a formula from a set of formulas. In case of Column 2, this lower level derivation is denoted by the relation $\vdash$, which is involved in the next level of derivation, denoted by the presence of a horizontal line. From the same perspective, one can notice that the structural rules are usually explicitly mentioned in case of sequent calculus presentation; this is not the case in the context of natural deduction presentation as the notion of derivation chain itself implicitly contains the features of overlap, monotonicity and cut. These derivations are basically metameta-level notions. The notion of derivation in case of Column 3 is similar as in the case for Column 2. Later, we will see that derivation in the context of GCT is also similar like Column 2, i.e. from a set of sequents to a sequent (cf. Examples 4.10 and 4.11).

Theorem 4.1 *Any derivation following the proof system of column 1 can be translated to a derivation following the proof system of column 2, and vice versa.*

Proof Column 1 to Column 2:
We first transform each form of rule of column 1 to a corresponding form of rule of column 2. The conversion scheme is as follows:

Column 1		Column 2
$\dfrac{\alpha, \quad \beta}{\gamma}$	$\leadsto$	$\alpha, \beta \vdash \gamma$
$\begin{matrix}(\alpha)\\ \vdots \\ \dfrac{\beta}{\gamma}\end{matrix}$	$\leadsto$	$\dfrac{X, \alpha \vdash \beta}{X \vdash \gamma}$

Hence, the transformed version of each rule of the column 1 of Table 4.2 in the form of a column 2 is as follows:

(&-I) transforms to: $\alpha, \beta \vdash \alpha \& \beta$.

(&-E) transforms to: $\alpha \& \beta \vdash \alpha, \ \alpha \& \beta \vdash \beta$.

($\underline{\vee}$-I) transforms to: $\alpha \vdash \alpha \underline{\vee} \beta, \ \beta \vdash \alpha \underline{\vee} \beta$.

($\underline{\vee}$-E) transforms to: $\dfrac{X, \alpha \vdash \gamma, \quad Y, \beta \vdash \gamma.}{X, Y, \alpha \underline{\vee} \beta \vdash \gamma}$

($\supset$-I) transforms to: $\dfrac{X, \alpha \vdash \beta\,.}{X \vdash \alpha \supset \beta}$

($\supset$-E) transforms to: $\alpha, \alpha \supset \beta \vdash \beta$.

($\neg$-I) transforms to: $\dfrac{X, \alpha \vdash \beta}{X \vdash \neg\alpha}$ for all β.

($\neg$-E) transforms to: $\alpha, \neg\alpha \vdash \beta$ for any β.

Now, with the help of these transformed version of rules, we want to derive the rules (i.e. left/right rules) given in column 2 of the above table.

Table 4.2 Different forms of non-structural rules

Connective	Natural deduction	Sequent calculus	New presentation
& (Conjunction)	(&-I) $\dfrac{\alpha \quad \beta}{\alpha\&\beta}$	(&-right) $\dfrac{X \vdash \alpha \quad X \vdash \beta}{X \vdash \alpha\&\beta}$	(&-introduction) $\dfrac{X \vdash \alpha \quad X \vdash \beta}{X \vdash \alpha\&\beta}$
	(&-E) $\dfrac{\alpha\&\beta}{\alpha}$ $\dfrac{\alpha\&\beta}{\beta}$	(&-left) $\dfrac{X, \alpha \vdash \gamma}{X, \alpha\&\beta \vdash \gamma}$ $\dfrac{X, \beta \vdash \gamma}{X, \alpha\&\beta \vdash \gamma}$	(&-Elimination) $\dfrac{X \vdash \alpha\&\beta}{X \vdash \alpha}$ $\dfrac{X \vdash \alpha\&\beta}{X \vdash \beta}$
$\vee$ (Disjunction)	($\underline{\vee}$-I) $\dfrac{\alpha}{\alpha\underline{\vee}\beta}$ $\dfrac{\beta}{\alpha\underline{\vee}\beta}$	($\underline{\vee}$-right) $\dfrac{X \vdash \alpha}{X \vdash \alpha\underline{\vee}\beta}$ $\dfrac{X \vdash \beta}{X \vdash \alpha\underline{\vee}\beta}$	($\underline{\vee}$-Introduction) $\dfrac{X \vdash \alpha}{X \vdash \alpha\underline{\vee}\beta}$ $\dfrac{X \vdash \beta}{X \vdash \alpha\underline{\vee}\beta}$
	($\underline{\vee}$-E) $\dfrac{\alpha\underline{\vee}\beta \quad \begin{matrix}[\alpha]\\ \vdots \\ \gamma\end{matrix} \quad \begin{matrix}[\beta]\\ \vdots \\ \gamma\end{matrix}}{\gamma}$	($\underline{\vee}$-left) $\dfrac{X,\alpha \vdash \gamma \quad Y,\beta \vdash \gamma}{X,Y,\alpha\underline{\vee}\beta \vdash \gamma}$	($\underline{\vee}$-Elimination) $\dfrac{X \vdash \alpha\underline{\vee}\beta \quad \alpha \vdash \gamma \quad \beta \vdash \gamma}{X \vdash \gamma}$
$\supset$ (Implication)	($\supset$-I) $\dfrac{\begin{matrix}[\alpha]\\ \vdots \\ \beta\end{matrix}}{\alpha \supset \beta}$	($\supset$-right) $\dfrac{X,\alpha \vdash \beta}{X \vdash \alpha \supset \beta}$	($\supset$-Introduction) $\dfrac{X,\alpha \vdash \beta}{X \vdash \alpha \supset \beta}$
	($\supset$-E) $\dfrac{\alpha \quad \alpha \supset \beta}{\beta}$	($\supset$-left) $\dfrac{X \vdash \alpha \quad Y,\beta \vdash \gamma}{X,Y,\alpha \supset \beta \vdash \gamma}$	($\supset$-Elimination) $\dfrac{X \vdash \alpha \quad Y \vdash \beta \supset \gamma}{X,Y \vdash \gamma}$
$\neg$ (Negation)	($\neg$-I) $\dfrac{\begin{matrix}[\alpha]\\ \vdots \\ \beta \text{ for all } \beta\end{matrix}}{\neg\alpha}$	($\neg$-right) $\dfrac{X,\alpha \vdash \beta \text{ for all } \beta}{X \vdash \neg\alpha}$	($\neg$-Introduction) $\dfrac{X,\alpha \vdash \beta \text{ for all } \beta}{X \vdash \neg\alpha}$
	($\neg$-E) $\dfrac{\alpha \quad \neg\alpha}{\beta \text{ for any } \beta}$	($\neg$-left) $\dfrac{X \vdash \alpha}{X, \neg\alpha \vdash \beta \text{ for any } \beta}$	($\neg$-Elimination) $\dfrac{X \vdash \alpha \quad Y \vdash \neg\alpha}{X \cup Y \vdash \beta \text{ for any } \beta}$

(&-right)

$$\begin{array}{c} X \vdash \alpha \quad Y \vdash \beta \\ \hline X, Y \vdash \alpha \quad X, Y \vdash \beta \\ \hline \alpha, \beta \vdash \alpha \& \beta \\ \hline X, Y \vdash \alpha \& \beta \end{array}$$

(From assumption using monotonicity)
(Transformed version of &-I added in the chain)
(From second and third steps using cut)

(&-left)

$$\begin{array}{c} X,\ \alpha \vdash \gamma \\ \hline \alpha \& \beta \vdash \alpha \\ \hline X,\ \alpha \& \beta \vdash \gamma \end{array}$$

(Transformed version of &-E added in the chain)
(From first and second steps using cut)

($\underline{\vee}$-right)

$$\begin{array}{c} X \vdash \alpha \\ \hline \alpha \vdash \alpha \underline{\vee} \beta \\ \hline X,\ \alpha \vdash \alpha \underline{\vee} \beta \end{array}$$

(Transformed version of $\underline{\vee}$-I added in the chain)
(From first and second steps using cut)

($\underline{\vee}$-left)

$$\begin{array}{c} X,\ \alpha \vdash \gamma \quad Y,\ \beta \vdash \gamma \\ \hline X, Y,\ \alpha \underline{\vee} \beta \vdash \gamma \end{array}$$

(Using transformed version of $\underline{\vee}$-E)

($\supset$-right)

$$\begin{array}{c} X,\ \alpha \vdash \beta \\ \hline X \vdash \alpha \supset \beta \end{array}$$

(Using transformed version of ($\supset$-I))

($\supset$-left)

$$\begin{array}{c} X \vdash \alpha \quad Y,\ \beta \vdash \gamma \\ \hline \alpha,\ \alpha \supset \beta \vdash \beta \\ \hline Y,\ \alpha,\ \alpha \supset \beta \vdash \gamma \\ \hline X, Y\ \alpha \supset \beta \vdash \gamma \end{array}$$

(transformed version of ($\supset$-E) added in the chain)
(Using cut on first and second steps)
(Using cut on first and third steps)

($\neg$-right)

$$\begin{array}{c} X,\ \alpha \vdash \beta \ all\ \beta \\ \hline X \vdash \neg \alpha \end{array}$$

(Using transformed version of ($\neg$-I))

($\neg$-left)

$$\begin{array}{c} X \vdash \alpha \\ \hline \alpha,\ \neg \alpha \vdash \beta \ any\ \beta \\ \hline X,\ \neg \alpha \vdash \beta \end{array}$$

(Using transformed version of ($\neg$-E))
(Using cut on first and second steps)

Column 2 to Column 1:
Now, in the same way, we transform each rule of column 2 to a rule of column 1.

Column 2		Column 1
$\dfrac{X, \alpha, \vdash \beta}{X, \gamma \vdash \delta}$	$\leadsto$	$\dfrac{\begin{array}{cc} & (\alpha) \\ & \vdots \\ \gamma & \beta \end{array}}{\delta.}$
$\dfrac{X \vdash \alpha \quad Y \vdash \beta}{X, Y \vdash \gamma}$	$\leadsto$	$\dfrac{\alpha \quad \beta}{\gamma.}$

The type of the rules available in column 2 is any one of the forms given above or their combination. Hence, the transformed version of each rule of the column 2 in column 1 is as follows:

(&-right) transforms to: $\dfrac{\alpha \quad \beta.}{\alpha \& \beta}$

(&-left) transforms to:
$$\dfrac{\begin{array}{cc} & (\alpha) \\ & \vdots \\ \alpha \& \beta & \gamma\,. \end{array}}{\gamma}$$

($\underline{\vee}$-right) transforms to: $\dfrac{\alpha}{\alpha \underline{\vee} \beta}$

($\underline{\vee}$-left) transforms to:
$$\dfrac{\begin{array}{ccc} & (\alpha) & (\beta) \\ & \vdots & \vdots \\ \alpha \underline{\vee} \beta & \gamma & \gamma \end{array}}{\gamma}$$

($\supset$-right) transforms to:
$$\dfrac{\begin{array}{c} (\alpha) \\ \vdots \\ \beta\;. \end{array}}{\alpha \supset \beta}$$

(⊃-left) transforms to:

$$\frac{\alpha \supset \beta \quad \alpha \quad \begin{matrix}(\beta)\\ \vdots \\ \gamma\end{matrix}}{\gamma}.$$

(¬-right) transforms to:

$$\frac{\begin{matrix}(\alpha)\\ \vdots \\ \beta\end{matrix}}{\neg\alpha}.$$

(¬-left) transforms to:

$$\frac{\alpha \quad \neg\alpha}{\beta}.$$

Now using these transformed version of the rules of column 2, we want to derive the rules (i.e. I/E rules) of column 1 of the above table.

(&-I): It is immediately obtained from the transformed version of (&-right).

(&-E):

$$\frac{\alpha \& \beta \quad \begin{matrix}(\alpha)\\ \vdots \\ \alpha\end{matrix}}{\alpha}$$

α is derived directly from assumption (α)

($\underline{\vee}$-I): It is immediately obtained from the transformed version of ($\underline{\vee}$-right).

($\underline{\vee}$-E): It is immediately obtained from the transformed version of ($\underline{\vee}$-left).

(⊃-I): is immediately obtained from the transformed version of (⊃-right).

(⊃-E):

$$\frac{\alpha \supset \beta \quad \alpha \quad \begin{matrix}(\beta)\\ \vdots \\ \beta\end{matrix}}{\beta}$$

(β is derived directly from assumption β)

(¬-I): It is immediately obtained from the transformed version of (¬-right).

(¬-E): It is immediately obtained from the transformed version of (¬-left).

□

Theorem 4.2 *1(i) Overlap, Cut and &-Elimination imply &-left.*
(ii) Overlap, Cut and &-left imply &-Elimination.
2(i) Cut and $\underline{\vee}$-Elimination imply $\underline{\vee}$-left.
(ii) Cut and $\underline{\vee}$-left imply $\underline{\vee}$-Elimination.
3(i) Overlap, Cut and ⊃-Elimination imply ⊃-left.
(ii) Overlap, Weakening, Cut and ⊃-left imply ⊃-Elimination.
4(i) Overlap and ¬-Elimination imply ¬-left.
(ii) Cut and ¬-left imply ¬-Elimination.

Proof 1(i) Let (&-Elimination) hold, and $X, \alpha \vdash \gamma$.

$$\begin{array}{cl} X, \alpha \vdash \gamma & \\ \hline \alpha \wedge \beta \vdash \alpha \wedge \beta & \text{(New sequent added using overlap)} \\ \hline \alpha \wedge \beta \vdash \alpha & \text{(From step 2 using \&-Elimination)} \\ \hline X, \alpha \wedge \beta \vdash \gamma & \text{(Using cut on first and third steps)} \end{array}$$

(ii) Let (&-left) hold, and $X \vdash \alpha \wedge \beta$.

$$\begin{array}{cl} X \vdash \alpha \wedge \beta & \\ \hline \alpha \vdash \alpha & \text{(New sequent added using overlap)} \\ \hline \alpha \wedge \beta \vdash \alpha & \text{(From step 2 using \&-Left)} \\ \hline X \vdash \alpha & \text{(Using cut on first and third steps)} \end{array}$$

2(i) Let ($\underline{\vee}$-Elimination) holds, and both $X, \alpha \vdash \gamma$ and $Y, \beta \vdash \gamma$.

$$\begin{array}{cl} X, \alpha \vdash \gamma \quad Y, \beta \vdash \gamma & \\ \hline \alpha \underline{\vee} \beta \vdash \alpha \underline{\vee} \beta & \text{(New sequent added using overlap)} \\ \hline X, Y, \alpha \underline{\vee} \beta \vdash \gamma & \text{(Using } \underline{\vee}\text{-Elimination on first and second steps)} \end{array}$$

(ii) Let ($\underline{\vee}$-left) hold. Let us assume $X \vdash \alpha \underline{\vee} \beta$, as well as $Y, \alpha \vdash \gamma$ and $Z, \beta \vdash \gamma$ hold.

$$\begin{array}{cl} X \vdash \alpha \underline{\vee} \beta \quad Y, \alpha \vdash \gamma \quad Z, \beta \vdash \gamma & \\ \hline Y, Z, \alpha \underline{\vee} \beta \vdash \gamma & \text{(From step 1 using } \underline{\vee}\text{-Left)} \\ \hline X, Y, Z \vdash \gamma & \text{(Using cut on first and second steps)} \end{array}$$

3(i) Let ($\supset$-Elimination) hold. Also let $X \vdash \alpha$ and $Y, \beta \vdash \gamma$.

$$\begin{array}{cl} X \vdash \alpha \quad Y, \beta \vdash \gamma & \\ \hline \alpha \supset \beta \vdash \alpha \supset \beta & \text{(New sequent added using overlap)} \\ \hline X, \alpha \supset \beta \vdash \beta & \text{(From 1st and 2nd steps using } \supset\text{-Elimination)} \\ \hline X, Y, \alpha \supset \beta \vdash \gamma & \text{(Using cut on first and third steps)} \end{array}$$

(ii) Let us assume ($\supset$-left) holds, and also $X \vdash \alpha$ and $Y \vdash \alpha \supset \beta$ hold.

$$\begin{array}{cl} X \vdash \alpha \quad Y \vdash \alpha \supset \beta & \\ \hline \beta \vdash \beta & \text{(New sequent added using overlap)} \\ \hline Y, \beta \vdash \beta & \text{(Using monotonicity on step 2)} \\ \hline X, Y, \alpha \supset \beta \vdash \beta & \text{(Using } \supset\text{-Left on first and third steps)} \\ \hline X, Y \vdash \beta & \text{(Using cut on first and fourth steps)} \end{array}$$

4(i) Let ($\neg$-Elimination) hold, and $X \vdash \alpha$.

$$\begin{array}{cl} X \vdash \alpha & \\ \hline \neg\alpha \vdash \neg\alpha & \text{(New sequent added using overlap)} \\ \hline X, \neg\alpha \vdash \beta & \text{(From first and second steps using } \neg\text{-Elimination)} \end{array}$$

(ii) Let ($\neg$-left) hold, and $X \vdash \alpha$, as well as $Y \vdash \neg\alpha$.

$$\dfrac{\dfrac{X \vdash \alpha \quad Y \vdash \neg\alpha}{X, \neg\alpha \vdash \beta}}{X, Y \vdash \beta} \quad \begin{array}{l}\text{(From first step using } \neg\text{-Left)} \\ \text{(Using cut on first and second steps)}\end{array}$$

□

Note 4.2 Intuitionistic logic along with the law of excluded middle gives rise to the system of classical logic. In Gentzen's natural deduction presentation, simply adding $\alpha\underline{\vee}\neg\alpha$ as an axiom with NJ the system NK is constructed. On the other hand, LK has been proved (Gentzen 1969) to be equivalent to the system LJ along with the primitive rule $\vdash \alpha\underline{\vee}\neg\alpha$. The above table represents different presentations for intuitionistic logic. To obtain classical logic from the presentation in column 3, we will propose the following changes:
(i) Instead of $\neg$-Introduction we shall take

$$(R\neg)\text{:}\ \dfrac{X \cup \{\alpha\} \vdash \beta \ X \cup \{\neg\alpha\} \vdash \beta}{X \vdash \beta} \qquad \text{(rule by cases)}$$

(ii) and instead of $\neg$-Elimination we shall take the rule $\{\alpha, \neg\alpha\} \vdash \beta$ for any β.

Proposition 4.1 *($R\neg$) implies $\neg$-Introduction.*

Proof Let us consider that $X, \alpha \vdash \beta$ holds for all β. So, in particular, $X, \alpha \vdash \neg\alpha$.

$$\dfrac{\dfrac{X, \alpha \vdash \beta\ for\ all \beta}{X, \alpha \vdash \neg\alpha \quad X, \neg\alpha \vdash \neg\alpha}}{X \vdash \neg\alpha} \quad \begin{array}{l}\text{(New sequent added using overlap)} \\ \text{(Using } R\neg \text{ on step 2)}\end{array}$$

□

Proposition 4.2 *($R\neg$) is equivalent to $\vdash \alpha\underline{\vee}\neg\alpha$.*

Proof Let us consider that $(R\neg)$ holds. Then by overlap

$$\dfrac{\dfrac{\alpha \vdash \alpha \quad \neg\alpha \vdash \neg\alpha}{\alpha \vdash \alpha\underline{\vee}\neg\alpha \quad \neg\alpha \vdash \alpha\underline{\vee}\neg\alpha}}{\vdash \alpha\underline{\vee}\neg\alpha} \quad \begin{array}{l}\text{(Using } \underline{\vee}\text{-Introduction on step 1)} \\ \text{(Using } R\neg \text{ on step 2)}\end{array}$$

On the other hand, let us assume that $\vdash \alpha\underline{\vee}\neg\alpha$ holds. Then we assume $X, \alpha \vdash \beta$ and $X, \neg\alpha \vdash \beta$.

$$\dfrac{\dfrac{\dfrac{X, \alpha \vdash \beta \quad X, \neg\alpha \vdash \beta}{\vdash \alpha\underline{\vee}\neg\alpha}}{X \vdash \alpha\underline{\vee}\neg\alpha}}{X \vdash \beta} \quad \begin{array}{l}\text{(New sequent added as an assumption)} \\ \text{(Using monotonicity on second step)} \\ \text{(From first and third steps using} \underline{\vee}\text{-Elimination)}\end{array}$$

□

Proposition 4.3 *The rule $\{\alpha, \neg\alpha\} \vdash \beta$, for any β is equivalent to the rule $\neg$-Elimination.*

Proof Let $\{\alpha, \neg\alpha\} \vdash \beta$ for all β. Let us consider $X \vdash \alpha$ and $Y \vdash \neg\alpha$ holds. So,

$$\cfrac{\cfrac{X \vdash \alpha \quad Y \vdash \neg\alpha}{X, Y \vdash \alpha \quad X, Y \vdash \neg\alpha}\ \text{(Using monotonicity on step 1)}}{\cfrac{\alpha, \neg\alpha \vdash \beta}{X, Y \vdash \beta}\ \text{(Using cut on second and third steps)}}\ \text{(New sequent added as an assumption)}$$

On the other hand, in ($\neg$-Elimination), if we assume X as $\{\alpha\}$ and Y as $\{\neg\alpha\}$, then by overlap and cut $\{\alpha, \neg\alpha\} \vdash \beta$ follows. □

So, introduction, elimination rules for &, $\vee$, $\supset$ along with $(R\neg)$ and $\{\alpha, \neg\alpha\} \vdash \beta$ for any β, known as explosiveness condition, give rise to classical logic. Later, we will see a generalized version of all these rules. The rules for explosiveness condition and $(R\neg)$ are presented as (GC4) and (GC5^M), respectively, in the graded context.

We will need a rewriting of the rules in order to adopt them within our framework. As mentioned in Chap. 1, $X \vdash \alpha$ is a meta-logical statement with the predicate $\vdash$ belonging to the meta-language. The line separating upper and lower sequents, that may be read as '**implies**', represents a binary predication of the metameta-language. So a typical rule, say monotonicity, should, in fact, be written as

'$X \vdash \alpha$' **implies** '$X \cup Y \vdash \alpha$' ...(A)

'$X \vdash \alpha$' and '$X \cup Y \vdash \alpha$' are, respectively, the names of the meta-language-sentences $X \vdash \alpha$ and $X \cup Y \vdash \alpha$ and the expression (A) is a well-formed sentence of the metameta-language. The intention behind accepting the monotonicity rule is to accept the above sentence to be true. In other words, the intension is to accept that truth value of '$X \vdash \alpha$' $\leq$ truth value of '$X \cup Y \vdash \alpha$' for all X, Y and α. This is to keep in mind that at present we are discussing classical logic in which the metameta-sentences are also classical with truth values 0 and 1, endowed with natural ordering. If gr('$X \vdash \alpha$') denotes truth value of '$X \vdash \alpha$', the form of monotonicity turns out to be $gr(\text{'}X \vdash \alpha\text{'}) \leq gr(\text{'}X \cup Y \vdash \alpha\text{'})$ or $gr(\text{'}X \vdash \alpha\text{'}) \leq gr(\text{'}Y \vdash \alpha\text{'})$ where $X \subseteq Y$ ...(B)

Similar would be the forms of other rules too. The quotation mark will be dropped henceforth for not making the representation cumbersome. For an illustration, the forms of introduction and elimination rules for conjunction are presented below.

$gr(X \vdash \alpha \textit{ and } X \vdash \beta) \leq gr(X \vdash \alpha\&\beta)$. ...(C)

$gr(X \vdash \alpha\&\beta) \leq gr(X \vdash \alpha)$ and $gr(X \vdash \alpha\&\beta) \leq gr(X \vdash \beta)$. ...(D)

It should be transparent that in the &-Introduction rule, & is the conjunction of the object language whereas '*and*' used to the left of $\leq$ sign in (C) is the meta-linguistic conjunction since it conjoins two meta-linguistic sentences $X \vdash \alpha$ and $X \vdash \beta$. (A), (B), (C) and (D) are all metameta-level sentences. Discerning levels of languages is the most vital point of the theory of graded consequence. In the presentation of classical logic, however, this important issue is not paid attention to.

In the theory of graded consequence, the relation of consequence being graded, the statement 'α is a consequence of X', now written as '$X \mid\!\sim \alpha$' in order to make a distinction from the crisp consequence relation $\vdash$, which may receive grades other than the top (1) or the least (0). That means $gr(X \mid\!\sim \alpha)$ is now a value in a suitable algebraic structure which is generally different from the algebraic structure for the object language semantics. In classical logic, as we have observed while discussing rules (B), (C) and (D), truth structure for all the levels are the same, viz. the two-point Boolean lattice $\{1, 0\}$ and the logical connectives are computed by the standard truth tables. So, the role of the two levels of algebraic structures and their interrelation behind the presence of a logical rule do not get counted properly. In the theory of graded consequence exploration of the special conditions or relationships between the algebraic structures, required for these rules to hold, is the main focus of the investigation. That is, given a rule of the types mentioned in the table involving a connective # (# represents any one of the object language connectives), we shall explore conditions to be imposed on the operator computing # and its interrelation with the algebraic counterparts of the concerned meta-linguistic connectives. Some or all of these conditions together fix the algebraic structures for the object and meta-languages. Conversely, we shall also observe that after fixing certain structures one is able to determine whether the graded counterparts of some specific or all of the logical rules hold. Thereby, this study makes explicit the hidden semantic considerations behind taking the rules of proof theories. The discussion regarding the sufficient and necessary conditions for having the inequalities, determining the rules, is presented in the following section.

4.2 Proof Theory of GCT: Links Between Object and Meta-level Algebras

Hilbert-type proof theory based on axioms and rules, in the context of graded consequence, has been discussed in Chap. 2. In fact, a few toy examples are also developed there, but basically the principles for constructing any such proof system are laid down. Here we proceed towards a natural-deduction-type proof theory generalizing the rules in column 3 of Table 4.2, presented in Sect. 4.1. When some specific object language is taken, demands to its logical connectives also crop up. These demands are stated in the non-structural rules, specifically the introduction and elimination rules. It is not obligatory that all the rules of classical logic must hold. For instance, in the intuitionistic logic, the law $\vdash \neg\neg\alpha \supset \alpha$ does not hold, whereas in paraconsistent logic (Arruda and Ayda 1989) this law is accepted but the law $\{\alpha, \neg\alpha\} \vdash \beta$ is dropped. We shall show in this section, how these acceptance and non-acceptance are related with the algebraic operations at the object level and the meta-level in the present context.

Given any particular connective, say &, of object language, in order to investigate whether or not $gr(X \mid\!\sim \alpha\&\beta) \leq gr(X \mid\!\sim \alpha)$ is present in a logical system we will study the necessary conditions which eventually emerge from the rule, and the conditions which suffice to have this. That is, we will study

(i) the algebraic meanings or properties of object-level connective '&' and
(ii) the mutual interrelation of the object and the meta-level connectives

to get the above rule in a logic with graded consequence.

In doing so, the entire study is done from semantic angle. That is, we start with a collection of fuzzy subsets, say $\{T_i\}_{i\in I}$, which are assumed to be truth functional and classical value preserving when restricted to the values 0 and 1.

Note 4.3 It should be noted that, while dealing with graded counterparts of structural rules we need not specify the nature of fuzzy subsets $\{T_i\}_{i\in I}$ generating the consequence relation. But when the object-level connectives come into the scenario we shall stick to those fuzzy subsets which satisfy the truth functionality condition. Moreover, to retain classical logic as a special case of the logical rules in the graded context we need to choose those truth functional fuzzy subsets $\{T_i\}_{i\in I}$ which behave classically in {0, 1}-valued structure. To explicate the necessity for such a consideration let us take an example.

Let $\mid\!\sim$ be a fuzzy relation from $P(F)$ to F, defined by $gr(X \mid\!\sim \alpha) = 1$ if $\alpha \in X$, and is equal to 0, otherwise. It can be shown that the fuzzy relation defined above satisfies all the conditions, viz. (GC1), (GC2) and (GC3). So, being a graded consequence relation, there exists a collection of fuzzy subsets such that the fuzzy relation $\mid\!\approx$ generated from that collection turns out to be the relation $\mid\!\sim$ given above. Following the technique of constructing such a collection as described in the proof of Theorem 2.1, we find $\{T_Y\}_{Y\in P(F)}$, defined by $T_Y(\beta) = 1$ if $\beta \in Y$ and 0 otherwise, is the collection generating the above $\mid\!\sim$. So, for the function $T_{\{\beta\}}$, $T_{\{\beta\}}(\beta) = 1$, but $T_{\{\beta\}}(\alpha \supset \beta) = 0$. That is, the collection contains valuation functions which are neither truth functional nor preserving classical valuation. Unless these two conditions are imposed on the set $\{T_i\}_{i\in I}$ some of the desired properties of (graded) consequence relation with logical connectives in the object level will not be available—and these two are quite justified conditions. Thus, when the focus is on the study of the semantic background of rules concerning object-level connectives, a graded consequence relation generated by a collection $\{T_i\}_{i\in I}$ without the above-mentioned conditions is to be abandoned.

So, from now onwards by any collection $\{T_i\}_{i\in I}$ we shall mean a collection of fuzzy subsets which are truth functional and classical value preserving.

Besides, we need to specify structures for both object-level and meta-level languages. That $(L, \wedge, \vee, *_m, \rightarrow_m, 0, 1)$, a complete residuated lattice needed for the meta-level language is already stated. As whatever be the algebraic structure, it is always based on a complete lattice structure $(L, \wedge, \vee)$, instead of displaying the operations corresponding to the order relation we, from now onwards, shall use $(L, *_m, \rightarrow_m, 0, 1)$ for denoting the meta-level algebraic structure. The object-level structure, which is L endowed with some operations (may be some or all of $*_o$, $\rightarrow_o$, $\oplus$, $\neg_o$), will come to the picture as and when a particular connective with its specific rules appear as our subject of study. We will use $*_o$, $\rightarrow_o$, $\oplus_o$ and $\neg_o$ as the operators for object-level connectives conjunction &, implication $\supset$, disjunction $\veebar$ and negation $\neg$, respectively. We shall denote the object-level algebra by L_o. As discussed in the previous paragraph, these operators need to be functions preserving classical valuations. But no other condition will be imposed on them, rather the properties of these operators will emerge gradually from the necessary and sufficient conditions of graded counterpart of rules involving these connectives. That is, there is no reason to assume that $*_o$, $\to_o$, $\oplus$ and $\neg_o$ have standard properties except on the set $\{0, 1\}$. We shall present examples later.

In graded context, we shall now explore (Dutta 2011a, b; Dutta and Chakraborty 2009a, b, 2017) the necessity and sufficiency conditions of some proof theoretic rules whose presence is traditionally sought for. In this connection, the necessary and sufficient conditions for both introduction, elimination rules for $\supset$ as well as for the transitivity condition of $\supset$ will be presented. On the other hand, in case of & we will study introduction rule (or right rule), left rule and elimination rule. For $\underline{\vee}$, we will concentrate on the introduction rule (or right rule) as well as the left rule and elimination rule of the connective. Besides, one more property of $\underline{\vee}$ will be studied; we shall denote the property by $P[\underline{\vee}]$ [cf. Theorem 4.12]. Classically, $\underline{\vee}$-Introduction, Overlap and Cut together imply the two-valued version of $P[\underline{\vee}]$. That is, in two-valued context $P[\underline{\vee}]$ is implied by $\underline{\vee}$-Introduction. Here, in graded context we will see that though the sufficient conditions for $P[\underline{\vee}]$ and $\underline{\vee}$-Introduction are exactly the same, the necessary conditions remain different. In classical case, both the necessary conditions are equivalent, but in graded context the equivalence does not hold. The necessary and sufficient conditions for (GC4) and (GC^{M}5), i.e. the many-valued version of the rules, viz. explosiveness condition and ($R\neg$) are already discussed in Chap. 3. Here we shall explore the graded version of $\neg$-Introduction.

Before entering into the study how does the relationships between different algebraic structures reveal the presence or absence of logical rules in a logic of graded consequence, let us present the basic definitions of some algebraic structures, which would be used as examples of object-level or meta-level algebraic structures in this chapter as well as some of the latter chapters. For meta-level algebraic structures, we shall always use different kinds of complete residuated lattices, some of the examples of which are discussed in Chap. 2. There are some algebraic structures different from residuated lattices but having many of the overlapping properties.

Definition 4.1 (*cf. Troelstra* 1992) $\langle L, \wedge, \vee, \perp, \to, *, 1\rangle$ is an intuitionistic linear algebra (IL algebra) if and only if

(i) $\langle L\wedge, \vee, \perp\rangle$ is a lattice with least element $\perp$,
(ii) $\langle L * 1\rangle$ is a commutative monoid with identity 1,
(iii) if $x \leq x'$, $y \leq y'$ then (a) $x * y \leq x' * y'$ and
(b) $x \to y \leq x \to y'$ for all $x, y \in L$,
(iv) $x * y \leq z$ if and only if $x \leq y \to z$ for all $x, y, z \in L$.

Definition 4.2 (*cf. Troelstra* 1992) An IL algebra with a specified constant say 0 (zero) is called an IL algebra with zero or ILZ algebra.

Note 4.4 In an ILZ algebra usually a negation $\neg$ is defined as $\neg x = x \to 0$, and a binary operation $\oplus$ is defined as $x \oplus y = \neg(\neg x * \neg y)$.

Definition 4.3 (*cf. Troelstra* 1992) An ILZ algebra is a classical linear algebra (CL algebra) if and only if $\neg\neg x = x$ for all $x \in L$.

Note 4.5 In a CL algebra, one can find $x \oplus y = \neg x \to y$ (Sen 2000).

Definition 4.4 (*cf. Hájek* 1998) A residuated lattice $(L, \wedge, \vee, *, \to, 0, 1)$ is a basic logic algebra (BL-algebra) if and only if the following two identities hold for all $x, y \in L$.

- $x \wedge y = x * (x \to y)$,
- $(x \to y) \vee (y \to x) = 1$.

Definition 4.5 An MV-algebra is a BL-algebra in which the identity, namely, $x = ((x \to 0) \to 0)$ is valid.

Note 4.6 $[0, 1]$ with truth functions of Łukasiewicz logic forms an MV-algebra and this is known as standard MV-algebra.

Definition 4.6 A BL-algebra satisfying the identity $x * x = x$ is called a Gödel algebra or in short G-algebra.

Note 4.7 A Heyting algebra is a residuated lattice where the lattice meet itself is the monoidal operation. A G-algebra is a Heyting algebra satisfying the prelinearity condition, viz. $(x \to y) \vee (y \to x) = 1$.

4.2.1 Implication in the Object Language

Let us consider '$\supset$' as an object-level implication. We now explore the necessary and sufficient conditions for the graded counterpart of the rules concerning $\supset$. For example, the deduction theorem (in classical context) is the statement: $X \cup \{\alpha\} \vdash \beta$ implies $X \vdash \alpha \supset \beta$. This is a metameta-level assertion. Its counterpart in the present context of GCT would be $gr(X \cup \{\alpha\} \mid\!\sim \beta) \leq gr(X \mid\!\sim \alpha \supset \beta)$. Now we determine the necessary and sufficient condition for this inequality to hold, keeping in mind that $\mid\!\sim = \models\!\!\!\!\approx_{\{T_i\}_{i\in I}}$ for some collection $\{T_i\}_{i\in I}$.

Theorem 4.3 *Let* $(L, \to_o)$ *and* $(L, *_m, \to_m, 0, 1)$ *be the object- and meta-level algebraic structures.*

(i) If $gr(X \cup \{\alpha\} \models\!\!\!\!\approx_{\{T_i\}_{i\in I}} \beta) \leq gr(X \models\!\!\!\!\approx_{\{T_i\}_{i\in I}} \alpha \supset \beta)$ *holds for all collections* $\{T_i\}_{i\in I}$, *then* $a \to_m b \leq a \to_o b$ *holds for any* $a, b \in L$. (Necessary condition for graded DT or $\supset$-Introduction rule.)

(ii) If for any $a, b \in L$, $a \to_m b \leq a \to_o b$ *holds, then for any* $\{T_i\}_{i\in I}$, $gr(X \cup \{\alpha\} \models\!\!\!\!\approx_{\{T_i\}_{i\in I}} \beta) \leq gr(X \models\!\!\!\!\approx_{\{T_i\}_{i\in I}} \alpha \supset \beta)$ *holds.* (Sufficient condition for graded DT or $\supset$-Introduction rule.)

Proof (i) Let us consider any collection $\{T_i\}_{i\in I}$. Then by representation theorem we know that the collection generates $\models\!\!\!\!\approx_{\{T_i\}_{i\in I}}$, a graded consequence relation. So, $gr(X \cup \{\alpha\} \models\!\!\!\!\approx_{\{T_i\}_{i\in I}} \beta) \leq gr(X \models\!\!\!\!\approx_{\{T_i\}_{i\in I}} \alpha \supset \beta)$ for any $X, Y \subseteq F$ and $\alpha, \beta \in F$.

In particular, for $X = \phi$, $gr(\{\alpha\} \models\!\!\!\!\approx_{\{T_i\}_{i\in I}} \beta) \leq gr(\models\!\!\!\!\approx_{\{T_i\}_{i\in I}} \alpha \supset \beta)$.

That is, $inf_i\{T_i(\alpha) \rightarrow_m T_i(\beta)\} \leq inf_i\{T_i(\alpha) \rightarrow_o T_i(\beta)\} \leq T_i(\alpha) \rightarrow_o T_i(\beta)$ for any i.

As for any $\{T_i\}_{i\in I}$, $gr(X \cup \{\alpha\} \approx_{\{T_i\}_{i\in I}} \beta) \leq gr(X \approx_{\{T_i\}_{i\in I}} \alpha \supset \beta)$ holds, let us consider $\{T_i\}_{i\in I}$ to be a singleton set consisting of T_i only.

Then $T_i(\alpha) \rightarrow_m T_i(\beta) \leq T_i(\alpha) \rightarrow_o T_i(\beta)$.

The above is true for any possible T_i. As, given any value $a \in L$, an arbitrary propositional variable can assume this value under some T_i, we can say that T_i can assume any possible value from the value set L. Hence $a \rightarrow_m b \leq a \rightarrow_o b$ for any $a, b \in L$.

(ii) Now $gr(X \cup \{\alpha\} \approx_{\{T_i\}_{i\in I}} \beta)$

$= inf_i\{(inf_{\gamma\in X} T_i(\gamma) \wedge T_i(\alpha)) \rightarrow_m T_i(\beta)\}$

$\leq inf_i\{(inf_{\gamma\in X} T_i(\gamma) *_m T_i(\alpha)) \rightarrow_m T_i(\beta)\}$ [as $\rightarrow_m$ is antitone at 1st variable]

$= inf_i\{inf_{\gamma\in X} T_i(\gamma) \rightarrow_m (T_i(\alpha) \rightarrow_m T_i(\beta))\}$ [as $(a *_m b) \rightarrow_m c = a \rightarrow_m (b \rightarrow_m c)$]

$\leq inf_i\{inf_{\gamma\in X} T_i(\gamma) \rightarrow_m (T_i(\alpha) \rightarrow_o T_i(\beta))\}$ [$a \rightarrow_m b \leq a \rightarrow_o b$, and $\rightarrow_m$ is isotone at 2nd variable]

$= gr(X \approx_{\{T_i\}_{i\in I}} \alpha \supset \beta)$. □

Proofs of the subsequent theorems follow basically the same pattern. So, we have included only the sketches of some of them indicating the nodal and distinctive features.

Theorem 4.4 *Let $(L, \rightarrow_o)$ and $(L, *_m, \rightarrow_m, 0, 1)$ be the object- and meta-level algebraic structures.*

(i) If $gr(X \approx_{\{T_i\}_{i\in I}} \alpha) *_m gr(Y \approx_{\{T_i\}_{i\in I}} \alpha \supset \beta) \leq gr(X \cup Y \approx_{\{T_i\}_{i\in I}} \beta)$ holds for any $\{T_i\}_{i\in I}$, then $a \rightarrow_o b \leq a \rightarrow_m b$ holds for any $a, b \in L$. (Necessary condition for graded Modus Ponens (MP) or ⊃-Elimination rule.)

(ii) If for any $a, b \in L$, $a \wedge (a \rightarrow_o b) \leq b$ hold, then for any collection of fuzzy sets $\{T_i\}_{i\in I}$, $gr(X \approx_{\{T_i\}_{i\in I}} \alpha) *_m gr(Y \approx_{\{T_i\}_{i\in I}} \alpha \supset \beta) \leq gr(X \cup Y \approx_{\{T_i\}_{i\in I}} \beta)$ holds. (Sufficient condition for graded Modus Ponens (MP) or ⊃-Elimination rule.)

Theorem 4.5 *Let $(L, \rightarrow_o)$ and $(L, *_m, \rightarrow_m, 0, 1)$ be the object- and meta-level algebraic structures.*

(i) If $gr(X \approx_{\{T_i\}_{i\in I}} \alpha \supset \beta) *_m gr(Y \approx_{\{T_i\}_{i\in I}} \beta \supset \gamma) \leq gr(X \cup Y \approx_{\{T_i\}_{i\in I}} \alpha \supset \gamma)$ holds for all collections $\{T_i\}_{i\in I}$, then for any $a, b, c \in L$, $(a \rightarrow_o b) *_m (b \rightarrow_o c) \leq (a \rightarrow_o c)$ holds. (Necessary condition for graded transitivity.)

(ii) If for any $a, b, c \in L$, $(a \rightarrow_o b) \wedge (b \rightarrow_o c) \leq (a \rightarrow_o c)$ holds, then for $\{T_i\}_{i\in I}$, $gr(X \approx_{\{T_i\}_{i\in I}} \alpha \supset \beta) *_m gr(Y \approx_{\{T_i\}_{i\in I}} \beta \supset \gamma) \leq gr(X \cup Y \approx_{\{T_i\}_{i\in I}} \alpha \supset \gamma)$ holds. (Sufficient condition for graded transitivity.)

Proof (i) Let for any graded consequence relation $\approx_{\{T_i\}_{i\in I}}$, $X, Y \subseteq F$ and $\alpha, \beta \in F$ $gr(X \approx_{\{T_i\}_{i\in I}} \alpha \supset \beta) *_m gr(Y \approx_{\{T_i\}_{i\in I}} \beta \supset \gamma) \leq gr(X \cup Y \approx_{\{T_i\}_{i\in I}} \alpha \supset \gamma)$ hold.

In particular, for $X = Y = \phi$,

$gr(\approx_{\{T_i\}_{i\in I}} \alpha \supset \beta) *_m gr(\approx_{\{T_i\}_{i\in I}} \beta \supset \gamma) \leq gr(\approx_{\{T_i\}_{i\in I}} \alpha \supset \gamma)$.

I.e. $inf_i\{T_i(\alpha) \rightarrow_o T_i(\beta)\} *_m inf_i\{T_i(\beta) \rightarrow_o T_i(\gamma)\} \leq inf_i\{T_i(\alpha) \rightarrow_o T_i(\gamma)\}$.

As any $\{T_i\}_{i \in I}$ generates a graded consequence relation satisfying the above inequality, in particular for a singleton collection $\{T_i\}$, $\approx_{\{T_i\}}$ also satisfies the graded rule transitivity.

Hence $\{T_i(\alpha) \rightarrow_o T_i(\beta)\} *_m \{T_i(\beta) \rightarrow_o T_i(\gamma)\} \leq \{T_i(\alpha) \rightarrow_o T_i(\gamma)\}$.

As T_is can assume any value from L, for any $a, b, c \in L$,

$(a \rightarrow_o b) *_m (b \rightarrow_o c) \leq (a \rightarrow_o c)$.

(ii) $gr(X \approx_{\{T_i\}_{i\in I}} \alpha \supset \beta) *_m gr(Y \approx_{\{T_i\}_{i\in I}} \beta \supset \gamma)$

$= inf_i\{inf_{\gamma \in X} T_i(\gamma) \rightarrow_m T_i(\alpha \supset \beta)\} *_m inf_i\{inf_{\gamma \in Y} T_i(\gamma) \rightarrow_m T_i(\beta \supset \gamma)\}$

$\leq inf_i[\{inf_{\gamma \in X} T_i(\gamma) \rightarrow_m T_i(\alpha \supset \beta)\} *_m \{inf_{\gamma \in Y} T_i(\gamma) \rightarrow_m T_i(\beta \supset \gamma)\}]$

$\leq inf_i[\{inf_{\gamma \in X \cup Y} T_i(\gamma) \rightarrow_m T_i(\alpha \supset \beta)\} *_m \{inf_{\gamma \in X \cup Y} T_i(\gamma) \rightarrow_m T_i(\beta \supset \gamma)\}]$

$\leq inf_i\{inf_{\gamma \in X \cup Y} T_i(\gamma) \rightarrow_m (T_i(\alpha \supset \beta) \wedge T_i(\beta \supset \gamma))\}$

$= inf_i\{inf_{\gamma \in X \cup Y} T_i(\gamma) \rightarrow_m (T_i(\alpha) \rightarrow_o T_i(\beta)) \wedge (T_i(\beta) \rightarrow_o T_i(\gamma))\}$

$\leq inf_i\{inf_{\gamma \in X \cup Y} T_i(\gamma) \rightarrow_m (T_i(\alpha) \rightarrow_o T_i(\gamma))\}$

$= inf_i\{inf_{\gamma \in X \cup Y} T_i(\gamma) \rightarrow_m T_i(\alpha \supset \gamma)\}$

$= gr(X \cup Y \approx_{\{T_i\}_{i\in I}} \alpha \supset \gamma)$. □

Example 4.1 1. In classical logic $\rightarrow_o$, the object-level implication and $\rightarrow_m$, the meta-level implication are the same. So DT holds as a trivial consequence.

2. As a non-trivial example, let us consider [0, 1] endowed with Gödel adjoint pair as the meta-level algebra, and Łukasiewicz implication for computing $\supset$. Then these structures for object and meta-level satisfy $a \rightarrow_m b \leq a \rightarrow_o b$, and hence DT holds for any $\approx_{\{T_i\}_{i\in I}}$ relative to the above-mentioned pair of algebraic structures.
3. For the standard implication (object level) of classical and intuitionistic logic $a \wedge (a \rightarrow_0 b) \leq b$ holds. So, MP is valid in classical as well as intuitionistic logic.
4. Assuming Gödel algebra for object-level structure one can immediately get $(a \rightarrow_o b) \wedge (b \rightarrow_o c) \leq (a \rightarrow_o c)$. So, Gödel algebra for object level and any complete residuated lattice for meta-level can be considered as a model of graded transitivity rule.
5. Let us consider $(L, *_o, \rightarrow_o)$ as the object-level algebraic structure. In some IL algebras (Sen 2000; Troelstra 1992), one can find a multiplicative conjunction operator $*_o$ for which $a \wedge b \leq a *_o b$ as well as $a *_o (a \rightarrow_o b) \leq b$ hold (see Examples 4.5, 4.7). Hence $a \wedge (a \rightarrow_o b) \leq b$ holds. So, there may be several cases, assuming some special IL algebras as object-level structure and a complete residuated lattice as the meta-level structure, where the graded rule MP can be obtained. Also, in these IL algebras $a \wedge b \leq a *_o b$ and $(a \rightarrow_o b) *_o (b \rightarrow_o c) \leq (a \rightarrow_o c)$, and hence $(a \rightarrow_o b) \wedge (b \rightarrow_o c) \leq (a \rightarrow_o c)$ are available. So, considering those IL algebras as object-level structure and any complete residuated lattice as the meta-level structure, graded transitivity rule can be obtained.

Note 4.8 (1) Apart from the sufficient condition considered in Theorem 4.4, $a \rightarrow_o b \leq a \rightarrow_m b$ also can be a sufficient condition for obtaining MP provided Gödel algebra is there as the meta-structure.

(2) Presence of MP and GC1 together implies the presence of converse of DT, i.e. $gr(X \mid\!\sim \alpha \supset \beta) \leq gr(X \cup \{\alpha\} \mid\!\sim \beta)$. On the other hand, converse of DT and (GC3) together implies MP. Hence, in the examples of logics with graded consequence where graded counterpart of MP is obtained, because of the presence of (GC1), converse of DT also can be obtained.

4.2.2 Conjunction in the Object Language

Let us consider '&' as an object-level conjunction, and explore the rules concerning & from the graded perspective.

Theorem 4.6 *Let* $(L, *_o)$ *and* $(L, *_m, \rightarrow_m, 0, 1)$ *be the object- and meta-level algebraic structures.*

(i) If for any $\{T_i\}_{i\in I}$, $gr(X \cup \{\alpha\} \mid\!\approx_{\{T_i\}_{i\in I}} \gamma) \leq gr(X \cup \{\alpha\&\beta\} \mid\!\approx_{\{T_i\}_{i\in I}} \gamma)$, *then for any* $a, b \in L$, $a \rightarrow_m c \leq (a *_o b) \rightarrow_m c$. (Necessary condition for &-left.)

(ii) If for $a, b \in L$, $a *_o b \leq a$, $gr(X \cup \{\alpha\} \mid\!\approx_{\{T_i\}_{i\in I}} \gamma) \leq gr(X \cup \{\alpha\&\beta\} \mid\!\approx_{\{T_i\}_{i\in I}} \gamma)$ *holds for any* $\{T_i\}_{i\in I}$. (Sufficient condition for &-left.)

Proof (i) Let for any graded consequence relation $\mid\!\approx_{\{T_i\}_{i\in I}}$, for any $X \subseteq F$ and $\alpha, \beta \in F$ $gr(X \cup \{\alpha\} \mid\!\approx_{\{T_i\}_{i\in I}} \gamma) \leq gr(X \cup \{\alpha\&\beta\} \mid\!\approx_{\{T_i\}_{i\in I}} \gamma)$ hold.

In particular, for $X = \phi$, $gr(\{\alpha\} \mid\!\approx_{\{T_i\}_{i\in I}} \gamma) \leq gr(\{\alpha\&\beta\} \mid\!\approx_{\{T_i\}_{i\in I}} \gamma)$.

I.e. $inf_i\{T_i(\alpha) \rightarrow_m T_i(\gamma)\} \leq inf_i\{(T_i(\alpha) *_o T_i(\beta)) \rightarrow_m T_i(\gamma)\}$.

As the above inequality holds for any $\{T_i\}_{i\in I}$, in particular, for a singleton collection containing T_i only, we have $\{T_i(\alpha) \rightarrow_m T_i(\gamma)\} \leq \{(T_i(\alpha) *_o T_i(\beta)) \rightarrow_m T_i(\gamma)\}$.

As $T_i(\alpha)$, $T_i(\beta)$, $T_i(\gamma)$ can assume any value of $a, b, c \in L$, the above inequality reduces to $a \rightarrow_m c \leq (a *_o b) \rightarrow_m c$.

(ii) $gr(X \cup \{\alpha\} \mid\!\approx_{\{T_i\}_{i\in I}} \gamma) = inf_i\{inf_{\gamma\in X}(T_i(\gamma) \wedge T_i(\alpha)) \rightarrow_m T_i(\gamma)\}$

$\leq inf_i\{inf_{\gamma\in X}(T_i(\gamma) \wedge (T_i(\alpha) *_o T_i(\beta))) \rightarrow_m T_i(\gamma)\}$ (by $a *_o b \leq a$ and (r5))

$= inf_i\{inf_{\gamma\in X}(T_i(\gamma) \wedge T_i(\alpha\&\beta)) \rightarrow_m T_i(\gamma)\}$

$= gr(X \cup \{\alpha\&\beta\} \mid\!\approx_{\{T_i\}_{i\in I}} \gamma)$. □

Theorem 4.7 *Let* $(L, *_o)$ *and* $(L, *_m, \rightarrow_m, 0, 1)$ *be the object- and meta-level algebraic structures.*

Then (i) if $gr(X \mid\!\approx_{\{T_i\}_{i\in I}} \alpha\&\beta) \leq gr(X \mid\!\approx_{\{T_i\}_{i\in I}} \alpha)$ *holds for all collections* $\{T_i\}_{i\in I}$, *then for any* $a, b \in L$, $a *_o b \leq a$ *holds.* (Necessary condition for &-Elimination.)

(ii) If for $a, b \in L$, $a *_o b \leq a$, *then* $gr(X \mid\!\approx_{\{T_i\}_{i\in I}} \alpha\&\beta) \leq gr(X \mid\!\approx_{\{T_i\}_{i\in I}} \alpha)$ *holds.* (Sufficient condition for &-Elimination.)

Proof Proof for (i) follows the similar line of argument as above. We only give an outline for the proof of (ii).

$gr(X \approx_{\{T_i\}_{i\in I}} \alpha\&\beta) = inf_i\{inf_{\gamma\in X}T_i(\gamma) \to_m T_i(\alpha\&\beta)\}$.
$= inf_i\{inf_{\gamma\in X}T_i(\gamma) \to_m (T_i(\alpha) *_o T_i(\beta))\} \leq inf_i\{inf_{\gamma\in X}T_i(\gamma) \to_m T_i(\alpha)\}$
$= gr(X \approx_{\{T_i\}_{i\in I}} \alpha)$. □

Theorem 4.8 *Let* $(L, *_o)$ *and* $(L, *_m, \to_m, 0, 1)$ *be the object- and meta-level algebraic structures.*

Then (i) if $gr(X \approx_{\{T_i\}_{i\in I}} \alpha) *_m gr(Y \approx_{\{T_i\}_{i\in I}} \beta) \leq gr(X \cup Y \approx_{\{T_i\}_{i\in I}} \alpha\&\beta)$ *holds for all collections* $\{T_i\}_{i\in I}$, *then* $a *_m b \leq a *_o b$ *holds for any* $a, b \in L$. (Necessary condition for &-Introduction or &-right.)

(ii) If $a \wedge b \leq a *_o b$ *for any* $a, b \in L$, *then for any* $\approx_{\{T_i\}_{i\in I}}$, *generated from* $\{T_i\}_{i\in I}$, $gr(X \approx_{\{T_i\}_{i\in I}} \alpha) *_m gr(Y \approx_{\{T_i\}_{i\in I}} \beta) \leq gr(X \cup Y \approx_{\{T_i\}_{i\in I}} \alpha\&\beta)$ *holds.* (Sufficient condition for &-Introduction or &-right.)

Proof (i) Let for any graded consequence relation $\approx_{\{T_i\}_{i\in I}}$, for any $X, Y \subseteq F$ and $\alpha, \beta \in F$, $gr(X \approx_{\{T_i\}_{i\in I}} \alpha) *_m gr(Y \approx_{\{T_i\}_{i\in I}} \beta) \leq gr(X \cup Y \approx_{\{T_i\}_{i\in I}} \alpha\&\beta)$ hold.

In particular, for $X = Y = \phi$, $gr(\approx_{\{T_i\}_{i\in I}} \alpha) *_m gr(\approx_{\{T_i\}_{i\in I}} \beta) \leq gr(\approx_{\{T_i\}_{i\in I}} \alpha\&\beta)$.
I.e. $inf_i\{T_i(\alpha)\} *_m inf_i\{T_i(\beta)\} \leq inf_i\{T_i(\alpha\&\beta)\}$.
In particular, for a singleton collection, the above inequality reduces to
$T_i(\alpha) *_m T_i(\beta) \leq T_i(\alpha\&\beta)$. I.e. $T_i(\alpha) *_m T_i(\beta) \leq T_i(\alpha) *_o T_i(\beta)$.
As T_i's can assume any value, for any $a, b \in L$, $a *_m b \leq a *_o b$ holds.

(ii) $gr(X \approx_{\{T_i\}_{i\in I}} \alpha) *_m gr(Y \approx_{\{T_i\}_{i\in I}} \beta)$
$= inf_i\{inf_{\gamma\in X}T_i(\gamma) \to_m T_i(\alpha)\} *_m inf_i\{inf_{\gamma\in Y}T_i(\gamma) \to_m T_i(\beta)\}$.
$\leq inf_i[\{inf_{\gamma\in X}T_i(\gamma) \to_m T_i(\alpha)\} *_m \{inf_{\gamma\in Y}T_i(\gamma) \to_m T_i(\beta)\}]$
$\leq inf_i[\{inf_{\gamma\in X\cup Y}T_i(\gamma) \to_m T_i(\alpha)\} *_m \{inf_{\gamma\in X\cup Y}T_i(\gamma) \to_m T_i(\beta)\}]$ [by (r5)]
$\leq inf_i\{inf_{\gamma\in X\cup Y}T_i(\gamma) \to_m (T_i(\alpha) \wedge T_i(\beta))\}$
[as $(a \to_m b) *_m (a \to_m c) \leq a \to_m (b \wedge c)$]
$\leq inf_i\{inf_{\gamma\in X\cup Y}T_i(\gamma) \to_m (T_i(\alpha) *_o T_i(\beta))\}$ [Since $a \wedge b \leq a *_o b$]
$= gr(X \cup Y \approx_{\{T_i\}_{i\in I}} \alpha\&\beta)$. □

Example 4.2 1. Usually, all the standard operators $*$ for conjunction (in [0,1] these are known as t-norms and more generally, called multiplicative conjunction) satisfy $a * b \leq a \wedge b \leq a, b$. So, in particular, for $*_o = \wedge$ and for any complete residuated lattice at the meta-level, all the above-mentioned necessary and sufficient conditions for graded rules of & hold.

2. Based on any complete residuated lattice for meta-level structure, the graded counterpart of &-left and &-Elimination can be obtained considering any standard operator for conjunction for object-level connective &.
3. Examples for the graded rule &-Introduction can be found by considering some of the IL algebras (see Examples 4.5, 4.7), where $a \wedge b \leq a *_o b$ holds, as the object-level structure.

4.2.3 Disjunction in the Object Language

Let us consider '$\underline{\vee}$' as an object-level disjunction and $\oplus_o$ as the algebraic operator for computing $\underline{\vee}$. Now, we will explore the rules concerning $\underline{\vee}$ from graded perspective.

Theorem 4.9 *Let $(L, \oplus_o)$ and $(L, *_m, \rightarrow_m, 0, 1)$ be the object- and meta-level algebraic structures.*

(i) If $gr(X \approx_{\{T_i\}_{i\in I}} \alpha) \leq gr(X \approx_{\{T_i\}_{i\in I}} \alpha\underline{\vee}\beta)$ holds for all collections $\{T_i\}_{i\in I}$, then for any $a, b \in L$, $a \leq a \oplus_o b$ holds. (Necessary condition for $\underline{\vee}$-Introduction or $\underline{\vee}$-right.)

(ii) If for $a, b \in L$, $a \leq a \oplus_o b$, then $gr(X \approx_{\{T_i\}_{i\in I}} \alpha) \leq gr(X \approx_{\{T_i\}_{i\in I}} \alpha\underline{\vee}\beta)$ holds. (Sufficient condition for $\underline{\vee}$-Introduction or $\underline{\vee}$-right.)

Theorem 4.10 *Let $(L, \oplus_o)$ and $(L, *_m, \rightarrow_m, 0, 1)$ be the object- and meta-level algebraic structures.*

(i) If $gr(X \cup \{\alpha\} \approx_{\{T_i\}_{i\in I}} \gamma) *_m gr(Y \cup \{\beta\} \approx_{\{T_i\}_{i\in I}} \gamma) \leq gr(X \cup Y \cup \{\alpha\underline{\vee}\beta\} \approx_{\{T_i\}_{i\in I}} \gamma)$ holds for all collections $\{T_i\}_{i\in I}$, then $(a \rightarrow_m c) *_m (b \rightarrow_m c) \leq ((a \oplus_o b) \rightarrow_m c)$ holds for any $a, b, c \in L$. (Necessary condition for $\underline{\vee}$-left.)

(ii) If $a \oplus_o b \leq a \vee b$ and $a \wedge (b \vee c) \leq (a \wedge b) \vee (a \wedge c)$ hold for any $a, b, c \in L$, then $gr(X \cup \{\alpha\} \approx_{\{T_i\}_{i\in I}} \gamma) *_m gr(Y \cup \{\beta\} \approx_{\{T_i\}_{i\in I}} \gamma) \leq gr(X \cup Y \cup \{\alpha\underline{\vee}\beta\} \approx_{\{T_i\}_{i\in I}} \gamma)$ holds. (Sufficient condition for $\underline{\vee}$-left.)

Theorem 4.11 *Let $(L, \oplus_o)$ and $(L, *_m, \rightarrow_m, 0, 1)$ be the object- and meta-level algebraic structures.*

(i) If $gr(X \approx_{\{T_i\}_{i\in I}} \alpha\underline{\vee}\beta) *_m gr(Y \cup \{\alpha\} \approx_{\{T_i\}_{i\in I}} \gamma) *_m gr(Z \cup \{\beta\} \approx_{\{T_i\}_{i\in I}} \gamma) \leq gr(X \cup Y \cup Z \approx_{\{T_i\}_{i\in I}} \gamma)$ holds for any collection of fuzzy sets $\{T_i\}_{i\in I}$, then $(a \rightarrow_m c) *_m (b \rightarrow_m c) \leq (a \oplus_o b) \rightarrow_m c$ holds for any $a, b, c \in L$. (Necessary condition for $\underline{\vee}$-Elimination.)

(ii) If $a \oplus_o b \leq a \vee b$ and $a \wedge (b \vee c) \leq (a \wedge b) \vee (a \wedge c)$ hold for $a, b, c \in L$, then $gr(X \approx_{\{T_i\}_{i\in I}} \alpha\underline{\vee}\beta) *_m gr(Y \cup \{\alpha\} \approx_{\{T_i\}_{i\in I}} \gamma) *_m gr(Z \cup \{\beta\} \approx_{\{T_i\}_{i\in I}} \gamma)$
$\leq gr(X \cup Y \cup Z \approx_{\{T_i\}_{i\in I}} \gamma)$ holds for any $\approx_{\{T_i\}_{i\in I}}$. (Sufficient condition for $\underline{\vee}$-Elimination.)

Proof (i) Assume for any graded consequence relation $\approx_{\{T_i\}_{i\in I}}$,

$$gr(X \approx_{\{T_i\}_{i\in I}} \alpha\underline{\vee}\beta) *_m gr(Y \cup \{\alpha\} \approx_{\{T_i\}_{i\in I}} \gamma) *_m gr(Z \cup \{\beta\} \approx_{\{T_i\}_{i\in I}} \gamma) \leq gr(X \cup Y \cup Z \approx_{\{T_i\}_{i\in I}} \gamma) \text{ holds.}$$

The above inequality holds for any X, Y, Z. In particular, for $X = Y = Z = \phi$, $gr(\approx_{\{T_i\}_{i\in I}} \alpha\underline{\vee}\beta) *_m gr(\{\alpha\} \approx_{\{T_i\}_{i\in I}} \gamma) *_m gr(\{\beta\} \approx_{\{T_i\}_{i\in I}} \gamma) \leq gr(\approx_{\{T_i\}_{i\in I}} \gamma)$.

$inf_i\{T_i(\alpha\underline{\vee}\beta)\} *_m inf_i\{T_i(\alpha) \rightarrow_m T_i(\gamma)\} *_m inf_i\{T_i(\beta) \rightarrow_m T_i(\gamma)\} \leq inf_i T_i(\gamma)$.

In particular, for a singleton collection $\{T_i\}$ the above inequality reduces to

$\{T_i(\alpha\underline{\vee}\beta)\} *_m \{T_i(\alpha) \rightarrow_m T_i(\gamma)\} *_m \{T_i(\beta) \rightarrow_m T_i(\gamma)\} \leq T_i(\gamma)$.

$\{T_i(\alpha) \oplus_o T_i(\beta)\} *_m \{T_i(\alpha) \rightarrow_m T_i(\gamma)\} *_m \{T_i(\beta) \rightarrow_m T_i(\gamma)\} \leq T_i(\gamma)$.

Hence $\{T_i(\alpha) \rightarrow_m T_i(\gamma)\} *_m \{T_i(\beta) \rightarrow_m T_i(\gamma)\} \leq \{T_i(\alpha) \oplus_o T_i(\beta)\} \rightarrow_m T_i(\gamma)$.

As T_is can assume any value, $(a \rightarrow_m c) *_m (b \rightarrow_m c) \leq (a \oplus_o b) \rightarrow_m c$, for $a, b, c \in L$.

(ii) $gr(X \approx_{\{T_i\}_{i\in I}} \alpha\underline{\vee}\beta) *_m gr(Y \cup \{\alpha\} \approx_{\{T_i\}_{i\in I}} \gamma) *_m gr(Z \cup \{\beta\} \approx_{\{T_i\}_{i\in I}} \gamma)$
$\leq inf_i[\{inf_{\delta\in X} T_i(\delta) \to_m T_i(\alpha\underline{\vee}\beta)\} *_m \{(inf_{\delta\in Y} T_i(\delta) \wedge T_i(\alpha)) \to_m T_i(\gamma)\} *_m$
$\{(inf_{\delta\in Z} T_i(\delta) \wedge T_i(\beta)) \to_m T_i(\gamma)\}]$
$\leq inf_i[\{inf_{\delta\in X\cup Y\cup Z} T_i(\delta) \to_m T_i(\alpha\underline{\vee}\beta)\} *_m \{(inf_{\delta\in X\cup Y\cup Z} T_i(\delta) \wedge T_i(\alpha))$
$\to_m T_i(\gamma)\} *_m \{(inf_{\delta\in X\cup Y\cup Z} T_i(\delta) \wedge T_i(\beta)) \to_m T_i(\gamma)\}]$
$\leq inf_i[\{inf_{\delta\in X\cup Y\cup Z} T_i(\delta) \to_m (T_i(\alpha\underline{\vee}\beta))\} *_m$
$\{inf_{\delta\in X\cup Y\cup Z} T_i(\delta) \wedge (T_i(\alpha) \vee T_i(\beta)) \to_m T_i(\gamma)\}]$
[by $(a \to_m c) *_m (b \to_m c) \leq (a \vee b) \to_m c$ and $a \wedge (b \vee c) \leq (a \wedge b) \vee (a \wedge c)$]
$= inf_i[\{inf_{\delta\in X\cup Y\cup Z} T_i(\delta) \to_m (T_i(\alpha) \oplus_o T_i(\beta))\} *_m$
$\{inf_{\delta\in X\cup Y\cup Z} T_i(\delta) \wedge (T_i(\alpha) \vee T_i(\beta)) \to_m T_i(\gamma)\}]$
$\leq inf_i[\{inf_{\delta\in X\cup Y\cup Z} T_i(\delta) \to_m (T_i(\alpha) \oplus_o T_i(\beta))\} *_m$
$\{(inf_{\delta\in X\cup Y\cup Z} T_i(\delta) \wedge (T_i(\alpha) \oplus T_i(\beta))) \to_m T_i(\gamma)\}]$ [Since $a \oplus_o b \leq a \vee b$]
$\leq inf_i\{inf_{\delta\in X\cup Y\cup Z} T_i(\delta) \to_m T_i(\gamma)\}$ [by (r6) listed in Sect. 2.1]
$= gr(X \cup Y \cup Z \approx_{\{T_i\}_{i\in I}} \gamma)$ □.

Theorem 4.12 *Let $(L, \oplus_o)$ and $(L, *_m, \to_m, 0, 1)$ be the object- and meta-level algebraic structures.*

(i) If $gr(X \cup \{\alpha\underline{\vee}\beta\} \approx_{\{T_i\}_{i\in I}} \gamma) \leq gr(X \cup \{\alpha\} \approx_{\{T_i\}_{i\in I}} \gamma)$ for all collections $\{T_i\}_{i\in I}$, then for any $a, b, c \in L$, $(a \oplus_o b) \to_m c \leq a \to_m c$. (Necessary condition for *$P[\underline{\vee}]$.)*

(ii) If for $a, b \in L$, $a \leq a \oplus_o b$, $gr(X \cup \{\alpha\underline{\vee}\beta\} \approx_{\{T_i\}_{i\in I}} \gamma) \leq gr(X \cup \{\alpha\} \approx_{\{T_i\}_{i\in I}} \gamma)$ holds for any $\{T_i\}_{i\in I}$. (Sufficient condition for *$P[\underline{\vee}]$.)*

Proof Let us show the proof outline of (ii).
$gr(X \cup \{\alpha\underline{\vee}\beta\} \approx_{\{T_i\}_{i\in I}} \gamma) = inf_i\{(inf_{\delta\in X} T_i(\delta) \wedge T_i(\alpha\underline{\vee}\beta)) \to_m T_i(\gamma)\}$.
$= inf_i\{(inf_{\delta\in X} T_i(\delta) \wedge (T_i(\alpha) \oplus_o T_i(\beta))) \to_m T_i(\gamma)\}$
$\leq inf_i\{(inf_{\delta\in X} T_i(\delta) \wedge T_i(\alpha)) \to_m T_i(\gamma)\}$ [as $a \leq a \oplus_o b$ and $\to_m$ satisfies (r5)]
$= gr(X \cup \{\alpha\} \approx_{\{T_i\}_{i\in I}} \gamma)$. □

Example 4.3 1. In BL-algebra (Hájek 1998), $(a \wedge (b \vee c)) = (a \wedge b) \vee (a \wedge c)$ and
$(a \to_m c) *_m (b \to_m c) \leq ((a \vee b) \to_m c)$ hold for $\vee$, the lattice join and $(*_m, \to_m)$, the adjoint pair. [0, 1] with respect to any standard t-norm forms a residuated lattice as well as BL-algebra. Besides, in a residuated lattice $(a \to_m b) *_m (b \to_m c) \leq (a \to_m c)$ holds. In a linear structure, $(a \to_m b) *_m (b \to_m c) \leq (a \to_m c)$ implies $(a \to_m b) *_m ((a \wedge b) \to_m c) \leq (a \to_m c)$. So considering [0, 1] endowed with any standard t-norm as meta-level structure and $\oplus_o = \vee$, the lattice joins as an operator for object-level disjunction, one can obtain all the rules for the connective $\underline{\vee}$.

2. In any residuated lattice $(a \to_m c) *_m (b \to_m c) \leq (a \vee b) \to_m c$ holds. Usual interpretation of $\underline{\vee}$ (i.e. any standard co-norm of [0, 1] or a multiplicative disjunction of a general structure) say, $\oplus_o$ does not allow $a \oplus_o b \leq a \vee b$. But in some of the CL algebras (Sen 2000; Troelstra 1992), one can find $\oplus$ as an operator for

$\underline{\vee}$ where $a \oplus_o b \leq a \vee b$ holds (see Example 4.7). So, assuming some of the CL algebras at object level may serve the purpose of finding models for graded rule $\underline{\vee}$-left as well as $\underline{\vee}$-Elimination.

Note 4.9 For convenience of presentation in some cases, viz. &-left, &-Elimination, $\underline{\vee}$-Introduction and $P[\underline{\vee}]$, only one form of the rules is given. In case of &, $gr(X \cup \{\beta\} \approx_{\{T_i\}_{i\in I}} \gamma) \leq gr(X \cup \{\alpha\&\beta\} \approx_{\{T_i\}_{i\in I}} \gamma)$ and $gr(X \approx_{\{T_i\}_{i\in I}} \alpha\&\beta) \leq gr(X \approx_{\{T_i\}_{i\in I}} \beta)$ are, respectively, the other forms of &-left and &-Elimination. If commutativity of $*_o$ is not assumed, then to get hold of both the versions of &-left and &-Elimination one needs to impose $a *_o b \leq a, b$ for all $a, b \in L$, to the object-level algebra. On the other hand, both the versions of &-left are satisfied only if $a \rightarrow_m c \leq (a *_o b) \rightarrow_m c$ and $b \rightarrow_m c \leq (a *_o b) \rightarrow_m c$ hold for all $a, b, c \in L$. Similarly, corresponding modifications are required in case of $\underline{\vee}$-Introduction and $P[\underline{\vee}]$ also, if one prefers not to assume $\oplus$ as a commutative operator.

Before exploring the graded rules for $\neg$, let us construct some examples for graded consequence relation having some of the rules together.

Example 4.4 Let $L = \{0, a, b, f, 1\}$ be the value set and $(L, *_o, \rightarrow_o, 0, 1)$ form an ILZ algebra (Sen 2000; Troelstra 1992) with zero element f, least element 0 and top as well as multiplicative identity 1. This is the structure for the object language. For the meta-language, let us take the Heyting algebra $(L, \wedge, \rightarrow_m, 0, 1)$ formed out of the order relation given in the diagram. The tables for all the operators are given below:

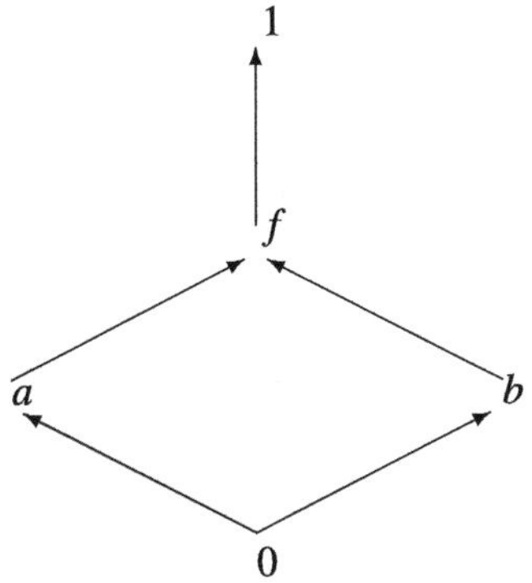

$*_o$	0	a	b	f	1
0	0	0	0	0	0
a	0	0	0	0	a
b	0	0	0	0	b
f	0	0	0	0	f
1	0	a	b	f	1

$\to_o$	0	a	b	f	1
0	1	1	1	1	1
a	f	1	f	1	1
b	f	f	1	1	1
f	f	f	f	1	1
1	0	a	b	f	1

$\to_m$	0	a	b	f	1
0	1	1	1	1	1
a	b	1	b	1	1
b	a	a	1	1	1
f	0	a	b	1	1
1	0	a	b	f	1

In this example, $x *_o y \leq x \wedge y \leq x, y$ and $x \to_m y \leq x \to_o y$ hold. So one can obtain the graded counterpart of DT, &-left and &-Elimination.

Example 4.5 Let $L = \{0, e, a, b, 1\}$ be the value set. The object-level structure is an IL algebra endowed with the operations $*_0$, $\to_o$ and the multiplicative identity element e. 1 and 0 are the respective top and least element of the lattice. The meta-level structure is $(L, \wedge, \to_m, 0, 1)$, a Heyting algebra. The definition of the operators are given below:

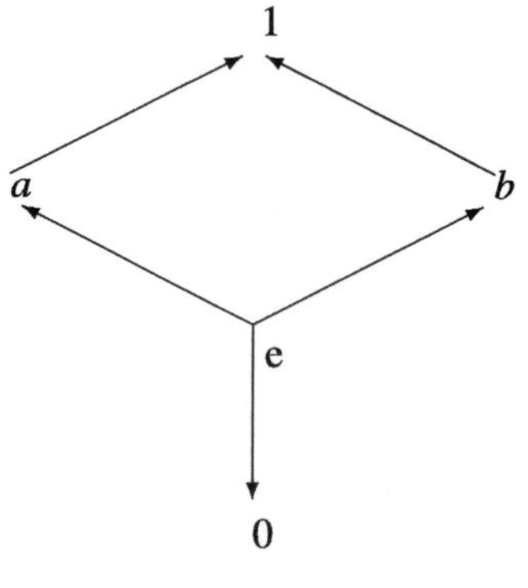

$*_o$	0	e	a	b	1
0	0	0	0	0	0
e	0	e	a	b	1
a	0	a	a	1	1
b	0	b	1	b	1
1	0	1	1	1	1

$\to_o$	0	e	a	b	1
0	1	1	1	1	1
e	0	e	a	b	1
a	0	0	a	0	1
b	0	0	0	b	1
1	0	0	0	0	1

$\to_m$	0	e	a	b	1
0	1	1	1	1	1
e	0	1	1	1	1
a	0	b	1	b	1
b	0	a	a	1	1
1	0	e	a	b	1

In this example, $x \wedge y \leq x *_o y$ and $x \to_o y \leq x \to_m y$ for all $x, y \in L$ hold. Now, as in an IL algebra $x *_o (x \to_o y) \leq b$ and $(x \to_o y) *_o (y \to_o z) \leq (x \to_o z)$ hold, one can immediately obtain $x \wedge (x \to_o y) \leq y$ as well as $(x \to_o y) \wedge (y \to_o z) \leq (x \to_o z)$ in the object-level structure. Hence, this case can be considered as a model for the graded rules &- Introduction, MP and transitivity. Besides, graded counterpart of converse of DT also can be obtained (see the Note 4.8).

4.2.4 Negation in the Object Language

In Sect. 4.1, we have seen that in presence of the rules for $\underline{\vee}$, $(R\neg)$ turns out to be equivalent to $\vdash \alpha \underline{\vee} \neg\alpha$. So, to have the graded version of intuitionistic logic we should consider ($\neg$-Introduction) instead of $(R\neg)$. We will study the necessary and sufficient conditions for the graded counterpart of ($\neg$-Introduction), i.e.

$inf_\beta\, gr(X \cup \{\alpha\} \mid\!\sim \beta) \leq gr(X \mid\!\sim \neg\alpha)$. The rule $\{\alpha, \neg\alpha\} \vdash \beta$ for all β, which is a special case of $\neg$-Elimination, is the two-valued version of (GC4). The necessary and sufficient conditions of (GC4) are already discussed in Chap. 3.

Theorem 4.13 *Let* $(L, \neg_o)$ *and* $(L, *_m, \to_m, 0, 1)$ *be the object- and meta-level algebraic structures. If* $inf_\beta\, gr(X \cup \{\alpha\} \approx\!\!\!|_{\{T_i\}_{i\in I}}\, \beta) \leq gr(X \approx\!\!\!|_{\{T_i\}_{i\in I}}\, \neg\alpha)$ *holds for all collections* $\{T_i\}_{i\in I}$*, then* $a \to_m 0 \leq \neg_o a$ *holds for any* $a \in L$. (Necessary condition for $\neg$-Introduction.)

Proof Let $inf_\beta\, gr(X \cup \{\alpha\} \approx\!\!\!|_{\{T_i\}_{i\in I}}\, \beta) \leq gr(X \approx\!\!\!|_{\{T_i\}_{i\in I}}\, \neg\alpha)$ hold for any $\approx\!\!\!|_{\{T_i\}_{i\in I}}$, X, and α.

Hence, in particular, for $X = \phi$, $inf_\beta\, gr(\{\alpha\} \approx\!\!\!|_{\{T_i\}_{i\in I}}\, \beta) \leq gr(\approx\!\!\!|_{\{T_i\}_{i\in I}}\, \neg\alpha)$ holds. That is, $inf_\beta[inf_i\{T_i(\alpha) \to_m T_i(\beta)\}] \leq inf_i\{\neg_o T_i(\alpha)\}$.

So, in particular, for a singleton collection $\{T_j\}$, $inf_\beta[\{T_j(\alpha) \rightarrow_m T_j(\beta)\}] \leq \{\neg_o T_j(\alpha)\}$.

Therefore $[\{T_j(\alpha) \rightarrow_m inf_\beta T_j(\beta)\}] \leq \{\neg_o T_j(\alpha)\}$. [by property (r7)] …(1)

(1) holds for any T_j. So, if we consider all possible T_js then

$inf_j[inf_\beta T_j(\beta)] \leq inf_\beta T_j(\beta)$ holds for any j.

Hence $T_j(\alpha) \rightarrow_m inf_j[inf_\beta T_j(\beta)] \leq T_j(\alpha) \rightarrow_m inf_\beta T_j(\beta) \leq \neg_o T_j(\alpha)$.

Considering all possible T_j's, $inf_j[inf_\beta T_j(\beta)]$ becomes 0. Then the above inequality reduces to $T_j(\alpha) \rightarrow_m 0 \leq \neg_o T_j(\alpha)$.

Hence, as T_j can assume any value $a \in L$, the necessary condition turns out to be $a \rightarrow_m 0 \leq \neg_o a$ for any $a \in L$. □

Theorem 4.14 *If there is an element* c $(\neq 1) \in L$, *such that for any* a $(\neq 1) \in L$, $a \rightarrow_m c \leq \neg_o a$, *then for any* $\approx_{\{T_i\}_{i\in I}}$ *defined over the structure* $(L, *_m, \rightarrow_m, 0, 1)$, $inf_\beta\, gr(X \cup \{\alpha\} \approx_{\{T_i\}_{i\in I}} \beta) \leq gr(X \approx_{\{T_i\}_{i\in I}} \neg\alpha)$ *holds.* (Sufficient condition for $\neg$-Introduction.)

Proof $inf_\beta\, gr(X \cup \{\alpha\} \approx_{\{T_i\}_{i\in I}} \beta)$

$= inf_\beta[inf_i\{(inf_{\gamma\in X} T_i(\gamma) \wedge T_i(\alpha)) \rightarrow_m T_i(\beta)\}]$.

$\leq inf_\beta[inf_i\{(inf_{\gamma\in X} T_i(\gamma) *_m T_i(\alpha)) \rightarrow_m T_i(\beta)\}]$

$= inf_i\{(inf_{\gamma\in X} T_i(\gamma) *_m T_i(\alpha)) \rightarrow_m inf_\beta T_i(\beta)\}$ [property (r7) of Sect. 2.1]

$= inf_i\{(inf_{\gamma\in X} T_i(\gamma) *_m T_i(\alpha)) \rightarrow_m inf_\beta T_i(\beta) : T_i(\alpha) \neq 1\} \wedge$
$inf_i\{(inf_{\gamma\in X} T_i(\gamma) *_m T_i(\alpha)) \rightarrow_m inf_\beta T_i(\beta) : T_i(\alpha) = 1\}$

$= inf_i\big\{(inf_{\gamma\in X} T_i(\gamma) *_m T_i(\alpha)) \rightarrow_m [inf_\beta\{T_i(\beta) : T_i(\beta) \leq c\}] : T_i(\alpha) \neq 1\big\} \wedge$
$inf_i\{inf_{\gamma\in X} T_i(\gamma) \rightarrow_m inf_\beta T_i(\beta) : T_i(\alpha) = 1\}$

$\leq inf_i\big\{(inf_{\gamma\in X} T_i(\gamma) *_m T_i(\alpha)) \rightarrow_m c : T_i(\alpha) \neq 1\big\} \wedge$
$inf_i\{inf_{\gamma\in X} T_i(\gamma) \rightarrow_m inf_\beta T_i(\beta) : T_i(\alpha) = 1\}$

$= inf_i\big\{inf_{\gamma\in X} T_i(\gamma) \rightarrow_m (T_i(\alpha) \rightarrow_m c) : T_i(\alpha) \neq 1\big\} \wedge$
$inf_i\{inf_{\gamma\in X} T_i(\gamma) \rightarrow_m inf_\beta T_i(\beta) : T_i(\alpha) = 1\}$

$\leq inf_i\big\{inf_{\gamma\in X} T_i(\gamma) \rightarrow_m \neg_o T_i(\alpha) : T_i(\alpha) \neq 1\big\} \wedge$
$inf_i\{inf_{\gamma\in X} T_i(\gamma) \rightarrow_m inf_\beta T_i(\beta) : T_i(\alpha) = 1\}$ [by $a(\neq 1) \in L$, $a \rightarrow_m c \leq \neg_o a$]

$\leq inf_i\big\{inf_{\gamma\in X} T_i(\gamma) \rightarrow_m \neg_o T_i(\alpha) : T_i(\alpha) \neq 1\big\} \wedge$
$inf_i\{inf_{\gamma\in X} T_i(\gamma) \rightarrow_m \neg_o T_i(\alpha) : T_i(\alpha) = 1\} = gr(X \approx_{\{T_i\}_{i\in I}} \neg\alpha)$. □

Example 4.6 In two-valued context, $c = 0$, and for $a \neq 1$, i.e. $a = 0, 0 \rightarrow_m 0 = 1$. So, as $\neg_o 0 = 1$ (since $\neg_o$ preserves classical values), the sufficient condition for $\neg$-Introduction holds. Hence, we get $\neg$-Introduction in two-valued case. To have a non-trivial example, let us consider $L = \{0, \frac{1}{2}, 1\}$ with Łukasiewicz adjoint pair at meta-level algebra. $\neg$ is defined by $\neg_o a = 0$ for $a = 1$ and 1 otherwise. Then for $c = \frac{1}{2}$ one can verify the sufficient conditions for graded $\neg$-Introduction.

Note 4.10 1. In Gentzen (1969), the sequent calculus presentation for intuitionistic logic differs from the sequent calculus presentation for classical logic by considering at most one formula at the right-hand side of $\vdash$. In the sequent calculus for classical logic, the right-hand side of $\vdash$ may contain more than one formula and this actually helps in deriving $\vdash \alpha\underline{\vee}\neg\alpha$ in the logic by using the rules ($\neg$-right(G)), ($\vee$-right), (Exchange right) and (Contraction right). These are redundant in our context where exactly one formula is considered in the right-hand side of $\vdash$. The rule ($R\neg$) ensures the presence of $\vdash \alpha\underline{\vee}\neg\alpha$ (by Proposition 4.2) in two-valued case.

2. So, in the graded context (GC1) to (GC3) along with the graded version of (&-Introduction), (&-Elimination), ($\underline{\vee}$-Introduction), ($\underline{\vee}$-Elimination), ($\supset$-Introduction), ($\supset$-Elimination), (GC4) and (GC5) generalize the classical sequent calculus presentation. On the other hand, replacing (GC5) by the graded counterpart of ($\neg$-Introduction) one can obtain the generalized version for intuitionistic logic.

Before entering into the next section, let us present some non-trivial examples where some of the rules concerning all the object language connectives are obtained.

Example 4.7 Let $L = \{0, f, a, b, e, 1\}$ be the value set. L along with the operators $*_o$, $\rightarrow_o$, $\neg_o$ (defined below) forms a CL algebra with the multiplicative identity element e, the least element 0, the top element 1 and the zero element f. Meta-level structure is $(L, \wedge, \rightarrow_m, 0, 1)$, a Heyting algebra.

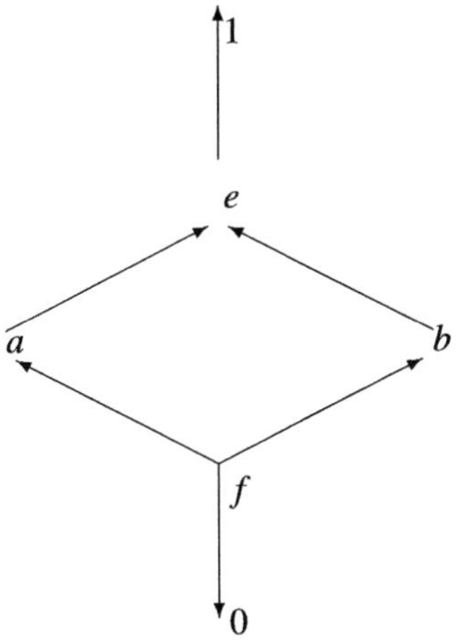

$*_o$	0	f	a	b	e	1
0	0	0	0	0	0	0
f	0	f	f	f	f	1
a	0	f	a	f	a	1
b	0	f	f	b	b	1
e	0	f	a	b	e	1
1	0	1	1	1	1	1

$\to_o$	0	f	a	b	e	1
0	1	1	1	1	1	1
f	0	e	e	e	e	1
a	0	b	e	b	e	1
b	0	a	a	e	e	1
e	0	f	a	b	e	1
1	0	0	0	0	0	1

$\neg_o$ and $\oplus_o$ are defined by – $\neg_o x = x \to_o f$ and $x \oplus_o y = \neg_o x \to_o y$.

	$\neg_o$
0	1
f	e
a	b
b	a
e	f
1	0

$\oplus_o$	0	f	a	b	e	1
0	0	0	0	0	0	1
f	0	f	a	b	e	1
a	0	a	a	e	e	1
b	0	b	e	b	e	1
e	0	e	e	e	e	1
1	1	1	1	1	1	1

$\to_m$	0	f	a	b	e	1
0	1	1	1	1	1	1
f	0	1	1	1	1	1
a	0	b	1	b	1	1
b	0	a	a	1	1	1
e	0	f	a	b	1	1
1	0	f	a	b	e	1

In this example, for all $x, y \in L$, $x \wedge y \leq x *_o y$, $x \oplus_o y \leq x \vee y$ and $x \to_o y \leq x \to_m y$ hold. Hence, as discussed in Example 4.5, one can obtain $x \wedge (x \to_o y) \leq y$ as well as $(x \to_o y) \wedge (y \to_o z) \leq (x \to_o z)$ in the object-level structure. Besides these, $x \wedge (y \vee z) \leq (x \wedge y) \vee (x \wedge z)$ and $((x \wedge z) \to_m y) \wedge (x \to_m z) \leq x \to_m y$ also hold. So with respect to these structures for object level and meta-level one can obtain the graded counterparts for &-Introduction, $\underline{\vee}$-left, $\underline{\vee}$-Elimination, MP and transitivity rule. And the presence of MP ensures the presence of converse of DT too. Regarding rules for $\neg$, it can be easily checked from the diagram of the lattice, and the table for $\neg$ that e satisfies all the condition of 'c' of Theorem 3.8. Hence, this example also can be considered as a model for GC5. Here e is not the topmost element of the lattice. On the other hand, for 0, the least element of the lattice $x \rightarrow_m 0 \leq \neg_o x$ holds for all $x \neq 1$, the topmost element of the lattice. So, the graded rule $\neg$-Introduction also can be verified here.

Example 4.8 Let us consider the same diagram and object-level structure as in Example 4.7. The meta-level structure is defined by the adjoint pair, viz. $(*_m, \rightarrow_m)$ given below:

$*_m$	0	f	a	b	e	1
0	0	0	0	0	0	0
f	0	0	0	0	0	f
a	0	0	a	0	a	a
b	0	0	0	b	b	b
e	0	0	a	b	e	1
1	0	f	a	b	e	1

$\rightarrow_m$	0	f	a	b	e	1
0	1	1	1	1	1	1
f	e	1	1	1	1	1
a	b	b	1	b	1	1
b	a	a	a	1	1	1
e	f	f	a	b	e	1
1	0	f	a	b	e	1

As the object-level structure is the same as that of Example 4.7, we have $x \wedge y \leq x *_o y$ and $x \oplus_o y \leq x \vee y$ for all $x, y \in L$. So, as in the above example, $x \wedge (x \rightarrow_o y) \leq y$, $(x \rightarrow_o y) \wedge (y \rightarrow_o z) \leq (x \rightarrow_o z)$ and $x \wedge (y \vee z) \leq (x \wedge y) \vee (x \wedge z)$ are available here also. Besides, with respect to the $\rightarrow_m$ defined above, $x \rightarrow_o y \leq x \rightarrow_m y$ holds. So, in this example, the graded rules transitivity, MP, converse of DT, &-Introduction, $\underline{\vee}$-left hold. Regarding the rules for $\neg$, here we can see $x \vee \neg_o x \geq e$ holds, but e does not satisfy the condition (iii) of Theorem 3.8. On the other hand, for $c = 0$, $x \rightarrow_m 0 \leq \neg_o x$ holds for all $x \neq 1$. So, one can find the graded counterpart of $\neg$-Introduction here.

From the above discussion and examples, it is clear that the choice of a pair of algebraic structures (L_o, L_m) over the same lattice L determines the availability of the logical rules. And hence a specific logic with graded notion of consequence is determined. Availability of some rules is completely determined by the interrelation between L_o and L_m, that is, the properties of the algebraic structures, irrespective of the specific value set. This observation plays the key role. Sometimes, for an arbitrarily chosen pair of algebraic structures, say (L_o, L_m), representing respective classes of algebras, some rules do not hold, but with respect to some specific value set L, and sometimes some specific collection $\{T_i\}_{i \in I}$, those rules may hold with respect to the same pair (L_o, L_m). As an illustration, we refer to Example 4.9 of Sect. 4.3

where the prefixed pair of algebraic structures is $(O_{Łukasiewicz}, M_{Gödel})$. In order to ensure the availability of GC5, apart from the above pair of structures, considering [0, 1] as the value set helps to get the sufficient condition of GC5.

4.3 Examples of Logics with Graded Notion of Consequence

Let us now concentrate on the logic building based on the meta-theory of GCT. While building a logic of graded consequence a specific object language containing some or all of the connectives $\neg, \supset, \&, \underline{\vee}$, and perhaps a few more, needs to be fixed. For the time being the focus is only on the propositional fragment of a language. Once the object language is specified, corresponding object-level algebraic structure is formed; a set L endowed with the respective operators $\neg_o$, $\rightarrow_o$, $*_o$, $\oplus_o$ for the connectives forms the object-level algebraic structure. Meta-level algebraic structure can be any arbitrarily chosen complete residuated lattice. The availability of rules corresponding to each connective is determined by the interrelation between the object- and meta-level algebraic structures that may be called L_o and L_m, respectively. The suffixes $_m$ and $_o$ are used to differentiate the respective operators (and hence structures) of the meta-language and the object language. Thus, the scheme for generating different logics with graded notion of consequence is as follows.

A collection $\{T_i\}_{i \in I}$ assigning values to the atomic formulae is considered. Depending on user's choice and necessity, different connectives in the object level are assumed. Hence, based on the user's intended meaning of the linguistic connectives of the object language, the object-level algebraic structure L_o emerges. The properties of the object-level algebraic structure as well as the meta-level algebraic structure along with their interrelations give shape to a particular logic with graded notion of consequence.

Let us consider an example to see how this scheme for generating logics of graded consequence works.

Example 4.9 We are going to construct a logic with graded consequence where the object-level logic is Łukasiewicz and the meta-level logic is Gödel. This logic will be denoted as $(O_{Łukasiewicz}, M_{Gödel})$. We consider the corresponding algebraic structures on a base set L. For the object level, we take Wajsberg algebra corresponding to Wajsberg axioms of Łukasiewicz logic. That is, L is endowed with an implication operation $\rightarrow_o$ corresponding to the Wajsberg's primitive logical connective $\supset$ and the 0-ary operator 0, already present in L corresponding to falsum $\perp$. Then by defining $\neg_o$, $*_o$, $\oplus_o$, $\wedge$ and $\vee$ in L by

- $\neg_o a = a \rightarrow_o 0$,
- $a *_o b = \neg_o(a \rightarrow_o \neg_o b)$,
- $a \oplus_o b = \neg_o a \rightarrow_o b$,
- $a \wedge b = a *_o (a \rightarrow_o b)$, and
- $a \vee b = (a \rightarrow_o b) \rightarrow_o b$,

we get an MV-algebra $(L, \wedge, \vee, *_o, \rightarrow_o, 0, 1)$. $(L, \wedge, \vee, 0, 1)$ turns out to be a bounded lattice. We take this lattice also complete.

Now the meta-level algebra $L_m = (L, \wedge, \vee, \rightarrow_m, 0, 1)$ is a complete lattice with another implication operator $\rightarrow_m$ such that $(\wedge, \rightarrow_m)$ forms the adjoint pair. This is the algebra for Gödel logic, in fact, a complete Heyting algebra.

Let us list the properties of the concerned algebraic structures.

Some properties of L_o:

(i) $(*_o, \rightarrow_o)$ is the adjoint pair.

(ii) $(L, *_o, 1)$ is a commutative monoid.

(iii) $a *_o \neg_o a = 0$, but $a \wedge \neg_o a = 0$ and $a \vee \neg_o a = 1$ do not hold.

(iv) $b \leq a \rightarrow_o b$.

(v) $a *_o (a \rightarrow_o b) \leq b$

(vi) $(a \rightarrow_o b) *_o (b \rightarrow_o c) \leq (a \rightarrow_o c)$.

(vii) $(a \rightarrow_o b) \wedge (b \rightarrow_o c) \leq (a \rightarrow_o c)$ does not hold in general.

(viii) $a *_o b \leq a \wedge b \leq a, b$.

(ix) $a, b \leq a \vee b \leq a \oplus_o b$.

(x) $\neg_o \neg_o a = a$, $a \rightarrow_o b = \neg_o b \rightarrow_o \neg_o a$.

Some interrelation between L_o and L_m.

(xi) $a \rightarrow_m b \leq a \rightarrow_o b$ for all $a, b \in L$.

(xii) $a \wedge (b \vee c) \leq (a \wedge b) \vee (a \wedge c)$ (this holds in BL-algebra, and hence holds in MV-algebra as well as Heyting algebra).

(xiii) $(a \rightarrow_m c) \wedge (b \rightarrow_m c) \leq (a \vee b) \rightarrow_m c$.

(xiv) $(a \rightarrow_m c) \wedge (b \rightarrow_m c) \leq (a \oplus_o b) \rightarrow_m c$ does not hold in general.

To establish the claims (vii) and (xiv), we have respective counterexamples below:

- In $[0, 1]$, $(.7 \rightarrow_o .3) \wedge (.3 \rightarrow_o .1) = .6 > (.7 \rightarrow_o .1) = .4$.
- In $[0, 1]$, $(.2 \rightarrow_m .3) \wedge (.1 \rightarrow_m .3) = 1 > (.2 \oplus_o .1) \rightarrow_m .3) = .8 \rightarrow_m .3 = .3$.

We take any collection $\{T_i\}_{i \in I}$ of valuations from the set of wffs F to $O_{Łukasiewicz}$ obeying the restrictions, and then define the semantic consequence relation $\approx_{\{T_i\}_{i \in I}}$ as in Definition 2.3 (B) using $M_{Gödel}$ as the meta-level structure. This constitutes the logic $(F, \approx_{\{T_i\}_{i \in I}})$.

The properties of the respective algebraic structures fixed for $(O_{Łukasiewicz}, M_{Gödel})$ immediately give us, following the discussion of Sect. 4.2, the proof theoretic nature of the logic. From Theorem 4.3, it can be noticed that the property (xi) gives the sufficient condition for DT. From property (viii), as given in Theorems 4.6 and 4.7, we can ensure the presence of &-E and &-L. Property (ix) satisfies the sufficient condition for $\underline{\vee}$-I given in Theorem 4.9. From the definition of $\neg_o$ and property (xi), we can have $\neg_o a = a \rightarrow_o 0 \geq a \rightarrow_m 0$. This shows that 0 plays the role of c given in Theorem 4.14, and this is sufficient to ensure the presence of $\neg$-I. So, in the proof theory of $(O_{Łukasiewicz}, M_{Gödel})$ we have rules like DT, &-E, &-L, $\underline{\vee}$-I and $\neg$-I. For GC5, if we restrict the value set to $[0, 1]$, then from the definition of $\neg_o$ it can be noticed that $a \wedge \neg_o a \geq \frac{1}{2}$, and $a \wedge \frac{1}{2} = a$ for all $a \leq \frac{1}{2}$. Hence, $\frac{1}{2}$ satisfies all the conditions of c mentioned in Theorem 3.8. So we can claim that GC5 holds, with the value of $c = \frac{1}{2}$, for the same logic $(O_{Łukasiewicz}, M_{Gödel})$ when $L = [0, 1]$.

This kind of construction may be extended to other structures. We fix different algebraic structures for different levels of logic and study the interrelations between the operators corresponding to the object- and meta-level algebraic structures. This process leads towards generating various logics based on GCT. At the end of Sect. 4.3.1, Table 4.3, obtained as an outcome of such investigation, shall be presented.

4.3.1 Many-Valued Logics as a Special Case of GCT

Let us concentrate on the many-valued logics which are generated from BL (Hájek 1998), the logic where the intersection of all tautologies generated from all t-norms constitutes the set of axioms. In Hájek (1998), Peter Hájek has shown that the well-known many-valued logics like Łukasiwicz logic, Product logic (also known as Goguen logic), Gödel logic, etc. can be found as an extension of BL. The set of formulae of these logics is usually mapped to the unit interval [0, 1] or to a subset of [0, 1]. Various t-norms defined over [0, 1] actually correspond to the strong conjunctions of many-valued logics (Hájek et al. 1996). In many-valued logics, usually we see two kinds of conjunctions; one is known as weak/additive conjunction and is computed by lattice meet and the other is known as strong/multiplicative conjunction and is computed by the t-norm. In many-valued context, generally, the classical laws generated from bi-valence do not get satisfied by weak conjunction. Whereas, t-norms corresponding to the strong conjunctions satisfy almost all properties of classical conjunction. So these t-norm-based many-valued logics become the centre of interest. Below we first present a brief overview of these many-valued logics as given in Hájek (1998).

Let $*$ be any continuous t-norm, i.e. a function from $[0, 1] \times [0, 1]$ to $[0, 1]$ such that $*$ is associative, commutative and non-decreasing on both the arguments, and 1 is the identity with respect to $*$. Three such well-known t-norms are as follows:

(Łukasiewicz t-norm) $a * b = \max(0, a + b - 1)$.
(Gödel t-norm) $a * b = \min(a, b)$.
(Product t-norm) $a * b = a.b$ (product of reals).

Given any t-norm $*$ there is an operation $\rightarrow$, called the residuum with respect to $*$, from $[0, 1] \times [0, 1]$ to $[0, 1]$. The residua of the above t-norms are such that $a \rightarrow b = 1$ for $a \leq b$, and for $a > b$ are as follows:

(Łukasiewicz implication) $a \rightarrow b = \min(1, 1 - a + b)$.
Gödel implication) $a \rightarrow b = b$.
(Goguen implication) $a \rightarrow b = \frac{b}{a}$.

Let a propositional language with basic propositional variables, and primitive connectives & and $\supset$ be chosen. Now considering propositional variables to assume values from [0, 1], and fixing a t-norm $*$ and its corresponding residuum $\rightarrow$ as truth functions for & and $\supset$, different propositional calculi based on $*$, denoted as $PC(*)$, are generated. Basic many-valued logic, denoted as BL, is the propositional calculus where the basic set of axioms are those which are tautologies under any t-norm and corresponding implication operation.

It is relevant to mention that a monoidal composition operation $*$ of a residuated lattice L (cf. Definition 2.1) is a generalization of what we call t-norm over [0, 1]. The usual way of getting the residuum operation $\rightarrow$ corresponding to a monoidal composition $*$ is given by $a \rightarrow b = \sup\{z \in L : a * z \leq b\}$.

Definition 4.7 Given a propositional language with primitive connectives & and $\supset$, BL is axiomatized by the following axioms and the rule modus ponens:

(A1) $(\alpha \supset \beta) \supset (\beta \supset \gamma) \supset (\alpha \supset \gamma))$.
(A2) $(\alpha \& \beta) \supset \alpha$.
(A3) $(\alpha \& \beta) \supset (\beta \& \alpha)$.
(A4) $(\alpha \& (\alpha \supset \beta)) \supset (\beta \& (\beta \supset \alpha))$.
(A5a) $(\alpha \supset (\beta \supset \gamma)) \supset ((\alpha \& \beta) \supset \gamma))$.
(A5b) $((\alpha \& \beta) \supset \gamma)) \supset (\alpha \supset (\beta \supset \gamma))$.
(A6) $(\alpha \supset \beta) \supset \gamma) \supset (((\beta \supset \alpha) \supset \gamma)) \supset \gamma)$.
(A7) $\overline{0} \supset \alpha$.

Łukasiewicz propositional logic axiomatizes the tautologies with respect to the residuated lattice on [0, 1] determined by Łuksiewiecz t-norm , and its corresponding Hilbert-type axiomatization is just an extension of BL.

Definition 4.8 Łukasiewicz propositional logic is an extension of BL with the following additional axiom concerning $\neg$, an unary connective defined as $\neg\alpha = \alpha \supset \overline{0}$.

$(\neg\neg)$ $\neg\neg\alpha \supset \alpha$.

Product logic axiomatizes the tautologies with respect to the residuated lattice on [0, 1] determined by the product t-norm, and is axiomatized as an extension of BL with the following additional axioms.

Definition 4.9 Axioms for product logic are those of BL-axioms and the following axioms Π_1, Π_2 concerning $\neg$ and $\wedge$, defined by $\neg\alpha = \alpha \supset \overline{0}$ and $\alpha \wedge \beta = \alpha \& (\alpha \supset \beta)$, respectively.

$(\Pi 1)$ $\neg\neg\gamma \supset ((\alpha \& \gamma \supset \beta \& \gamma) \supset (\alpha \supset \beta))$,
$(\Pi 2)$ $(\alpha \wedge \neg\alpha) \supset \overline{0}$.

Gödel logic is obtained by axiomatizing the tautologies with respect to the residuated lattice on [0, 1] considering lattice meet operation as the t-norm, and its corresponding axiom system is an extension of BL.

Definition 4.10 Gödel logic is obtained by extending the BL-axioms with the following axiom:

(G) $\alpha \supset (\alpha \& \alpha)$.

It will be shown that the t-norm-based many-valued logics can be obtained as a special case of the scheme proposed in GCT. In fact, the fifth, sixth and the seventh rows of Table 4.3 correspond to the product logic, Gödel logic and the Łukasiewicz logic, respectively. In order to obtain a many-valued logic, say ML under the scheme as shown in Example 4.9, we have to consider $\langle [0, 1], \wedge, \rightarrow_c, 0, 1\rangle$ as the meta-level algebraic structure, denoted as $M_{(\wedge, \rightarrow_c)}$, where $\rightarrow_c$ is a crisp implication defined by $a \rightarrow_c b = 1$ if $a \leq b$ and 0, otherwise. In literature $\rightarrow_c$ is known as Gaines–Rescher implication (Klir and Yuan 2006). If ML is, in particular, Łukasiewicz logic, then correspondingly we shall choose $(O_{Łukasiewicz}, M_{(\wedge, \rightarrow_c)})$ as the pair of algebraic structures for the object language and meta-language, respectively, where $O_{Łukasiewicz}$ is as in Example 4.9. It is to be noted here that the algebraic structure $M_{(\wedge, \rightarrow_c)} = \langle [0, 1], \wedge, \vee, \rightarrow_c, 0, 1\rangle$ is not a residuated lattice structure, but it satisfies the required properties (cf. Note 2.1) which are needed to establish that the semantic consequence of many-valued logics is a graded consequence relation (cf. Theorem 2.1).

Below the notion of semantic consequence of many-valued logic is presented following the definition of semantic consequence of GCT. It will be shown that the meta-linguistic implication, present in the defining sentence of $\approx\!\!|$ given in Definition 2.3, needs to be computed by $\rightarrow_c$ in order to capture the notion of semantic consequence of many-valued logics. Moreover, it will be shown that such a notion of consequence turns out to be a graded consequence relation.

Let us recall the designated-set-based notion of semantic consequence of many-valued logics (Bolc and Borowik 1992). A designated set D is generally a proper subset of value set not containing the least element 0 and containing the greatest element 1. In most cases, the singleton set $\{1\}$ is taken as D, but there are examples where it is not a singleton.

Definition 4.11 Let D be the designated set of the value set [0, 1], and $\{T_i\}_{i\in I}$ be the set of all valuations satisfying property (TC) based on some algebra over [0, 1]. Then $X \models_{ML} \alpha$ holds if and only if for all T_i, $T_i(X) \subseteq D$ implies $T_i(\alpha) \in D$.

To obtain $\models_{ML}$, the semantic consequence of ML, as a special case of the notion of graded consequence relation we start with the collection of all possible fuzzy sets $\{T_i\}_{i\in I}$ from F to [0, 1], and construct a collection $\{T_i^D\}_{i\in I}$ where for each $i \in I$, $T_i^D(\alpha) = 1$ if $T_i(\alpha) \in D$, and 0 otherwise. Identifying the function T_i^D with the set it determines, the above definition reduces to—'for all T_i^D, $X \subseteq T_i^D$ implies $\alpha \in T_i^D$'. Then defining $gr(X \approx\!\!|_{\{T_i^D\}_{i\in I}} \alpha) = \inf_{i\in I}\{\inf_{x\in X} T_i^D(x) \rightarrow_c T_i^D(\alpha)\}$, it can be noticed that $X \models_{ML} \alpha$ iff $gr(X \approx\!\!|_{\{T_i^D\}_{i\in I}} \alpha) = 1$. It can be proved that $\approx\!\!|_{\{T_i^D\}_{i\in I}}$ is a graded consequence relation, i.e. $\approx\!\!|_{\{T_i^D\}_{i\in I}}$ satisfies (GC1)–(GC3).

To prove $\approx\!\!|_{\{T_i^D\}_{i\in I}}$ is a graded consequence relation, we need to establish that $\rightarrow_c$ can be a candidate for $\rightarrow_m$ with properties (r3), (r5), (r6) and (r7) (cf. Note 2.1). Moreover, we need to show that the meta-linguistic conjunction $*_m$ satisfies property (r2). In this regard, as mentioned earlier, we should note that many-valued logics do not properly distinguish between the operators for connectives of different layers of a logic. Since it has been our concern to identify the algebraic structures corresponding to the levels (object, meta and metameta) of logic, in case of many-valued logics also

Table 4.3 Logics of graded consequence generated from two algebraic structures

	DT	MP/DT_c	Tran Tran	&-I/&-R	&-E/&-L	$\underline{\vee}$-I/$\underline{\vee}$-R	$\underline{\vee}$-E/$\underline{\vee}$-L	¬-I/¬-R	GC5	GC4
$(O_{Gödel}, M_{Łukasiewicz})$	N	Y	Y	Y	Y	Y	Y	N	–	Y
$(O_{Łukasiewicz}, M_{Gödel})$	Y	N	N	N	Y	Y	N	Y	Y in [0, 1]	N
$(O_{Goguen}, M_{Gödel})$	Y	N	N	N	Y	Y	N	Y	–	Y $k = 1$
$(O_{Gödel}, M_{Goguen})$	N	Y	Y	Y	Y	Y	Y	N	–	Y $k = 1$
$(O_{Goguen}, M_{(\wedge, \rightarrow_c)})$ with $D = \{1\} \subseteq [0, 1]$ Product logic	N	Y	Y	Y	Y	Y	Y	N	N for $c = 1$	Y $k = 1$
$(O_{Gödel}, M_{(\wedge, \rightarrow_c)})$ with $D = \{1\} \subseteq [0, 1]$ ≡ Gödel logic	Y	Y	Y	Y	Y	Y	Y Y	Y	N for $c = 1$	Y $k = 1$
$(O_{Łukasiewicz}, M_{(\wedge, \rightarrow_c)})$ with $D = \{1\} \subseteq [0, 1]$ ≡ Łukasiewicz logic	N	Y	Y	Y	Y	Y	Y	N	N for $c = 1$	Y $k = 1$
with $D = \{1, \frac{1}{2}\}$, $L = \{0, \frac{1}{2}, 1\}$ Łukasiewicz 3-valued paraconsistent logic		N							Y $c = \frac{1}{2}$	N

we are led towards this study. Regarding meta-linguistic 'conjunction' of many-valued logics, there is a convention of understanding meta-linguistic conjunction as object-level conjunction or in terms of a connective representing lattice meet. In both the cases, the operators are t-norms which by definition satisfy $a \leq b$ implies $a \circ c \leq b \circ c$, the monotonicity property of t-norm. From this property, one can easily obtain the condition (r2). Now, the theorem below would establish that $\rightarrow_c$, along with any t-norm for $*_m$, has all the properties (r3), (r5), (r6), (r7), which are required in order to generate a graded consequence relation.

Theorem 4.15 *The crisp implication operation* $\rightarrow_c$ *and any t-norm* $*_m$ *satisfy the following properties:*

(1) $a \rightarrow_c a = 1$.
(2) $a \leq b$ *implies* $a \rightarrow_c d \geq b \rightarrow_c d$.
(3) $a \rightarrow_c b \geq ((a \wedge d) \rightarrow_c b) *_m (a \rightarrow_c d)$.
(4) $a \rightarrow_c inf_i b_i = inf_i (a \rightarrow_c b_i)$.

Proof In order to prove this theorem, let us first prove the isotone property of $\rightarrow_c$ in the second variable, i.e. $a \leq b$ implies $c \rightarrow_c a \leq c \rightarrow_c b$ in [0, 1]. For $a, b, c \in L$ and $a \leq b$, two cases arise. (Case-I) $c \leq a$ (Case-II) $a < c$.

Case-I Let $c \leq a$. Then $c \leq a \leq b$ and hence $c \rightarrow_c a = c \rightarrow_c b = 1$.

Case-II Let $a < c$. Then $c \rightarrow_c a = 0 \leq c \rightarrow_c b$.

(1) The definition of $\rightarrow_c$ itself verifies the condition $a \rightarrow_c a = 1$.

(2) Let $a \leq b$. Now two cases arise. (A) $a \leq c$ (B) $a > c$.

(A) If $a \leq c$ then $a \rightarrow_c c = 1 \geq b \rightarrow_c c$.

(B) If $a > c$, i.e. $c < a \leq b$ then both $a \rightarrow_c c = b \rightarrow_c c = 0$. Hence combining (A) and (B) we have $a \leq b$ implies $a \rightarrow_c c \geq b \rightarrow_c c$.

(3) To prove $a \rightarrow_c b \geq ((a \wedge c) \rightarrow_c b) *_m (a \rightarrow_c c)$ let us consider two cases.

Case - I $a \leq b$. Hence $a \rightarrow_c b = 1$.

$a \leq b$ implies $a \wedge c \leq b$. Hence $(a \wedge c) \rightarrow_c b = 1$.

Now either $a \rightarrow_c c = 0$ or $a \rightarrow_c c = 1$.

In both the cases, $a \rightarrow_c b \geq ((a \wedge c) \rightarrow_c b) *_m (a \rightarrow_c c)$.

Case -II $a > b$. Hence $a \rightarrow_c b = 0$.

Now either $a \wedge c \leq b$ or $a \wedge c \nleq b$.

If $a \wedge c \nleq b$ then $(a \wedge c) \rightarrow_c b = 0$ and hence $a \rightarrow_c b \geq ((a \wedge c) \rightarrow_c b) *_m (a \rightarrow_c c)$.

If $a \wedge c \leq b$ then as $a \nleq b$, $c \leq b$.

Hence $a \rightarrow_c c \leq a \rightarrow_c b$. [By isotone property of $\rightarrow_c$ in the second variable]

$a \rightarrow_c c = 0$.

Hence, combining all the cases we can conclude $a \rightarrow_c b \geq ((a \wedge c) \rightarrow_c b) *_m (a \rightarrow_c c)$.

(4) As $inf_i b_i \leq b_i$, $a \rightarrow_c inf_i b_i \leq a \rightarrow_c b_i$, for each i. [By isotone property of $\rightarrow_c$ in the second variable]

Hence $a \rightarrow_c inf_i b_i \leq inf_i (a \rightarrow_c b_i)$.

Now two cases arise. Case-I $a \leq inf_i b_i$ Case-II $a \nleq inf_i b_i$.

Case-I $a \leq inf_i b_i$ implies $a \to_c inf_i b_i = 1$.
Also $a \leq inf_i b_i \leq b_i$ for all i.
So, $a \to_c b_i = 1$ for all i.
Hence $inf_i(a \to_c b_i) = 1$.
Case -II $a \nleq inf_i b_i$ implies $a \to_c inf_i b_i = 0$.
Now $a \nleq inf_i b_i$ implies there is some i such that $b_i \leq a$ and hence $a \to_c b_i = 0$.
Hence $inf_i(a \to_c b_i) = 0$. □

Hence, Theorem 4.15 establishes that the implication operator $\to_c$ of many-valued logics turns out to be an implication operator $\to_m$, as required to generate a graded consequence relation with respect to $\{T_i^D\}_{i \in I}$.

Thus, as an outcome of the study made in Sects. 4.2, 4.3, and 4.3.1, we have the following table with different logics of graded consequence (Dutta and Chakraborty 2016). For any connective #, #-I, #-E, #-R, #-L, respectively, denote the graded counterparts of the introduction, elimination, right and left rule of the connective. Let DT, DT_c, MP, Trans be the abbreviations for the graded version of the deduction theorem, its converse, modus ponens and transitivity, respectively.

4.4 A Comparative Analysis

It should be clear by this time that the study of proof theory in the context of graded consequence has been carried out from the semantic angle, to be more specific from the angle of algebraic semantics. A proof theory from the semantic angle may seem to be strange. So a clarification is needed. The intention behind introduction of natural deduction systems and sequent calculi by Gentzen is well known. We have mentioned that in brief at the beginning of this chapter. It would however be helpful to consider Ono's paper (Ono 2003) as another point of reference. Like Gentzen, Ono also begins the presentation of sequent calculus with the comment that structural rules describe the roles of two meta-linguistic symbols, viz. 'comma' and 'implication'. In cases of LJ and LK, Ono remarks that with the help of the structural rules, viz. contraction-left, weakening-left, cut, &-left and &-right one can obtain the following logical principle: $\delta, \phi \vdash \psi$ if and only if $\delta \& \phi \vdash \psi$. This indicates that the usual practice in LJ, LK is to see ',' and object-level conjunction '&' equivalently. Similar is the attitude towards the meta-implication $\vdash$ and object-level implication $\supset$.

Now the question arises (Ono 2003), how would the ',' be treated in substructural logics where some of the structural rules remain absent. Ono presents a modified version of &-left in this context. Let us call this new object-level conjunction as $\&_H$ and the rules for $\&_H$ be given by

$$\frac{\Gamma, \alpha, \beta, \Sigma \vdash \gamma}{\Gamma, \alpha \&_H \beta, \Sigma \vdash \gamma} \ (\&_H\text{-left}) \qquad \frac{\Gamma \vdash \alpha \quad \Sigma \vdash \beta}{\Gamma, \Sigma \vdash \alpha \&_H \beta} \ (\&_H\text{-right}).$$

With these new rules for $\&_H$ and cut it has been proved in Ono (2003) that 'a sequent

$\alpha_1, \alpha_2, \ldots \alpha_m \vdash \beta$ is provable if and only if $\alpha_1 \&_H \alpha_2 \&_H \ldots \&_H \alpha_m \vdash \beta$ is provable'.

Thus, in this new context again, i.e. in the case of substructural logics also 'comma' and object-level conjunction are understood equivalently. Hence, as a result, the rule ($\supset$-right) implies 'if $\alpha \&_H \beta \vdash \gamma$ is provable then $\alpha \vdash \beta \supset \gamma$ is provable' and the rule ($\supset$-left) along with overlap and cut implies 'if $\alpha \vdash \beta \supset \gamma$ is provable then $\alpha \&_H \beta \vdash \gamma$ is provable'. So, the conditions imposed on the rules, viz. to see 'comma' and $\vdash$ equivalently with (or at least reflected in) $\&_H$ and $\supset$ are responsible for having $\alpha \&_H \beta \vdash \gamma$ if and only if $\alpha \vdash \beta \supset \gamma$. In algebraic term, the counter part of this property is given by $a *_o b \leq c$ if and only if $a \leq b \rightarrow_o c$ where $*_o$ and $\rightarrow_o$ are the operators for $\&_H$ and $\supset$, respectively. In fact, there is a hidden semantic content in the conditions on the rules. One wants to have the following:

if $\alpha_1, \alpha_2, \ldots \alpha_m \vdash \beta$ is 'true' then $\alpha_1 \&_H \alpha_2 \&_H \ldots \&_H \alpha_m \vdash \beta$ is 'true' and conversely. Or if $\alpha \&_H \beta \vdash \gamma$ is 'true' then $\alpha \vdash \beta \supset \gamma$ is 'true' and conversely. Because of such inherent semantic demands the syntactic rules are so framed and as a result we get the deduction theorem (DT) as well as the converse of DT in the syntax and the property of residuation in the algebra. All the syntactic rules taken in all sequent calculi are semantically sound.

However, the rules of substructural logics are 'sound' in a crisp sense. As an example, let us take the sequent system FLe. Let $\mathscr{L}$ be a logic over FLe in which the consequence relation $\vdash_{\mathscr{L}}$ is defined in the standard way. Now for any set Γ of wffs and a wff β, $\Gamma \vdash_{\mathscr{L}} \beta$ if and only if $1 \leq v(\gamma)$ holds for all $\gamma \in \Gamma$ implies $1 \leq v(\beta)$ holds for every valuation v on every FLe-algebra which validates $\mathscr{L}$ (Kowalski and Ono 2010). The proof of this claim rests upon soundness of the rules of FLe. For instance, the rule ($\Rightarrow \cdot$), viz.

$$\frac{\Gamma \Rightarrow \alpha, \quad \Delta \Rightarrow \beta}{\Gamma, \Delta \Rightarrow \alpha \cdot \beta}$$

is sound since if ($1 \leq v(\Gamma)$ implies $1 \leq v(\alpha)$) and ($1 \leq v(\Delta)$ implies $1 \leq v(\beta)$) then ($1 \leq v(\Gamma \cup \Delta)$ implies $1 \leq v(\alpha \cdot \beta)$), where $1 \leq v(\Gamma)$ stands for $1 \leq v(\gamma)$ for each $\gamma \in \Gamma$ and every valuation v.

Theory of graded consequence differs in both respects.

First, some logics may like to relax the conditions. For example, it may demand only one part of DT to hold. In other words, from the semantic standpoint, a logic may accept the metameta-conditional if $\alpha \& \beta \vdash \gamma$ then $\alpha \vdash \beta \supset \gamma$ as true but not the converse. From such a standpoint, in the corresponding algebra, the demand would be $a *_o b \leq c$ implies $a \leq b \rightarrow_o c$ but not the converse. Also, one may very legitimately accept that $\{\alpha, \beta\} \vdash \alpha$ but the object language may not contain a conjunction at all and hence $\alpha \& \beta \vdash \alpha$ would be meaningless. Thus, identifying meta-level connectives with those belonging to the object level is not essential to construct a logic. But some relationship between the connectives of two levels has to be accepted. Our present formulation allows for this space. In Sect. 4.2, conditions for various other relationships are obtained. A designer of a logic may wish to have one or more of them and should structure the truth value sets accordingly.

Second, sequents here are graded, their interpretations are many-valued meta-level sentences as discussed in Sect. 4.1: $\Gamma \vdash \alpha$ or $\Gamma \Rightarrow \alpha$ is interpreted as 'α follows from Γ' and this is not a crisp sentence in general. That is, from semantic angle, instead of understanding the above sentence in terms of $1 \leq v(\Gamma)$ implies $1 \leq v(\alpha)$, here we consider the grade of $\Gamma \mid\!\sim \alpha$ with respect to the collection $\{T_i\}_{i \in I}$ generating $\mid\!\sim$.

In this connection, the main departure of the present approach from the sequent calculi or hypersequent calculi developed for fuzzy logics is indicated below.

Sequents as mentioned above, in the context of graded consequence, are many-valued. While in the context of sequent/hypersequent calculus presentation for t-norm-based fuzzy logics (Metcalfe 2004; Metcalfe et al. 2009), sequents as well as rules are two-valued. To illustrate, let us consider a typical rule say

$$\frac{G_1 \mid H_1 \quad G_2 \mid H_2 \ldots\ldots G_{n-1} \mid H_{n-1} \quad G_n \mid H_n}{G \mid H}$$

in some hypersequent calculus where for each i $(i = 1, \ldots, n)$, G_i, H_i are multisets of sequents of the form $S^1_{m_1} \mid S^1_{m_2} \ldots \mid S^1_{m_i}$. That is, for each i, G_i is a multiset of the form $S^1_{m_1} \mid S^1_{m_2} \ldots \mid S^1_{m_i}$. Each S with super and subscripts are sequents of the type $\Gamma \Rightarrow \Delta$, Γ, Δ being finite multisets of wffs. Soundness of the rule is defined through steps. Some aggregation operators $\star_1$, $\star_2$ (may be the same in some cases) are defined such that for each pair of multisets Γ, Δ of wffs, and each valuation v, $v^{\star_1}(\Gamma)$ aggregates the values $\{v(\gamma) : \gamma \in \Gamma\}$ and $v^{\star_2}(\Delta)$ aggregates the values $\{v(\delta) : \delta \in \Delta\}$. The aggregation functions vary depending on variation of the logics. In fact, these functions take care of the meta-linguistic operator (,) present in both Γ and Δ. But no logical interpretation of these operators are offered in general (such as in the cases of Łukasiewicz logic and product logic). Now, a G with or without a subscript is valid, written as $\models G$ iff for each valuation v, $v^{\star_1}(\Gamma) \leq v^{\star_2}(\Delta)$ for some sequent $\Gamma \Rightarrow \Delta$ present in G. Then finally,

$$\frac{G_1 \mid H_1 \quad G_2 \mid H_2 \ldots\ldots G_{n-1} \mid H_{n-1} \quad G_n \mid H_n}{G \mid H}$$

is sound iff for each i $(i = 1, \ldots, n)$, if $\models G_i$ or $\models H_i$, then $\models G$ or $\models H$. In particular, if $n = 2$ and the hypersequents $G_1 \mid H_1$, $G_2 \mid H_2$ and $G \mid H$ contain just one sequent

each, say $\Gamma_1 \Rightarrow \Delta_1$, $\Gamma_2 \Rightarrow \Delta_2$ and $\Gamma \Rightarrow \Delta$, respectively, then the rule

$$\frac{G_1 \mid H_1 \quad G_2 \mid H_2}{G \mid H}$$

is sound iff $v^{\star_1}(\Gamma_1) \leq v^{\star_2}(\Delta_1)$ and $v^{\star_1}(\Gamma_2) \leq v^{\star_2}(\Delta_2)$ imply $v^{\star_1}(\Gamma) \leq v^{\star_2}(\Delta)$. The crispness of the procedure at all stages is obvious which does not change essentially even though a distinction between many-valued logics and fuzzy logics is made in the sense of Esteva et al. (2009) by incorporating names of all the truth values in [0, 1] within the language as was proposed in Hájek (1998).

On the other hand, in graded context, rules which are metameta-level sentences of the form '$\{\Gamma_1 \mid\!\sim \alpha_1, \ldots, \Gamma_n \mid\!\sim \alpha_n\}$ implies $\Delta \mid\!\sim \beta$' indicates a relation between two meta-level sentences, viz. ($\Gamma_1 \mid\!\sim \alpha_1$ **and** …**and** $\Gamma_n \mid\!\sim \alpha_n$) and $\Delta \mid\!\sim \beta$ which are generally many-valued. So, from semantic angle such a rule is sound iff the above metameta-level sentence is true and that is represented as $gr(\Gamma_1 \approx\!\!\!| \alpha_1) *_m \dots *_m gr(\Gamma_n \approx\!\!\!| \alpha_n) \le gr(\Delta \approx\!\!\!| \beta)$. This metameta-level sentence is crisp but the sequents $\Gamma_1 \mid\!\sim \alpha_1, \dots, \Gamma_n \mid\!\sim \alpha_n, \Delta \mid\!\sim \beta$ are not.

Sequents/hypersequents are meta-linguistic sentences, but in some cases (such as Gödel logic) (Metcalfe et al. 2009), these are equated with an object-level formula. But in the context of graded consequence, the entire study on proof theoretic rules maintains the distinction between object-level and meta-level connectives. A proof theoretic rule, $gr(\Gamma \approx\!\!\!| \alpha) \le gr(\Delta \approx\!\!\!| \beta)$, stated semantically, may entail some interrelation between object-level and meta-level connectives, but $\Gamma \mid\!\sim \alpha$ and/or $\Delta \mid\!\sim \beta$ cannot be equated generally with an object-level formula.

The present study reveals that

- behind the syntactic rules of existing proof theories there is a semantic consideration and
- the relationship between the object- and meta-level operations in the truth set plays a key role in the setting up of the proof theoretic rules.

The usefulness of this proof theoretic study of the logic of graded consequence lies in another aspect too. Apart from the traditional use of sequent calculi in which a proof tree begins with sequents of the form $\alpha \vdash \alpha$, it has been beneficial in some cases (Ghosh and Chakraborty 2004; Lehmaman and Magidor 1992) to consider a set of sequents as the given datum. Then the query is—which other sequents are derivable from this set by using the available rules? In the present case, sequents are graded. Thus, as datum, we have a set like $\{gr(X_1 \mid\!\sim \alpha_1) = a_1, \dots, gr(X_n \mid\!\sim \alpha_n) = a_n\}$, the interpretation being α_i follows from X_i to the extent a_i in a logic with levels stratified. The generating set $\{T_i\}_{i\in I}$ may not be known. One would now ask the question: Does $X \mid\!\sim \alpha$ hold for some X and α and if so, how is the $gr(X \mid\!\sim \alpha)$ related with the grades $a_1, a_2, \dots, a_n$? The available rules in the logic that depend upon available relation in the algebras of the two levels will give an answer. Let us present some specific examples of reaching at such a graded sequent from a set of graded sequents.

Example 4.10 Referring to Example 4.7, we have seen how the algebras of two levels determine the set of logical rules in the logic. Let as the datum we are given the following set: $\{gr(X \mid\!\sim \neg\alpha \supset \beta) = e,\ gr(Y \mid\!\sim \alpha \supset \gamma) = b,\ gr(Z \mid\!\sim \gamma \supset \delta) = a,\ gr(\{\beta\} \mid\!\sim \delta) = 1\}$.

Now the logic allows us to derive the following:

$gr(X \mid\!\sim \neg\alpha \supset \beta) \wedge gr(Y \mid\!\sim \alpha \supset \gamma) \wedge gr(Z \mid\!\sim \gamma \supset \delta) \wedge gr(\{\beta\} \mid\!\sim \delta)$
$\le gr(X \mid\!\sim \neg\alpha \supset \beta) \wedge gr(Y \cup Z \mid\!\sim \alpha \supset \delta) \wedge gr(\{\beta\} \mid\!\sim \delta)$ [transitivity]
$\le gr(X \mid\!\sim \neg\alpha \supset \beta) \wedge gr(Y \cup Z \cup \{\alpha\} \mid\!\sim \delta) \wedge gr(\{\beta\} \mid\!\sim \delta)$ [converse of DT]
$= gr(X \mid\!\sim \alpha \underline{\vee} \beta) \wedge gr(Y \cup Z \cup \{\alpha\} \mid\!\sim \delta) \wedge gr(\{\beta\} \mid\!\sim \delta)$ [definition of $\supset$]
$\le gr(X \cup Y \cup Z \mid\!\sim \delta)$. [$\underline{\vee}$-Elimination]

The grades of the sequents present in the datum imply that
$e \wedge b \wedge a \wedge 1\ (= f)$
$= gr(X \mid\!\sim \neg\alpha \supset \beta) \wedge gr(Y \mid\!\sim \alpha \supset \gamma) \wedge gr(Z \mid\!\sim \gamma \supset \delta) \wedge gr(\{\beta\} \mid\!\sim \delta)$

$\leq gr(X \cup Y \cup Z \mid\sim \delta)$.
i.e. $gr(X \cup Y \cup Z \mid\sim \delta) \geq f$.

Example 4.11 With respect to Example 4.8, let us start with the data consisting of the graded sequents

$gr(X \mid\sim \alpha \supset \beta) = e, gr(X \mid\sim \beta \supset \gamma) = a, gr(X, \neg\alpha \mid\sim \delta) = a, gr(\mid\sim \alpha\&\gamma \supset \delta) = e$.

Now we can have the following derivation:

$gr(X \mid\sim \alpha \supset \beta) *_m gr(X \mid\sim \beta \supset \gamma) *_m gr(X, \neg\alpha \mid\sim \delta) *_m gr(\mid\sim \alpha\&\gamma \supset \delta)$
$\leq gr(X \mid\sim \alpha \supset \gamma) *_m gr(X, \neg\alpha \mid\sim \delta) *_m gr(\mid\sim \alpha\&\gamma \supset \delta)$ [transitivity]
$\leq gr(X, \alpha \mid\sim \gamma) *_m gr(X, \neg\alpha \mid\sim \delta) *_m gr(\mid\sim \alpha\&\gamma \supset \delta)$ [converse of DT]
$\leq gr(X, \alpha \mid\sim \gamma) *_m gr(X, \alpha \mid\sim \alpha) *_m gr(X, \neg\alpha \mid\sim \delta) *_m gr(\mid\sim \alpha\&\gamma \supset \delta)$ [GC1]
$\leq gr(X, \alpha \mid\sim \alpha\&\gamma) *_m gr(X, \neg\alpha \mid\sim \delta) *_m gr(\mid\sim \alpha\&\gamma \supset \delta)$ [&-Introduction]
$\leq gr(X, \alpha \mid\sim \delta) *_m gr(X, \neg\alpha \mid\sim \delta)$ [MP]
$\leq gr(X, \alpha\underline{\vee}\neg\alpha \mid\sim \delta)$. [$\underline{\vee}$-left]

From the grades of the sequents, we have started with, the following can be obtained:

$a \leq gr(X, \alpha\underline{\vee}\neg\alpha \mid\sim \delta)$. ...(1)

Also, as $x \oplus \neg_o x \geq e$, for a $\mid\sim$ generated from any arbitrary collection $\{T_i\}_{i\in I}$, we have $e \leq gr(\mid\sim \alpha\underline{\vee}\neg_o\alpha)$. ...(2)

So, from (1) and (2) we obtain

$a *_m e \leq gr(X, \alpha\underline{\vee}\neg\alpha \mid\sim \delta) *_m gr(\mid\sim \alpha\underline{\vee}\neg_o\alpha)$

Hence $a \leq gr(X \mid\sim \delta)$. [by GC3]

References

Arruda, A.I.: Aspects of the historical development of paraconsistent logic. In: Priest, G., Routley, R., Norman, J. (eds.) Paraconsistent Logic: Essays on the Inconsistent, pp. 99–129. Philosophia Verlag, München, Handen, Wien (1989)

Barwise, J.: Introduction to first order logic. In: Barwise, J. (ed.) Handbook of Mathematical Logic. Studies in Logic and the Foundations of Mathematics, North Holland, Amsterdam (1982). ISBN 978-0-444-86388-1

Bolc, L., Borowik, P.: Many-valued Logics: Theoretical Foundations. Springer, Berlin (1992)

Dubois, D., Prade, H.: Possibilistic logic: a retrospective and prospective view. Fuzzy Sets Syst. **144**(1), 3–23 (2004)

Dutta S., Chakraborty, M.K.: Towards a proof theory of graded consequence: semantics behind syntax. In: Beziau, J-Y., Costa-Leite, A., D'Ottaviano, I.M.L. (orgs.) Aftermath of the Logical Paradise. Coleção CLE, vol. 81, pp. 7–50 (2017)

Dutta, S., Chakraborty, M.K.: Rule modus ponens vis-á-vis explosiveness condition in graded perspective. In: Lowen, Roubens (eds.) Proceedings of the International Conference on Rough Sets, Fuzzy Sets and Soft Computing, held on 5–7 November 2009 at Tripura University, pp. 271–284. SERIALS (2009a)

Dutta, S., Chakraborty, M.K.: Some proof theoretic results depending on context from the perspective of graded consequence. In Sakai, H., Chakraborty, M.K., Assanien, A.E.H., Ślejak, D., Zhu, W. (eds.) RSFDGrC 2009: Proceedings of the 12th International Conference on Rough Sets, Fuzzy Sets, Data Mining, and Granular Computing, volume 5908 of LNAI, pp. 144–151. Springer, Berlin (2009b)

Dutta, S.: Graded Consequence as the Foundation of Fuzzy Logic, Doctoral Thesis, University of Calcutta (2011a)

Dutta, S.: Introduction, elimination rule for $\neg$ and $\supset$: a study from graded context. In: Kuznetsov, S.O., et al. (eds.) PReMI 2011. LNCS, vol. 6744, pp. 80–85. Springer, Berlin (2011b)

Dutta, S., Chakraborty, M.K.: The role of metalanguage in graded logical approaches. Fuzzy Sets Syst. **298**, 238–250 (2016)

Esteva, F., Godo, L., Noguera, C.: First order t-norm based fuzzy logics with truth-constants: distinguished semantics and completeness properties. Ann. Pure Appl. Log. **161**(2), 185–202 (2009)

Gentzen, G.: Investigations into logical deductions. In: Szabo, M.E. (ed.) Collected papers of G. Gentzen, pp. 68–131. North Holland Publications, Amsterdam (1969)

Ghosh, S., Chakraborty, M.K.: Non-monotonic proof systems: algebraic foundations. Fundam. Inform. **59**, 39–65 (2004)

Hájek, P., Godo, L., Esteva, F.: A complete many-valued logic with product conjunction. Arch. Math. Log. **35**, 191208 (1996)

Hájek, P.: Metamathematics of Fuzzy Logic. Kluwer Academic Publishers, Dordrecht (1998)

Katsura, S.: Nagarjuna and the tetralemma. In: Silk, J.A. (ed.) Buddhist Studies: The Legacy of Godjin M. Nagas. Motilal Banarsidass Pvt. Ltd., Delhi (2008)

Klir, G.J., Yuan, B.: Fuzzy Sets And Fuzzy Logic: Theory and Applications. Prentice Hall of India Private Limited, New Delhi (2006)

Kowalski, T., Ono, H.: Fuzzy logics from substructural perspective. Fuzzy Sets Syst. **161**, 301–310 (2010)

Lehmaman, D., Magidor, M.: What does a conditional knowledge base entail? Artif. Intell. **55**, 1–60 (1992)

Lemmon, E.J.: Beginning Logic, Nelson (1965)

Metcalfe, G.: Proof Theory for Propositional Fuzzy Logics. PhD thesis, University of London (2004)

Metcalfe, G., Olivetti, N., Gabbay, D.: Proof Theory for Fuzzy Logics. Springer, Berlin (2009)

Ono, H.: Substructural logics and residuated lattices - an introduction. In: Hendricks, V.F., Malinowski, J. (eds.) Trends in Logic: 20, pp. 177–212. Kluwer Academic Publishers, Printed in the Netherlands (2003)

Sen, J.: Doctoral thesis on 'Some embeddings in linear logic and related issues', University of Calcutta, December 2000

Takeuti, G.: Proof Theory. North Holland, 2nd edn. (1987)

Troelstra, A.S.: Lectures on Linear Logic. Number 29, CSLI, Standford (1992)

Chapter 5
Meta-logical Notions Generalized: Graded Consequence with Fuzzy Set of Premises

Abstract In the classical context, the notion of consequence can be equivalently presented by consequence operator approach and consequence relation approach. Similar approaches are observed in fuzzy context too. Different proposals came up in order to bridge the gap between Pavelka's notion of fuzzy consequence operator and Chakraborty's notion of graded consequence relation in fuzzy context. In this regard, we shall discuss the notion of fuzzy consequence relation by Castro et al., and implicative consequence operator and implicative consequence relation by Rodríguez et al. We shall discuss some of the limitations of these approaches, and propose extensions of fuzzy consequence relation of Castro et al., and the notion of implicative consequence relation of Rodríguez et al, in a graded meta-language. In both the cases, the notion of graded consequence with fuzzy set of premises is considered.

5.1 Consequence Operator and Consequence Relation in the Context of Fuzzy Logics

It has been mentioned that in logic, the notion of consequence is understood either as an operator which maps every premise set X to its set of consequences $C(X)$ or as a binary relation $\vdash$ which consists of a set of pairs (X, α), indicating that α is a consequence of X. In classical two-valued context, these two approaches to the notion of consequence are equivalent (cf. Chap. 1). In fuzzy context, retaining this equivalence did not appear as straightforward as the classical. Below we shall present a brief history of the proposals for such studies in the chronological order of their emergence. The focus would be on the stage-by-stage development (Castro et al. 1994; Chakraborty 1995; Dutta and Chakraborty 2014, 2016; Pavelka 1979; Rodríguez et al. 2003) where researchers proposed their theories from different aspects of generalization of classical consequence along with an attempt of overcoming some of the previous gaps.

In this section and in Chap. 6, we shall present a detailed survey of different fuzzy logical approaches. We shall, however, observe that each approach lacks in one or more of the following points:

M. K. Chakraborty and S. Dutta, *Theory of Graded Consequence*, Logic in Asia: Studia Logica Library, https://doi.org/10.1007/978-981-13-8896-5_5

(i) A notion of 'degree of consequence' is introduced, but the notion of consistency remains crisp.
(ii) Though a notion of 'degree of consequence' is available, the process of derivation and the rules of inference are basically crisp.
(iii) Degrees are assigned by different ways to the notion of consequence, claimed as degree of consequence or provability degree or value of proof, but none of the degrees are computed by paying proper care to the underlying meta-logical expression defining the respective notions.

Pavelka in (1979) introduced the concept of consequence operator over fuzzy sets of formulae as a generalization of Tarski's (1956) notion of consequence operator. Chakraborty (1988, 1995), on the other hand, in 1987 introduced a notion of graded consequence relation generalizing the notion of classical consequence relation. Castro et al. (1994) defined a notion of fuzzy consequence relation in order to retain the classical equivalence between consequence operator, consequence relation in fuzzy context in 1994. Thus, a one-to-one correspondence between fuzzy closure operator of Pavelka and fuzzy consequence relation of Castro et al. is obtained, but other meta-logical notions like consistency remain two-valued. So, classical consequence–consistency correspondence is sacrificed here. On the other hand, introducing a notion of implicative consequence relation Rodríguez et al. (2003) made a good contribution, as it generalizes both Chakraborty's notion of graded consequence and fuzzy consequence relation of Castro et al. But they (Rodríguez et al. 2003) introduced two different notions of inclusion between fuzzy sets, leaving their interpretation unexplained. In the following sections, we shall attempt to make both of these loose ends meaningful from the perspective of building a well-defined graded meta-logic.

In Sect. 5.1.1, we first introduce Pavelka's notion of fuzzy consequence operator, and point out the differences with the notion of graded consequence relation. In Sect. 5.1.2, we discuss the approach taken by Castro et al. in order to bridge the gap between Pavelka's fuzzy consequence operator and Chakraborty's graded consequence relation. Then in Sect. 5.2, we shall present an extension of Castro's proposal by developing the complete meta-theory parallel to the work presented in Sect. 2.4 of Chap. 2 and Sect. 3.1 of Chap. 3. In Sect. 5.3, we shall discuss the approach taken by Rodríguez et al. in order to bridge the gaps among different notions of fuzzy consequence appeared in Castro et al. (1994), Chakraborty (1995), Rodríguez et al. (2003). Section 5.4 will contain a proposal to overcome the limitations of the approach presented in Sect. 5.3.

5.1.1 Pavelka's Fuzzy Consequence Operator and Chakraborty's Graded Consequence Relation

Pavelka in (1979) introduced the consequence operator as follows.

Definition 5.1 A consequence operator C in fuzzy context is a function from the set of all fuzzy subsets ($\mathscr{F}(F)$) over F to $\mathscr{F}(F)$ satisfying the following conditions:

1. $X \subseteq C(X)$,
2. if $X \subseteq Y$ then $C(X) \subseteq C(Y)$ and
3. $C(C(X)) = C(X)$,

where for any $A, B \in \mathscr{F}(F)$, $A \subseteq B$ means $A(\alpha) \leq B(\alpha)$ for all $\alpha \in F$, that is, the degree to which the formula α belongs to the fuzzy premise A is less or equal to the degree to which α belongs to the fuzzy premise B, where the membership degrees are taken from a complete lattice. C is called the fuzzy closure operator.

In 1987, on the other hand, Chakraborty (1988, 1995) introduced the concept of graded consequence relation as a generalization of the notion of classical consequence relation Gentzen (1969) in many-valued context (cf. Definition 2.2 of Chap. 2). From Definitions 5.1 and 2, it can be noticed that Pavelka's consequence operator and Chakraborty's graded consequence relation are two different notions.

(i) For the first one, the premise set is considered to be a fuzzy set over formulae, and for the other it is simply an ordinary set of premises.

(ii) In the first case, the notion of consequence is a crisp function which maps a fuzzy set of formulae to another fuzzy set of formulae. Whereas the notion of consequence in the second case is a graded notion, captured by a fuzzy relation from $P(F)$ to F.

(iii) Restricting the domain to the set of all ordinary sets of formulae, Gerla in (2001) has shown that given any graded consequence relation $\mid\!\sim$, one can obtain Pavelka's notion of consequence by defining $C(X)(\alpha) = gr(X \mid\!\sim \alpha)$. But the converse does not hold. Let us consider the following example (Gerla 2001), where the fuzzy relation $\mid\!\sim$ is defined in terms of Pavelka's C operator by the definition $gr(X \mid\!\sim \alpha) = C(X)(\alpha)$.
Let $F = \{\alpha_1, \alpha_2, \alpha_3, \alpha_4\}$ be the set of formulae, and $\{T_1, T_2\}$ be a collection of fuzzy sets over F with $T_1(\alpha_1) = T_1(\alpha_3) = 1$, $T_1(\alpha_2) = 0.7$, $T_1(\alpha_4) = 0.8$, and $T_2(\alpha_2) = T_2(\alpha_3) = T_2(\alpha_4) = 1$, $T_2(\alpha_2) = 0.9$. It can be shown that $\{T_1, T_2\}$ defines a fuzzy closure operator C where $C(X)(\alpha) = \inf_i\{T_i(\alpha) : X \subseteq T_i\}$. Let $X = \{\alpha_1, \alpha_3\}$ and $Z = \{\alpha_4\}$. Then $C(X)(\alpha_2) = 0.7$, $C(X \cup Z)(\alpha_2) = 0.9$ and $C(X)(\alpha_4) = 0.8$. Now defining $gr(X \mid\!\sim \alpha) = C(X)(\alpha)$, it can be observed that (GC3) is not satisfied as $\inf_{\beta \in Z} gr(X \mid\!\sim \beta) = gr(X \mid\!\sim \alpha_4) = 0.8$, $gr(X \cup Z \mid\!\sim \alpha_2) = 0.9$ and $gr(X \mid\!\sim \alpha_2) = 0.7$ imply $\inf_{\beta \in Z} gr(X \mid\!\sim \beta) \wedge gr(X \cup Z \mid\!\sim \alpha) \nleq gr(X \mid\!\sim \alpha)$. So, the restriction of Pavelka's closure operator C to the crisp sets X does not yield a graded consequence relation.

(iv) For any $X \in \mathscr{F}(F)$ and formula α, $C(X)(\alpha)$ is a value which may be read as the degree of belongingness of α to the set of consequences of X, i.e. $C(X)$. We shall analyse in the following chapter that $C(X)(\alpha)$ cannot be read as the degree of derivation/consequence of α from X. On the other hand, for $X \in P(F)$ and $\alpha \in F$, $gr(X \mid\!\sim \alpha)$ denotes the degree of derivation of α from X.

(v) While there is a matching between graded consequence relation and graded inconsistency (cf. Chap. 3), there is no such relationship between Pavelka's fuzzy consequence operator (closure operator) and inconsistency (or consistency). In fact in Pavelka (1979) consistency is a crisp notion defined by X is consistent if and only $C(X) \neq F$.

5.1.2 *Fuzzy Consequence Relation: Castro et al.*

There had been natural attempts to generalize graded consequence relation admitting fuzzy sets as premises and matching perfectly with Pavelka's consequence operator.

Castro et al. (1994) introduced a notion parallel to the notion of graded consequence relation with fuzzy sets of premises, and they came up with a notion of fuzzy consequence relation $f_c : \mathscr{F}(F) \times F \mapsto L$, where L is a complete bounded lattice defined as below.

Definition 5.2 A fuzzy consequence relation is a fuzzy relation $f_c : \mathscr{F}(F) \times F \mapsto L$ satisfying the following conditions (Castro et al. 1994):

(f_c1) $X(\alpha) \leq f_c(X, \alpha)$.
(f_c2) If $X \subseteq Y$ then $f_c(X, \alpha) \leq f_c(Y, \alpha)$.
(f_c3) If for all β, $Y(\beta) \leq f_c(X, \beta)$ then $f_c(X \cup Y, \alpha) \leq f_c(X, \alpha)$.

They proved that this notion of fuzzy consequence relation is equivalent to the notion of fuzzy closure operator due to Pavelka. It is to be noted that independently, in Dutta and Chakraborty (2014), the present authors have developed the same axiomatization for a graded consequence relation in the context of fuzzy sets of premises, and established the equivalence with Pavelka's notion of consequence. The development in Dutta and Chakraborty (2014) has been more extensive in that other meta-logical notions, viz. consistency/inconsistency has also been graded and the relationship between graded consequence relation and graded inconsistency has been established. This latter point is not touched upon in Castro et al. (1994). In the notation of GCT, $f_c(X, \alpha)$ is to be written as $gr(X \mid\!\sim \alpha)$; the only difference from Chap. 1 is that X here represents a fuzzy subset of wffs in general. In the following section, we deal with this extended notion of graded consequence.

5.2 Meta-Logic of Graded Consequence with Fuzzy Premises

Definition 5.3 A fuzzy relation $\mid\!\sim$ from $\mathscr{F}(F)$, the set of all fuzzy subsets over formulae to F giving value to every pair (X, α) from a value set L, is said to be a graded consequence relation if it satisfies the following conditions:

(GC$_f$1) $X(\alpha) \leq gr(X \mid\!\sim \alpha)$.
(GC$_f$2) If $X \subseteq Y$ then $gr(X \mid\!\sim \alpha) \leq gr(Y \mid\!\sim \alpha)$.
(GC$_f$3) If for all $\beta \in F$, $Z(\beta) \leq gr(X \mid\!\sim \beta)$ then $gr(X \cup Z \mid\!\sim \alpha) \leq gr(X \mid\!\sim \alpha)$.

We use the same name 'graded consequence relation' in the context of fuzzy sets also. In the present context, (L $\wedge$, $\vee$, 0, 1), the meta-level algebraic structure for a graded consequence relation, is assumed to be a complete lattice.

Theorem 5.1 *Pavelka's definition of consequence operator is equivalent to Definition 5.3 of graded consequence relation.*

Proof To prove this theorem, we need to establish that from Pavelka's consequence operator a graded consequence relation can be obtained and conversely from a graded consequence relation a Pavelka's consequence operator can be derived.
Pavelka's notion of consequence to graded consequence relation:
Definition: Let C: $\mathscr{F}(F) \mapsto \mathscr{F}(F)$ be a consequence operator satisfying Pavelka's axioms. A fuzzy relation $|\!\sim$ is defined by $gr(X \mid\!\sim \alpha) = C(X)(\alpha)$.
Claim: The fuzzy conclusion relation $|\!\sim$ is a graded consequence relation.
(GC_f1) By (1) of Pavelka's axioms $X(\alpha) \leq C(X)(\alpha)$ for any α.
Hence, $X(\alpha) \leq C(X)(\alpha) = gr(X \mid\!\sim \alpha)$, for any α.
(GC_f2) Let $X \subseteq Y$, then $gr(X \mid\!\sim \alpha) = C(X)(\alpha)$
$\leq C(Y)(\alpha)$ [By (2) of Definition 5.1]
$= gr(Y \mid\!\sim \alpha)$.
(GC_f3) Let $Z(\beta) \leq gr(X \mid\!\sim \beta) = C(X)(\beta)$ for all β.
$C(Z)(\beta) \leq C(C(X))(\beta)$. [By (2) of Definition 5.1]
$C(Z)(\beta) \leq C(X)(\beta)$ for all β. ...(I) [By (3) of Definition 5.1]
Now $(X \cup Z)(\alpha) = X(\alpha) \vee Z(\alpha) \leq C(X)(\alpha) \vee C(Z)(\alpha)$. [By (1) of Definition 5.1]
$= C(X)(\alpha)$. [By (I)]
Hence, $C(X \cup Z)(\alpha) \leq C(C(X))(\alpha) = C(X)(\alpha)$. [By (2) and (3) of Definition 5.1]
Hence, $gr(X \cup Z \mid\!\sim \alpha) \leq gr(X \mid\!\sim \alpha)$. [By definition of $|\!\sim$].
Graded consequence relation to Pavelka's notion of consequence operator:
Definition: Let $|\!\sim$ be a graded consequence relation in the sense of Definition 5.3. Define C: $\mathscr{F}(F) \mapsto \mathscr{F}(F)$ by $C(X)(\alpha) = gr(X \mid\!\sim \alpha)$.
Claim: The consequence operator C defined above satisfies all three axioms proposed by Pavelka.
(1) $C(X)(\alpha) = gr(X \mid\!\sim \alpha) \geq X(\alpha)$ for all α. [By (GC_f1)]
Hence $X \subseteq C(X)$.
(2) Let $X \subseteq Y$, where X, Y are two fuzzy subsets of F.
Then $gr(X \mid\!\sim \alpha) \leq gr(Y \mid\!\sim \alpha)$ for all α. [By (GC_f2)]
Hence, by definition of C, $C(X)(\alpha) \leq C(Y)(\alpha)$ for all α.
Therefore $C(X) \subseteq C(Y)$.
(3) $C(C(X))(\alpha) = gr(C(X) \mid\!\sim \alpha) \geq C(X)(\alpha)$. [Using ($GC_f$1)]
Hence, $C(X)(\alpha) \leq C(C(X))(\alpha)$. ...(III)
Now by (GC_f3) we have, $gr(X \mid\!\sim \alpha) \geq gr(X \cup Z \mid\!\sim \alpha)$ when $Z(\beta) \leq gr(X \mid\!\sim \beta)$ for all β.
Now as $C(X)(\alpha) = gr(X \mid\!\sim \alpha)$ for all α, by (GC_f3), $gr(X \mid\!\sim \beta) \geq gr(X \cup C(X) \mid\!\sim \beta) = gr(C(X) \mid\!\sim \beta)$.
[By (GC_f1), $X(\beta) \leq gr(X \mid\!\sim \beta) = C(X)(\beta)$ for all β. Hence, $X \cup C(X) \subseteq C(X)$.]
Hence, $C(X)(\beta) \geq C(C(X))(\beta)$. ...(IV)
Combining (III) and (IV) we have $C(C(X)) = C(X)$. □

Note 5.1 Theorem 5.1 establishes that the axioms proposed for the consequence operator in the fuzzy context and the axioms proposed for the graded consequence relation with fuzzy sets of premises are interdefinable. Besides, the defining criteria of one notion in terms of the other immediately imply that these two notions perfectly match to each other. That means, given a consequence operator of Pavelka if we define a graded consequence relation with fuzzy sets of premises and again in terms of the graded consequence relation if we generate a consequence operator in Pavelka's sense then the newly obtained notion of consequence operator coincides with the initial one. Similarly, the other direction can be obtained, i.e. one starts from a graded consequence relation, proceeds by defining a consequence operator in Pavelka's sense and then by defining a new graded consequence relation, thereby comes back to the earlier graded consequence relation.

The following theorem shows that given any collection of graded consequence relations their intersection, computed by infimum, turns out to be a graded consequence relation.

Theorem 5.2 *Let* $\{\mid\!\sim_i\}_{i\in I}$ *be a collection of graded consequence relations in the sense of Definition 5.3. Then* $\mid\!\sim$ *defined by* $gr(X \mid\!\sim \alpha) = \inf_{i\in I} gr(X \mid\!\sim_i \alpha)$ *is a graded consequence relation.*

Theorem 5.3 $\langle\{\mid\!\sim_i\}_{i\in I}, \cap, \cup\rangle$ *forms a lattice, where* $\mid\!\sim_i \cap \mid\!\sim_j = \inf\{\mid\!\sim_i, \mid\!\sim_j\}$ *and* $\mid\!\sim_i \cup \mid\!\sim_j = \cap_{k\in K} \mid\!\sim_k$*, where* $K = \{k \in I : \mid\!\sim_i, \mid\!\sim_j \subseteq \mid\!\sim_k\}$.

Proof For $\cap$, the proof is straightforward as 'inf' satisfies commutativity, associativity, absorption and idempotence.
Let $\mid\!\sim_1$ and $\mid\!\sim_2$ be two graded consequence relations and $\{\mid\!\sim_j\}_{j\in J}$ be the subcollection of $\{\mid\!\sim_i\}_{i\in I}$ such that $\mid\!\sim_1, \mid\!\sim_2 \subseteq \mid\!\sim_j$, for each $j \in J$. Let us call, $\cap_{j\in J} \mid\!\sim_j = \mid\!\sim_{12}$.
Therefore, by Theorem 5.2, $\mid\!\sim_{12}$ is a graded consequence relation.
From the definition of $\cup$, it is immediate that $\cup$ is commutative.
Let $\mid\!\sim_1, \mid\!\sim_2$ and $\mid\!\sim_3$ be graded consequence relations, and $\{\mid\!\sim_j\}_{j\in J}, \{\mid\!\sim_k\}_{k\in K}, \{\mid\!\sim_l\}_{l\in L}$ be the subcollections of $\{\mid\!\sim_i\}_{i\in I}$ such that $\mid\!\sim_1, \mid\!\sim_2 \subseteq \mid\!\sim_j$, for each $j \in J$, $\mid\!\sim_2, \mid\!\sim_3 \subseteq \mid\!\sim_k$, for each $k \in K$ and $\mid\!\sim_1, \mid\!\sim_3 \subseteq \mid\!\sim_l$, for each $l \in L$.
For $\mid\!\sim_1 \cup (\mid\!\sim_2 \cup \mid\!\sim_3)$, we know $(\mid\!\sim_2 \cup \mid\!\sim_3) = \mid\!\sim_{23} = \cap_{k\in K} \mid\!\sim_k$.
It can be easily proved that $\{\mid\!\sim_k\}_{k\in K}$ is the collection containing $\mid\!\sim_{23}$ also.
Let $\{\mid\!\sim_m\}_{m\in M}$ be the collection containing $\mid\!\sim_1, \mid\!\sim_2$ and $\mid\!\sim_3$.
Therefore, $\{\mid\!\sim_m\}_{m\in M} = \{\mid\!\sim_j\}_{j\in J} \bigcap \{\mid\!\sim_k\}_{k\in K} \bigcap \{\mid\!\sim_l\}_{l\in L}$, where $\bigcap$ represents the set theoretic intersection.
Hence, $\mid\!\sim_1 \cup (\mid\!\sim_2 \cup \mid\!\sim_3) = \mid\!\sim_{1(23)} = \cap\left[\{\mid\!\sim_j\}_{j\in J} \bigcap \{\mid\!\sim_k\}_{k\in K} \bigcap \{\mid\!\sim_l\}_{l\in L}\right] = \mid\!\sim_{(12)3}$.
Consider $\mid\!\sim_i \cup(\mid\!\sim_i \cap \mid\!\sim_j)$. We know, $\mid\!\sim_i \cap \mid\!\sim_j \subseteq \mid\!\sim_i$.
Therefore, if $\{\mid\!\sim_r\}_{r\in R}$ be the collection of graded consequence relations containing $\mid\!\sim_i$, then $\mid\!\sim_i \cup(\mid\!\sim_i \cap \mid\!\sim_j) = \cap_{r\in R} \mid\!\sim_r$. Also, it can be proved that $\mid\!\sim_i = \cap_{r\in R} \mid\!\sim_r$.
That is $\mid\!\sim_i \cup(\mid\!\sim_i \cap \mid\!\sim_j) = \mid\!\sim_i$. ...(A)
Now for $\mid\!\sim_i \cap(\mid\!\sim_i \cup \mid\!\sim_j)$, let $\{\mid\!\sim_s\}_{s\in S}$ be the collection containing $\mid\!\sim_i$ and $\mid\!\sim_j$.
Therefore $\mid\!\sim_i \cup \mid\!\sim_j = \cap_{s\in S} \mid\!\sim_s = \mid\!\sim_{ij}$. Hence, $\mid\!\sim_i \cap(\mid\!\sim_i \cup \mid\!\sim_j) = \mid\!\sim_i$. ...(B)

(A) and (B) prove that $\cup$ satisfies the law of absorption.
Also, $\cup$ satisfies the law of idempotence since $\mid\!\sim_i \cup \mid\!\sim_i = \cap_{r\in R} \mid\!\sim_r = \mid\!\sim_i$.
Hence, $\langle\{\mid\!\sim_i\}_{i\in I}, \cap, \cup\rangle$ forms a lattice. □

In this new context, the graded counterpart of semantic consequence is given by the underlying definition.

Definition 5.4 For a given collection $\{T_i\}_{i\in I}$ of fuzzy subsets assigning values from a complete lattice L to every formula of F, the graded semantic consequence, viz. $\approx_{\{T_i\}_{i\in I}}$ is defined as $gr(X \approx_{\{T_i\}_{i\in I}} \alpha) = inf_i\{T_i(\alpha) : X \subseteq T_i\}$.

Theorem 5.4 *The fuzzy relation* $\approx_{\{T_i\}_{i\in I}}$ *defined in Definition 5.4 is a graded consequence relation.*

Proof By Definition 5.4, for a collection $\{T_i\}_{i\in I}$, $\approx_{\{T_i\}_{i\in I}}$ is defined as $gr(X \approx_{\{T_i\}_{i\in I}} \alpha)$ $= inf_i\{T_i(\alpha) : X \subseteq T_i\}$.
For $X \subseteq T_i$ we have $X(\alpha) \leq T_i(\alpha)$ for all α. I.e. $X(\alpha) \leq inf_i\{T_i(\alpha) : X \subseteq T_i\}$.
Hence, $X(\alpha) \leq gr(X \approx_{\{T_i\}_{i\in I}} \alpha)$. ...(1)
Let $X \subseteq Y$, i.e. $X(\beta) \leq Y(\beta)$ for all β.
Now all those T_is for which $Y \subseteq T_i$ holds $X \subseteq T_i$ also holds.
I.e. $X(\beta) \leq Y(\beta) \leq T_i(\beta)$ for all β.
The number of T_is containing Y is less or equal to the number of T_is containing X.
Let $\{T_{i_j}\}_{j\in J}$ and $\{T_{i_k}\}_{k\in K}$ be the subcollections of $\{T_i\}_{i\in I}$ containing Y and X respectively. So, $\{T_{i_j}\}_{j\in J} \subseteq \{T_{i_k}\}_{k\in K}$.
I.e. $inf_{i\in I}\{T_i(\alpha) : X \subseteq T_i\} \leq inf_{i\in I}\{T_i(\alpha) : Y \subseteq T_i\}$.
Hence, $gr(X \approx_{\{T_i\}_{i\in I}} \alpha) \leq gr(Y \approx_{\{T_i\}_{i\in I}} \alpha)$ ...(2)
Let for all β, $Z(\beta) \leq gr(X \approx_{\{T_i\}_{i\in I}} \beta) = inf_i\{T_i(\beta) : X \subseteq T_i\}$.
That is, for all β, $Z(\beta) \leq T_i(\beta)$ hold for all such T_i which satisfy $X(\gamma) \leq T_i(\gamma)$ for all γ.
Let $\{T_{i_j}\}_{j\in J}$ be the subcollection which satisfies $X(\beta) \leq T_{i_j}(\beta)$ for all β, and hence satisfies $Z(\beta) \leq T_{i_j}(\beta)$.
Hence, for all $j \in J$, $X(\beta) \vee Z(\beta) \leq T_{i_j}(\beta)$, i.e. $(X \cup Z)(\beta) \leq T_{i_j}(\beta)$.
So, for the collection $\{T_{i_j}\}_{j\in J}$, for any γ, $X(\gamma) \leq T_{i_j}(\gamma)$ implies $(X \cup Z)(\gamma) \leq T_{i_j}(\gamma)$.
Let $\{T_{i_k}\}_{k\in K}$ be the collection satisfying $(X \cup Z)(\gamma) \leq T_{i_k}(\gamma)$.
Then $\{T_{i_j}\}_{j\in J} \subseteq \{T_{i_k}\}_{k\in K}$.
So, $inf_i\{T_i(\alpha) : (X \cup Z)(\gamma) \leq T_i(\gamma)\} \leq inf_i\{T_i(\alpha) : X(\gamma) \leq T_i(\gamma)\}$
That is, $gr(X \cup Z \approx_{\{T_i\}_{i\in I}} \alpha) \leq gr(X \approx_{\{T_i\}_{i\in I}} \alpha)$.
Hence, for all β, $Z(\beta) \leq gr(X \approx_{\{T_i\}_{i\in I}} \beta)$ implies $gr(X \cup Z \approx_{\{T_i\}_{i\in I}} \alpha) \leq gr(X \approx_{\{T_i\}_{i\in I}} \alpha)$ ...(3)
Combining (1), (2) and (3) we conclude that $\approx_{\{T_i\}_{i\in I}}$ is a graded consequence relation. □

Theorem 5.5 *If* $\mid\!\sim$ *is a graded consequence relation then there exists a family* $\{T_i\}_{i\in I}$ *of fuzzy subsets of* F *such that* $gr(X \mid\!\sim \alpha) = gr(X \approx_{\{T_i\}_{i\in I}} \alpha)$.

Proof Let us take the indexing set I as $\mathscr{F}(F)$, the collection of all fuzzy subsets over F.
Now we define the fuzzy subset T_X, for each $X \in \mathscr{F}(F)$ by $T_X(\alpha) = gr(X \mid\!\sim \alpha)$.
We want to prove that $gr(X \mid\!\sim \alpha) = inf_Y\{T_Y(\alpha) : X \subseteq T_Y\}$.
As from the first condition of $\mid\!\sim$ we know $X(\beta) \leq gr(X \mid\!\sim \beta) = T_X(\beta)$, it is obvious that T_X is a member of the collection $\{T_Y(\alpha) : X \subseteq T_Y\}$.
Our claim is that $T_X(\alpha) = inf_Y \{T_Y(\alpha) : X \subseteq T_Y\}$.
Now three cases arise.
(I) $X \subseteq T_X \subseteq T_Y$, (II)$X \subseteq T_Y \subseteq T_X$ and (III)$X \subseteq T_Y$, but T_X, T_Y are non-comparable.
Case-I If $X \subseteq T_X \subseteq T_Y$ then nothing to prove.
Case-II If $X \subseteq T_Y \subseteq T_X$, $X(\alpha) \leq T_Y(\alpha) = gr(Y \mid\!\sim \alpha)$.
This implies $gr(X \cup Y \mid\!\sim \beta) \leq gr(Y \mid\!\sim \beta)$ for all β. [By (GC$_f$3)]
That is, $T_{X\cup Y}(\beta) \leq T_Y(\beta)$ for all β. ...(1)
Also by (GC$_f$2) $gr(X \mid\!\sim \beta) \leq gr(X \cup Y \mid\!\sim \beta)$. I.e. $T_X(\beta) \leq T_{X\cup Y}(\beta)$ for all β. ...(2)
Hence, from (1) and (2) $T_X \subseteq T_Y$.

The above inequality together with $T_Y \subseteq T_X$ implies $T_X = T_Y$.
Case-III Let $X \subseteq T_Y$ and T_X, T_Y be non-comparable.
Now as $X \subseteq T_Y$, $X(\alpha) \leq T_Y(\alpha) = gr(Y \mid\!\sim \alpha)$ for all α.
Hence, by (GC$_f$3), $gr(X \cup Y \mid\!\sim \beta) \leq gr(Y \mid\!\sim \beta)$, for all β.
That is, $T_{X\cup Y}(\beta) \leq T_Y(\beta)$, for all β.
Again by (GC$_f$2) $gr(X \mid\!\sim \beta) \leq gr(X \cup Y \mid\!\sim \beta)$, i.e. $T_X(\beta) \leq T_{X\cup Y}(\beta)$, for all β.
Hence, $T_X(\beta) \leq T_{X\cup Y}(\beta) \leq T_Y(\beta)$, for all β.
This contradicts the fact that T_X, T_Y are non-comparable.

So, there cannot be any such T_Y containing X which is non-comparable with T_X.
Hence, combining all these three cases we have $inf_Y\{T_Y(\alpha) : X \subseteq T_Y\} = T_X(\alpha) = gr(X \mid\!\sim \alpha)$. □

Theorems 5.4 and 5.5 are representation theorems, parallel, respectively, to (i) and (ii) of Theorem 2.1, in Chap. 2.

From the definition of $\approx_{\{T_i\}_{i\in I}}$ (cf. Definition 5.4), it can be noticed that a complete lattice is the algebraic structure for the meta-language of a graded consequence relation with fuzzy set of premises. Whereas in the context of graded consequence relation with ordinary set of premises (cf. Chap. 2), the corresponding algebraic structure has to be a complete residuated lattice where $*_m$ and its residuum $\rightarrow_m$ compute the meta-linguistic 'and' and 'if-then', respectively. In the following theorem, we shall explore the relation between the notions of graded semantic consequence proposed in Chap. 2 and the present one. For convenience of presentation, we shall denote the notion of graded semantic consequence, defined in the present context, by $\approx_f$ and the earlier one by $\approx$, and in both the cases the semantic base being $\{T_i\}_{i\in I}$.

Theorem 5.6 *Given a collection of fuzzy subsets* $\{T_i\}_{i\in I}$, *let*
(I) $gr(X \approx_f \alpha) = inf_i\{T_i(\alpha) : X \subseteq T_i\}$, *where* $X \in \mathscr{F}(F)$ *and*

(II) $gr(X \approx\!\!\models \alpha) = inf_i\{inf_{\beta \in X}\ T_i(\beta) \rightarrow_m T_i(\alpha)\}$ *where* $X \in P(F)$.
If in (I) X *is taken as an ordinary set then* $gr(X \approx\!\!\models \alpha) \leq gr(X \approx\!\!\models_f \alpha)$.

Proof For an ordinary set X of formulae $X \subseteq T_i$ means $T_i(\beta) = 1$ for $\beta \in X$.
Now $inf_i\{inf_{\beta \in X}\ T_i(\beta) \rightarrow_m T_i(\alpha)\}$
$\leq inf_i\{inf_{\beta \in X}\ T_i(\beta) \rightarrow_m T_i(\alpha) : X \subseteq T_i\}$.
$= inf_i\{1 \rightarrow T_i(\alpha) : X \subseteq T_i\}$
$= inf_i\{T_i(\alpha) : X \subseteq T_i\}$
Hence, $gr(X \approx\!\!\models \alpha) \leq gr(X \approx\!\!\models_f \alpha)$. □

In the earlier chapters to define graded semantic consequence relation, a meta-level implication satisfying certain properties played the role of the truth function for meta-linguistic 'if-then' (cf. Note 2.1, Chap. 2). In the present case, while defining graded semantic consequence in the context of fuzzy sets of premises, no 'implication' operator has been used to compute the logical connective 'if-then' used in the meta-language. But it is clear that in the present context use of an implication operator $\Rightarrow$, defined by $a \Rightarrow b = 1$ if $a \leq b$ and 0 otherwise, serves the purpose, that is, defines the same graded consequence relation as in Definition 5.4.

5.2.1 Graded Inconsistency and Graded Consequence

Let us assume the presence of negation ($\neg$) in the object language. We assume no other logical connective in it since to understand inconsistency other logical connectives are not essential even in the case of classical logic. If the language is more expressive, negation ($\neg$) usually has some interaction with the other object language connectives, viz. conjunction, disjunction, implication, etc. But these interactions do not play any essential role in the definition of inconsistency. So, the axioms for graded consequence relation are extended with respect to the connective $\neg$. Let us propose the extended set of axioms for graded consequence relation as follows.

Definition 5.5 A graded consequence relation $\mid\!\sim$ is a fuzzy relation from $\mathscr{F}(F)$ to F satisfying following conditions:

(GC_f1) $X(\alpha) \leq gr(X \mid\!\sim \alpha)$.
(GC_f2) If $X \subseteq Y$ then $gr(X \mid\!\sim \alpha) \leq gr(Y \mid\!\sim \alpha)$.
(GC_f3) If $Z(\beta) \leq gr(X \mid\!\sim \beta)$ for all $\beta \in F$, then $gr(X \cup Z \mid\!\sim \alpha) \leq gr(X \mid\!\sim \alpha)$.
(GC_f4) There is some $k > 0$ such that for any $X \in \mathscr{F}(F)$, $inf_\beta gr(X \mid\!\sim \beta) \geq k$ where for some α, $\{\alpha, \neg\alpha\} \subseteq \mathrm{Supp}(X)$.
(GC_f5) $gr(X \cup \{\alpha\} \mid\!\sim \beta) \wedge gr(X \cup \{\neg\alpha\} \mid\!\sim \beta) \wedge c \leq gr(X \mid\!\sim \beta)$ for some $c > 0$.
(GC_f6) $gr(X \mid\!\sim \alpha) \leq inf_\beta\ gr(X \cup \{\neg\alpha\} \mid\!\sim \beta)$.
(GC_f7) $gr(X \cup \{\neg\alpha\} \mid\!\sim \alpha) \leq gr(X \mid\!\sim \alpha)$.

That (GC_f1), (GC_f2) and (GC_f3) are the graded version of reflexivity, monotonicity and cut is quite clear. (GC_f4) is the graded version of explosiveness condition, and it ensures that for any $X \in \mathscr{F}(F)$, if for some α, $\{\alpha, \neg\alpha\}$ is in the support of X then any β follows from X at least to a non-zero grade. (GC_f5) is the graded counterpart of reasoning by cases, which ensures if β follows from both $X \cup \{\alpha\}$ and $X \cup \{\neg\alpha\}$ at least to a non-zero degree $c > 0$, then β follows from X at least to the degree c as well. In the context of classical logic, we know, if X yields α, then from $X \cup \{\neg\alpha\}$ every β follows, and (GC_f6) is the graded version of the same. (GC_f7) generalizes the property that if α follows from $X \cup \{\neg\alpha\}$, then α follows from X.

On the other hand, let *INCONS* be a fuzzy subset over $\mathscr{F}(F)$ denoting the notion of graded inconsistency. The formal definition is as follows.

Definition 5.6 For any $X \in \mathscr{F}(F)$, $INCONS(X)$, denoting the degree to which X is inconsistent, is axiomatized by the following axioms:

(INC_f1) $X(\alpha) \leq INCONS(X \cup \{\neg\alpha\})$ for all α.
(INC_f2) If $X \subseteq Y$ then $INCONS(X) \leq INCONS(Y)$.
(INC_f3) If $Y(\alpha) \leq INCONS(X \cup \{\neg\alpha\})$ for all α, then $INCONS(X \cup Y) \leq INCONS(X)$.
(INC_f4) There is some $k > 0$ such that for any $X \in \mathscr{F}(F)$, $INCONS(X) \geq k$ where for some α, $\{\alpha, \neg\alpha\} \subseteq \mathrm{Supp}(X)$.
(INC_f5) $INCONS(X \cup \{\alpha\}) \wedge INCONS(X \cup \{\neg\alpha\}) \wedge c \leq INCONS(X)$ for all α, and some $c > 0$.

It can be noticed that the axioms proposed above generalize the basic features of the classical notion of inconsistency. As an instance let us consider (INC_f1) which generalizes the idea that if $\alpha \in X$, then $X \cup \{\neg\alpha\}$ is inconsistent. (INC_f2) is the graded version of the property, viz. if a set of formulae is inconsistent then its superset is also inconsistent. Similarly, (INC_f3) generalizes that for all $\alpha \in Y$, if $X \cup \{\neg\alpha\}$ is inconsistent and $X \cup Y$ is inconsistent as well, then X is inconsistent. (INC_f4) ensures that for any $X \in \mathscr{F}(F)$, containing for some α, $\{\alpha, \neg\alpha\}$ in its support, the degree of inconsistency of X must be non-zero. In classical context, this is always 1. (INC_f5) captures the idea that if $X \cup \{\alpha\}$ and $X \cup \{\neg\alpha\}$ both are inconsistent at least to a non-zero degree c, then so is X.

In classical logic, the notions of consequence and inconsistency are equivalent (Surma 1981) in the sense that assuming one as the primitive notion the other can be obtained. In Chap. 2, the same result has been established in the graded context where the notion of graded consequence is a fuzzy relation between $P(F)$ and F and the notion of graded inconsistency, i.e. *INCONS* is a fuzzy set over $P(F)$. The following theorems of this section are presented to study this consequence–inconsistency interconnection in the present context.

Definition 5.7 1. Let *INCONS* be a fuzzy subset over $\mathscr{F}(F)$ satisfying (INC_f1) to (INC_f6) axioms. We define $\mid\!\sim$ in the following way. $gr(X \mid\!\sim \alpha) = INCONS(X \cup \{\neg\alpha\})$.

2. Let $\mid\!\sim$ be a graded consequence relation satisfying ($GC_f 1$) to ($GC_f 6$) axioms. Let us define *INCONS*, a fuzzy subset over $\mathscr{F}(F)$, in the following way.
   $INCONS(X) = inf_\alpha\, gr(X \mid\!\sim \alpha)$.

Theorem 5.7 *Let INCONS, following Definition 5.6, be given, and* $\mid\!\sim$ *be defined following (1) of Definition 5.7. Then* $\mid\!\sim$ *satisfies the axioms (*$GC_f 1$*) to (*$GC_f 6$*).*

Proof ($GC_f 1$) and ($GC_f 2$) are straightforward from ($INC_f 1$) and ($INC_f 2$), respectively.

($GC_f 3$) Let $Z(\beta) \leq gr(X \mid\!\sim \beta)$, for all β.
$\leq INCONS(X \cup \{\neg\beta\})$, for all β.
$\leq INCONS(X \cup \{\neg\alpha\} \cup \{\neg\beta\})$, for all α, β.
Hence, by ($INC_n 3$), $INCONS(X \cup Z \cup \{\neg\alpha\}) \leq INCONS(X \cup \{\neg\alpha\})$.
Therefore, $gr(X \cup Z \mid\!\sim \alpha) \leq gr(X \mid\!\sim \alpha)$.

($GC_f 4$) Let $X \in \mathscr{F}(F)$ such that $\{\alpha, \neg\alpha\} \subseteq$ Supp(X). Then by ($INC_n 4$) there is some $k > 0$ such that $INCONS(X) \geq k > 0$. Hence, $INCONS(X \cup \{\neg\beta\}) \geq INCONS(X) \geq k > 0$ for any β.
That is, for any β, $gr(X \mid\!\sim \beta) \geq k > 0$. Hence, $\inf_\beta gr(X \mid\!\sim \beta) \geq k > 0$.

($GC_f 5$) can be obtained directly from ($INC_n 5$).
($GC_f 6$) is immediate from the definition of $\mid\!\sim$ and axiom ($INC_n 2$).
($GC_f 7$) By definition of $\mid\!\sim$, $gr(X \cup \{\neg\alpha\} \mid\!\sim \alpha) = INCONS(X \cup \{\neg\alpha\}) = gr(X \mid\!\sim \alpha)$. □

Theorem 5.8 *Let* $\mid\!\sim$*, following Definition 5.5, be given, and INCONS be defined following (2) of Definition 5.7. Then INCONS satisfies axioms (*$INC_f 1$*) to (*$INC_f 5$*).*

Proof ($INC_f 1$) $X(\alpha) \leq gr(X \mid\!\sim \alpha)$. [By ($GC_f 1$)]
$\leq inf_\beta\, gr(X \cup \{\neg\alpha\} \mid\!\sim \beta)$. [By ($GC_f 6$)]
$= INCONS(X \cup \{\neg\alpha\})$.
($INC_f 2$) is straightforward by using the axiom ($GC_f 2$).

($INC_f 3$) $Y(\alpha) \leq INCONS(X \cup \{\neg\alpha\})$ for all α.
$= inf_\beta\, gr(X \cup \{\neg\alpha\} \mid\!\sim \beta) \leq gr(X \cup \{\neg\alpha\} \mid\!\sim \beta)$ for all β.
In particular, $Y(\alpha) \leq gr(X \cup \{\neg\alpha\} \mid\!\sim \alpha) = gr(X \mid\!\sim \alpha)$ [by ($GC_f 7$)].
So, $Y(\alpha) \leq gr(X \mid\!\sim \alpha)$ for all α.
By ($GC_f 3$), $gr(X \cup Y \mid\!\sim \beta) \leq gr(X \mid\!\sim \beta)$ for all β.
$inf_\beta\, gr(X \cup Y \mid\!\sim \beta) \leq inf_\beta\, gr(X \mid\!\sim \beta)$.
Hence, $INCONS(X \cup Y) \leq INCONS(X)$.

($INC_f 4$) and ($INC_f 5$) can be proved directly by using ($GC_f 4$) and ($GC_f 5$), respectively. □

Theorem 5.9 *Let* $\mid\!\sim_1$ *be a graded consequence relation and INCONS be defined by* $INCONS(X) = inf_\alpha\, gr(X \mid\!\sim_1 \alpha)$*. Let* $\mid\!\sim_2$ *be a fuzzy relation from* $\mathscr{F}(F)$ *to* F*, defined by* $gr(X \mid\!\sim_2 \alpha) = INCONS(X \cup \{\neg\alpha\})$*. Then* $\mid\!\sim_1 = \mid\!\sim_2$*.*

Proof $gr(X \mid\!\sim_2 \alpha) = INCONS(X \cup \{\neg\alpha\})$.
$= inf_\beta\, gr(X \cup \{\neg\alpha\} \mid\!\sim_1 \beta)$. [By definition of *INCONS*]
$\leq gr(X \cup \{\neg\alpha\} \mid\!\sim_1 \beta)$ for any β.
In particular, for $\beta = \alpha$, $gr(X \mid\!\sim_2 \alpha) \leq gr(X \cup \{\neg\alpha\} \mid\!\sim_1 \alpha)$. ...(1)
Now by (GC_f7) we have $gr(X \cup \{\neg\alpha\} \mid\!\sim_1 \alpha) = gr(X \mid\!\sim_1 \alpha)$. ...(2)
So, by (1) and (2), we have $gr(X \mid\!\sim_2 \alpha) \leq gr(X \mid\!\sim_1 \alpha)$ ...(3)
$gr(X \mid\!\sim_1 \alpha) \leq inf_\beta\, gr(X \cup \{\neg\alpha\} \mid\!\sim_1 \beta)$ [By (GC_f6)]
$= INCONS(X \cup \{\neg\alpha\}) = gr(X \mid\!\sim_2 \alpha)$. ...(4)
By (3) and (4), we can conclude that $\mid\!\sim_1 = \mid\!\sim_2$. □

Theorem 5.10 *Let $INCONS_1$ satisfy all (INC_f1) to (INC_f5) axioms and $\mid\!\sim$ be defined by $gr(X \mid\!\sim \alpha) = INCONS_1(X \cup \{\neg\alpha\})$. Let $INCONS_2$ be a fuzzy set over $\mathscr{F}(F)$ defined by $INCONS_2(X) = inf_\alpha\, gr(X \mid\!\sim \alpha)$. Then $INCONS_1(X) \leq INCONS_2(X)$ for any $X \in \mathscr{F}(F)$.*

Proof $INCONS_2(X) = inf_\alpha\, gr(X \mid\!\sim \alpha)$. [By definition of $INCONS_2$]
$= inf_\alpha\, INCONS_1(X \cup \{\neg\alpha\})$. ...(1) [By definition of $INCONS_1$]
Again $INCONS_1(X) \leq INCONS_1(X \cup \{\neg\alpha\})$ for all α. [By (INC_f2)]
Hence, $INCONS_1(X) \leq inf_\alpha\, INCONS_1(X \cup \{\neg\alpha\}) = INCONS_2(X)$. [By (1)]
Hence, $INCONS_1(X) \leq INCONS_2(X)$ for any $X \in \mathscr{F}(F)$. □

Note 5.2 In classical logic, the notions of inconsistency and consequence are equivalent, and they perfectly match with each other. That is, given a consequence relation $\vdash_1$ defining *INCONS* in terms of $\vdash_1$, and again defining another consequence relation $\vdash_2$ in terms of this obtained notion *INCONS*, $\vdash_2$ turns out to be the same as $\vdash_1$. Similar is the case when one starts with *INCONS*, the classical notion of inconsistecy. The above theorems show, in the graded context, though one of the notions of consequence and inconsistency can be obtained in terms of the other (see Theorems 5.8 and 5.7) they do not perfectly match with each other (see Theorem 5.10). In this context, let us consider the following postulate (INC_f6) inf_α $INCONS(X \cup \{\neg\alpha\}) \leq INCONS(X)$.
From two-valued interpretation, it can be read as 'if for all α, $X \cup \{\neg\alpha\}$ is inconsistent, then X is inconsistent', which is true in the case of classical logic. We show that addition of this postulate to the already existing INC_f axioms gives the abovementioned perfect matching in this graded context too.

Theorem 5.11 *$\mid\!\sim$ with (GC_f1) to (GC_f6) generates a notion of INCONS which satisfies (INC_f1) to (INC_f6).*

Proof We already have proved that the notion of *INCONS* generated from $\mid\!\sim$ by the definition, viz. $INCONS(X) = inf_\alpha\, gr(X \mid\!\sim \alpha)$ satisfies (INC_f1) to (INC_f5) in Theorem 5.8. So, we only need to show that the definition of *INCONS* satisfies (INC_f6).
$INCONS(X \cup \{\neg\alpha\}) = inf_\beta\, gr(X \cup \{\neg\alpha\} \mid\!\sim \beta)$.
$\leq gr(X \cup \{\neg\alpha\} \mid\!\sim \beta)$ for any β.
In particular, for $\beta = \alpha$, $INCONS(X \cup \{\neg\alpha\}) \leq gr(X \cup \{\neg\alpha\} \mid\!\sim \alpha)$. ...(1)

By (GC_f7), we have $gr(X \cup \{\neg\alpha\} \mid\!\sim \alpha) = gr(X \mid\!\sim \alpha)$.
Therefore, (1) becomes $INCONS(X \cup \{\neg\alpha\}) \leq gr(X \mid\!\sim \alpha)$.
$inf_\alpha\ INCONS(X \cup \{\neg\alpha\}) \leq inf_\alpha\ gr(X \mid\!\sim \alpha) = INCONS(X)$. □

Theorem 5.12 *Let $INCONS_1$ satisfy all (INC_f1) to (INC_f6) axioms and $\mid\!\sim$ be defined by $gr(X \mid\!\sim \alpha) = INCONS_1(X \cup \{\neg\alpha\})$. Let $INCONS_2$ be a fuzzy set over $\mathscr{F}(F)$ defined by $INCONS_2(X) = inf_\alpha\ gr(X \mid\!\sim \alpha)$. Then $INCONS_1(X) = INCONS_2(X)$ for any $X \in \mathscr{F}(F)$.*

Proof In Theorem 5.10, it has already been proved that $INCONS_1(X) \leq INCONS_2(X)$. Here, we need to prove the other direction of the inequality.

$INCONS_2(X) = inf_\alpha\ gr(X \mid\!\sim \alpha)$. [By definition of $INCONS_2$]
$= inf_\alpha\ INCONS_1(X \cup \{\neg\alpha\})$. [By definition of $\mid\!\sim$]
$\leq INCONS_1(X)$. [By (INC_f6)] □

Note 5.3 In this section, while establishing that the two sets of axioms proposed for the graded notions of consequence and inconsistency are equivalent the graded notion of inconsistency has been defined by using the definition, viz. $INCONS(X) = \inf_\alpha gr(X \mid\!\sim \alpha)$. That is, in terms of semantics, the definition becomes $INCONS(X) = \inf_\alpha \inf_i\{T_i(\alpha)/X \subseteq T_i\}$. In the previous sections, starting from the definition of graded semantic consequence to proving representation theorem no restriction on the nature of fuzzy sets, i.e. T_i's constituting a graded consequence relation $\approx$ need to be imposed. But for *INCONS*, we need to exclude the presence of such T_i's which assign 1 to every formula of F; this restriction is imposed as if such a T_i exists then for a fuzzy set X from which every formula can be derived to the full extent (i.e. $INCONS(X) = 1$) will have T_i as its model. This goes against the classical result, viz. an inconsistent set does not have a model. We shall see in the next section, in Gerla (2001), Gerla also has imposed the same restriction while considering the semantic base.

5.2.2 Graded Consistency

As discussed in Sect. 5.2, we know that the proposed notion of graded consequence with fuzzy set of premises becomes equivalent to the notion Pavelka's fuzzy closure operator. Besides, the definition of graded semantic consequence (Definition 5.4) is exactly the same as that of semantic consequence proposed by Pavelka (1979). But while defining the notion of consistency, Pavelka (1979) made it two-valued by imposing the following condition.
X is consistent if $C(X) \neq F$, i.e. it is not that $C(X)(\alpha) = 1$ for all α.
So, although Pavelka's notion of consequence set is fuzzy, the notion of consistency is a crisp concept. Thus, in Pavelka's fuzzy logic, consequence and inconsistency are not commensurate to each other.

On the other hand, in Gerla (2001), Gerla has defined the notion of inconsistency as $Inc(X) = \inf\{\mathscr{C}(X)(\alpha)/\alpha \in F\}$, where $\mathscr{C}(X) = \cap\{T_i \in \{T_i\}_{i\in I}/X \subseteq T_i\}$ for any class $\{T_i\}_{i\in I}$ of fuzzy sets of formulae excluding the fuzzy set which maps every formula to 1 (Gerla 2001). Then, the notion of consistency, i.e. *Cons* which is a fuzzy set on $\mathscr{F}(F)$ has been defined as $Cons(X) = \sim Inc(X)$, where $\sim$ is an operator to compute the complement of the fuzzy set *Inc*.

Here, it can be easily noticed that the definition of $\mathscr{C}$ as proposed by Gerla is the same as the definition proposed by Pavelka and hence it coincides with our case too. But Gerla's definition of *Inc* and *Cons*, both graded, differ from those proposed in Pavelka. In our case, the definition of *INCONS* coincides with the definition of *Inc* of Gerla. But to define *Cons* Gerla (2001) has used an operator $\sim$, which has to be an operator for meta-level negation. Whereas throughout our development, in the context of graded consequence with fuzzy set of premises, we have not introduced any meta-level negation. Besides, Gerla's formalism also presumes the truth set as [0, 1]. In our case, this set is a general complete lattice where negation cannot be generally defined by a subtraction operation. It should also be mentioned that the definition of inconsistency defined in Sect. 5.2.1 is proposed by generalizing the syntactic notion of explosiveness condition. And consistency, on the other hand, is defined below from semantic angle. From syntactic angle, a set of formulae is consistent means there is some formula which does not follow from X. This needs a meta-level negation in the defining criterion which is not assumed.

With respect to the above discussion, the following definition of consistency may turn out to be satisfactory.

Definition 5.8 $CONS(X) = 1$ if $INCONS(X) = 0 = \sup_i \{\inf_{\alpha/X(\alpha)>0} T_i(\alpha)/X \subseteq T_i\}$, otherwise.

Note 5.4 Classically, a set of formulae X is said to be consistent if and only if it has a model, i.e. a valuation T such that $X \subseteq T$. This is equivalent to the statement: X is consistent if and only if for some truth valuation T every formula of X gets the truth value 'true'. In the present context, a fuzzy set of formulae either has a model T or does not, in the sense that either $X(\alpha) \leq T(\alpha)$ holds for all α, or not. That is, being contained in a model is a two-valued notion. Whereas truth of a formula is a matter of grade. So, in this graded context, characterizing consistency in terms of model or in terms of truth may not be equivalent. Hence, the expression for $CONS(X)$ in the present context reflects that X is consistent if and only if X has a model T such that every formula belonging to X is true under T.

Theorem 5.13 *If $INCONS(X) \neq 1$, then $INCONS(X) \leq CONS(X)$.*

Proof If $INCONS(X) = 0$ then by definition of *CONS* it is immediate.
Let $INCONS(X) \neq 0$. Now $\inf_\alpha T_i(\alpha) \leq \inf_{\alpha/X(\alpha)>0} T_i(\alpha)$. ...(1)
Therefore, for any T_i such that $X \subseteq T_i$ also (1) holds.
That is, $\inf_i \inf_\alpha\{T_i(\alpha)/X \subseteq T_i\} \leq \inf_{\alpha/X(\alpha)>0} T_i(\alpha)$ for any $X \subseteq T_i$
$\leq \sup_i \{\inf_{\alpha/X(\alpha)>0} T_i(\alpha)/X \subseteq T_i\}$.
Hence, $INCONS(X) \leq CONS(X)$. □

Proposition 5.1 *(i) If $INCONS(X) = 1$ then $CONS(X) = 0$.*

(ii) If $INCONS(X) \neq 1$ *then* $CONS(X) \neq 0$
(iii) $CONS(X) = 1$ *implies* $INCONS(X) \neq 1$.

Proof (i) According to Note 5.3 of Sect. 5.2.1, $INCONS(X) = 1$ implies there is no T_i such that $X \subseteq T_i$. Therefore, supremum of an empty collection is 0 and hence the result.
(ii) Let $INCONS(X) \neq 1$. then either $INCONS(X) = 0$ or $INCONS(X) \neq 0$.
For $INCONS(X) = 0$, $CONS(X) = 1$ $(\neq 0)$.
Suppose $INCONS(X) \neq 0, 1$, then by previous theorem $INCONS(X) \leq CONS(X)$ implies $CONS(X) \neq 0$.
(iii) Let $CONS(X) = 1$. If possible let, $INCONS(X) = 1$. Then by (i) $CONS(X) = 0$. This contradicts the assumption. □

Theorem 5.14 *If* $X \subseteq Y$ *then* $CONS(Y) \leq CONS(X)$.

Proof We have four cases, viz. (i) $INCONS(X) = 0$, (ii) $INCONS(Y) = 0$, (iii) $INCONS(X) = INCONS(Y) = 0$, (iv) $INCONS(X) \neq 0$ and $INCONS(Y) \neq 0$. The result is immediate in the first and the third cases. So, let us prove (ii) and (iv). (ii) As $INCONS(X) \leq INCONS(Y) = 0$, $CONS(X) = 1$ and hence the result. (iv) For $INCONS(X) \neq 0$ and $INCONS(Y) \neq 0$ two cases arise. (a) $INCONS(X) = 1$ or $INCONS(Y) = 1$, (b) $INCONS(X) \neq 1$, $INCONS(Y) \neq 1$. (a) If $INCONS(X) = 1$ then as $INCONS(X) \leq INCONS(Y)$, $INCONS(Y) = 1$ and hence by Proposition 5.1(i, ii), $CONS(Y) = 0$. Hence, the result is immediate.
Similarly, if $INCONS(Y) = 1$ or both of $INCONS(X)$ and $INCONS(Y)$ are 1, the result follows immediately.
(b) If $INCONS(X) \neq 1$ and $INCONS(Y) \neq 1$ then there are collections say, $\{T_i\}_{i \in I}$ and $\{T_j\}_{j \in J}$ such that $X \subseteq T_i$ for each $i \in I$, and $Y \subseteq T_j$, for each $j \in J$.

$$
\begin{aligned}
CONS(X) &= \sup_i \{\inf_{\alpha/X(\alpha)>0} T_i(\alpha)/X \subseteq T_i\} \\
&\geq \sup_i \{\inf_{\alpha/Y(\alpha)>0} T_i(\alpha)/X \subseteq T_i\} \quad [\text{Since } X(\alpha) \leq Y(\alpha) \text{ for all } \alpha] \\
&\geq \sup_j \{\inf_{\alpha/Y(\alpha)>0} T_i(\alpha)/Y \subseteq T_j\} \quad [\text{As } X \subseteq Y \subseteq T_j, \text{ for } j \in J] \\
&= CONS(Y).
\end{aligned}
$$
□

Theorem 5.15 *If for some* α, $gr(X \mid\!\sim \alpha) = 1$ *then*
(i) $CONS(X) \neq 0$ *implies* $CONS(X \cup \{\alpha\}) \neq 0$ *and*
(ii) $CONS(X) = CONS(X \cup \{\alpha\})$.

Proof (i) Let $CONS(X) \neq 0$. Then $CONS(X \cup \{\alpha\}) \leq CONS(X)$ implies either $CONS(X \cup \{\alpha\}) \neq 0$ or $CONS(X \cup \{\alpha\}) = 0$.
If $CONS(X \cup \{\alpha\}) = 0$ then by Proposition 5.1, $INCONS(X \cup \{\alpha\}) = 1$.
Therefore, $\inf_\beta gr(X \cup \{\alpha\} \mid\!\sim \beta) = 1$. That is, $gr(X \cup \{\alpha\} \mid\!\sim \beta) = 1$ for any β.
Now, $\{\alpha\}(\beta) \leq gr(X \mid\!\sim \beta)$ for all β. [Since $gr(X \mid\!\sim \alpha) = 1$]
Then by cut, $gr(X \cup \{\alpha\} \mid\!\sim \gamma) \leq gr(X \mid\!\sim \gamma)$ for all γ.
Hence, $gr(X \mid\!\sim \gamma) = 1$ for all γ. So, $\inf_\beta gr(X \mid\!\sim \beta) = 1$, i.e. $INCONS(X) = 1$.
Therefore, $CONS(X) = 0$. This contradicts the assumption.
Hence, $CONS(X \cup \{\alpha\}) \neq 0$.

(ii) To prove $CONS(X) = CONS(X \cup \{\alpha\})$, we need to consider four cases.
(I) $CONS(X) = 1$, (II) $CONS(X \cup \{\alpha\}) = 1$, (III) $CONS(X) = CONS(X \cup \{\alpha\}) = 1$ and (IV) $CONS(X) \neq 1$, $CONS(X \cup \{\alpha\}) \neq 1$.
(I) For $CONS(X) = 1$ either (Ia) $INCONS(X) = 0$ or (Ib) $INCONS(X) \neq 0$.
(Ia) Now as $\{\alpha\}(\beta) \leq gr(X \mid\!\sim \beta)$ for all β [since $gr(X \mid\!\sim \alpha) = 1$]
by cut, $gr(X \cup \{\alpha\} \mid\!\sim \gamma) \leq gr(X \mid\!\sim \gamma)$ for all γ.
Hence, $\inf_\gamma gr(X \cup \{\alpha\} \mid\!\sim \gamma) \leq \inf_\gamma gr(X \mid\!\sim \gamma) = INCONS(X) = 0$
Therefore, $INCONS(X \cup \{\alpha\}) = 0$. Then by definition 5.8 $CONS(X \cup \{\alpha\}) = 1$.
(Ib) Let $INCONS(X) \neq 0$. Then $INCONS(X \cup \{\alpha\}) \neq 0$.

$$CONS(X \cup \{\alpha\}) = \sup_i\{\inf_{\beta/(X\cup\{\alpha\})(\beta)>0} T_i(\beta) : X \cup \{\alpha\} \subseteq T_i\}.$$
$$= \sup_i\{\inf_{\beta/(X(\beta)>0} T_i(\beta) : X \cup \{\alpha\} \subseteq T_i\}. \quad [\text{Since } T_i(\alpha) = 1]$$
$$= \sup_i\{\inf_{\beta/(X(\beta)>0} T_i(\beta) : X \subseteq T_i\}. \quad [\text{As } X \subseteq X \cup \{\alpha\}, gr(X \mid\!\sim \alpha) = 1]$$
$$= CONS(X).$$

Cases (II) and (III) are immediate.
(IV) Here two cases arise—(a) $X(\alpha) > 0$, (b) $X(\alpha) = 0$.

(a) $$CONS(X) = \sup_i \{\inf_{\beta/X(\beta)>0,\ \beta\neq\alpha} T_i(\beta) \wedge T_i(\alpha)/X \subseteq T_i\} \quad [\text{Since } X(\alpha) > 0]$$
$$= \sup_i \{\inf_{\beta/(X\cup\{\alpha\})(\beta)>0} T_i(\beta)/X \subseteq T_i\} \ldots(1)$$

Let $\{T_i\}_{i\in I}$ and $\{T_j\}_{j\in J}$ be the collections of fuzzy sets containing X and $X \cup \{\alpha\}$, respectively. Therefore, $\{T_j\}_{j\in J} \subseteq \{T_i\}_{i\in I}$. On the other hand, for $T_i \in \{T_i\}_{i\in I}$, $X \cup \{\alpha\} \subseteq T_i$ since $gr(X \mid\!\sim \alpha) = 1$. Therefore, $\{T_i\}_{i\in I} = \{T_j\}_{j\in J}$.
Hence, (1) reduces to $CONS(X) = \sup_i \{\inf_{\beta/(X\cup\{\alpha\})(\beta)>0} T_i(\beta)/X \cup \{\alpha\} \subseteq T_i\}$
$= CONS(X \cup \{\alpha\})$.

(b) $$CONS(X \cup \{\alpha\}) = \sup_i \{\inf_{\beta/(X\cup\{\alpha\})(\beta)>0} T_i(\beta)/X \cup \{\alpha\} \subseteq T_i\}$$
$$= \sup_i \{\inf_{\beta/X(\beta)>0,\ \beta\neq\alpha} T_i(\beta)/X \cup \{\alpha\} \subseteq T_i\}. \quad [\text{Since } T_i(\alpha) = 1]$$
$$= CONS(X). \quad [\text{As } X(\alpha) = 0 \text{ and } \{T_i\}_{i\in I} = \{T_j\}_{j\in J}]$$
□

Theorem 5.16 *If for some* α, $CONS(X) \leq gr(X \mid\!\sim \alpha)$ *then* $CONS(X) \neq 1$ *or* $CONS(\{\alpha\}) = 1$.

Proof If $CONS(X) \neq 1$ we are already done. Let $CONS(X) = 1$.
Then by Proposition 5.1(iii), $INCONS(X) \neq 1$.
That is, there is some T_i such that $X \subseteq T_i$. Now as $1 = CONS(X) \leq gr(X \mid\!\sim \alpha)$, for $X \subseteq T_i$, $T_i(\alpha) = 1$. Hence, $CONS(\{\alpha\}) = \sup_i\{T_i(\alpha) : \{\alpha\} \subseteq T_i\} = 1$. □

5.2.3 Equivalence

In the context of classical logic, two sets of formulae are equivalent if whatever can be derived from the one set that can also be derived from the other set. This definition of equivalence between two sets of formulae has been generalized in Chap. 2, where the consequence relation is fuzzy but the sets are crisp. Here we define equivalence between two fuzzy sets of formulae and the consequence relation is fuzzy too.

Definition 5.9 Let X, Y be two fuzzy subsets of formulas. X is said to be equivalent to Y, denoted by $X \equiv Y$, if for all α, $gr(X \mid\!\sim \alpha) = gr(Y \mid\!\sim \alpha)$.

Theorem 5.17 *For any two fuzzy subsets X, Y of formulas $X \cup Y \equiv X$ if and only if $Y(\beta) \leq gr(X \mid\!\sim \beta)$ for all β.*

Proof Let $X \cup Y \equiv X$, i.e. $gr(X \cup Y \mid\!\sim \alpha) = gr(X \mid\!\sim \alpha)$.
Also, $(X \cup Y)(\alpha) \leq gr(X \cup Y \mid\!\sim \alpha)$. [By ($GC_f1$)]
$Y(\alpha) \leq (X \cup Y)(\alpha) \leq gr(X \cup Y \mid\!\sim \alpha) = gr(X \mid\!\sim \alpha)$.
Hence, $Y(\alpha) \leq gr(X \mid\!\sim \alpha)$.
Conversely, let $Y(\beta) \leq gr(X \mid\!\sim \beta)$ for all β.
Then by (GC_f3), $gr(X \cup Y \mid\!\sim \alpha) \leq gr(X \mid\!\sim \alpha)$. ...(1)
Hence, by (GC_f2), $gr(X \mid\!\sim \alpha) = gr(X \cup Y \mid\!\sim \alpha)$, which implies $X \cup Y \equiv X$. □

Theorem 5.18 *For any two fuzzy subsets X, Y of formulas $X \equiv Y$ if and only if $Y(\alpha) \leq gr(X \mid\!\sim \alpha)$ and $X(\alpha) \leq gr(Y \mid\!\sim \alpha)$.*

Proof Let $X \equiv Y$, i.e. $gr(X \mid\!\sim \alpha) = gr(Y \mid\!\sim \alpha)$. Then $X(\alpha) \leq gr(Y \mid\!\sim \alpha)$ and $Y(\alpha) \leq gr(X \mid\!\sim \alpha)$ are immediate by (GC_f1).
Conversely, let $Y(\alpha) \leq gr(X \mid\!\sim \alpha)$ and $X(\alpha) \leq gr(Y \mid\!\sim \alpha)$.
By Theorem 5.17, $X \cup Y \equiv X$ and $X \cup Y \equiv Y$. Hence $X \equiv Y$. □

Theorem 5.19 *If for fuzzy subsets X, Y, Z, W of formulas $X \equiv Y$ and $Z \equiv W$ then $X \cup Z \equiv Y \cup W$.*

Proof Let $X \equiv Y$ and $Z \equiv W$. Then $gr(X \mid\!\sim \alpha) = gr(Y \mid\!\sim \alpha)$, $gr(Z \mid\!\sim \alpha) = gr(W \mid\!\sim \alpha)$.
Now by (GC_f2), $gr(Y \mid\!\sim \alpha) = gr(X \mid\!\sim \alpha) \leq gr(X \cup Z \mid\!\sim \alpha)$. ...(1)
And similarly, $gr(W \mid\!\sim \alpha) \leq gr(X \cup Z \mid\!\sim \alpha)$. ...(2)
(1) and (2) imply $Y(\alpha) \vee W(\alpha) \leq gr(X \cup Z \mid\!\sim \alpha)$, i.e. $(Y \cup W)(\alpha) \leq gr(X \cup Z \mid\!\sim \alpha)$.
Similarly, we have $(X \cup Z)(\alpha) \leq gr(Y \cup W \mid\!\sim \alpha)$. Hence $X \cup Z \equiv Y \cup W$. □

Corollary 5.1 *For three fuzzy subsets X, Y, Z of formulas if $X \equiv Y$ then $X \cup Z \equiv Y \cup Z$.*

Corollary 5.2 *For two fuzzy subsets X, Y of formulas if $X \equiv Y$ then $X \cup Y \equiv X$ and $X \cup Y \equiv Y$.*

Theorem 5.20 *Let X, Y, Z be three fuzzy subsets of formulas such that $X \subseteq Y \subseteq Z$. Then $X \equiv Z$ implies $Y \equiv Z$.*

While considering a fuzzy set of premises, the following questions come into fore.
–What would be the interpretation of a fuzzy set of premises?
–What is the difference between membership degree of a formula in a fuzzy set of premises and its truth values?

Some researchers (Gottwald 1993, 1995) consider that the degree of membership of a formula in a fuzzy set of premises represents the degree of confidence of the formula as a premise while the latter may be taken as truth degree. How satisfactory is this interpretation is a matter of debate. But mathematically speaking, elegant equivalence theorems have been established, and at no point a structure richer than a complete lattice has been necessary for determining values of meta-level assertions. Besides, taking fuzzy sets as premises has been a practice in the research on fuzzy logic (Biacino and Gerla 1993, 1996, 1999; Dubois and Prade 1979; Esteva et al. 2007, 2009; Gaines 1977; Gerla 1994, 1996, 2001; Giles 1976; Gottwald 1993; Hájek 1998; Novak 1990; Pavelka 1979). So, the present approach turns out to be meaningful.

5.3 Implicative Consequence Operator and Consequence Relation

In Rodríguez et al. (2003), Rodríguez et al. made an extensive study on different approaches to capture the notion of consequence in the context of fuzzy logics. The gaps between ($C3$) due to Pavelka, (f_c3) due to Castro et al. and (GC3) due to Chakraborty were their main concern. They remarked

> ..., *it is worth noticing that fuzzy consequence relations as defined above, when restricted over crisp sets of formulas, becomes only a particular class of graded consequence relation. Namely, regarding the two versions of the fuzzy cut properties, (GC3) and (f_c3), it holds that for* $A, B \in \mathscr{P}(\mathscr{L})$, *if* $B(p) \leq f_c(A, p)$ *for all* $p \in \mathscr{L}$, *it is clear that* $\inf_{q \in B} f_c(A, q) = 1$.

It is to be noted that in Rodríguez et al. (2003) $\mathscr{L}$ is used to denote the set of all formulae of a language. The authors further introduced a pair of notions called implicative closure operator and its corresponding notion of implicative consequence relation. Their intention was to generalize the notion of graded consequence in the context of fuzzy sets of premises in such a way that it is based upon the notion of *graded inclusion* between fuzzy sets as well as it can be translated equivalently in terms of a consequence operator.

Definition 5.10 (Rodríguez et al. (2003)) An implicative closure operator is defined to be a mapping $\tilde{C}:\mathscr{F}(F) \mapsto \mathscr{F}(F)$ such that the following holds:

($\tilde{C}1$) $X \subseteq \tilde{C}(X)$.
($\tilde{C}2$) If $X \subseteq Y$, then $\tilde{C}(X) \subseteq \tilde{C}(Y)$.
($\tilde{C}3$) $[X \sqsubseteq_* \tilde{C}(Y)] * \tilde{C}(Y \cup X) \subseteq \tilde{C}(Y)$.

$[X \sqsubseteq_* \tilde{C}(Y)]$ is defined as $\inf_x [X(x) \rightarrow \tilde{C}(Y)(x)]$, and the values assigned to the formulae by fuzzy sets are members of a complete BL-algebra (cf. Chap. 4) with the residuated pair $(*, \rightarrow)$. The fuzzy set $X * Y$ is defined pointwise, i.e. $(X * Y)(x) = X(x) * Y(x)$.

Definition 5.11 (Rodríguez et al. (2003)) An implicative consequence relation, viz. I_c is defined to be a fuzzy relation between $\mathscr{F}(F)$ and F satisfying

(I_c1) $X(\alpha) \leq I_c(X, \alpha)$,
(I_c2) if $X \subseteq Y$, then $I_c(X, \alpha) \leq I_c(Y, \alpha)$, and

(I_c3) $[X \sqsubseteq_* \tilde{C}(Y)] * I_c(Y \cup X, \alpha) \leq I_c(Y, \alpha)$.

That I_c is equivalent to the notion of $\tilde{C}$ was then established Rodríguez et al. (2003) simply by defining $I_c(X, \alpha) = \tilde{C}(X)(\alpha)$. Moreover, the authors then showed that restricting the notion in the context of ordinary sets (I_c3) turns out to be (GC3).

The semantic notion of implicative consequence relation is also defined, in the same paper Rodríguez et al. (2003), as follows.

Definition 5.12 Given a collection of fuzzy sets $\{T_i\}_{i \in I}$ over formulae, the semantic counterpart of the implicative consequence relation is defined to be $I_c^{\{T_i\}_{i \in I}}(X, \alpha) = \inf_{i \in I}([X \sqsubseteq_* T_i] \to T_i(\alpha))$.

This definition generalizes the notion of graded semantic consequence (Definition 2.3), as restricting X to be an ordinary set the above definition coincides with the notion of graded semantic consequence (Definition 2.3). Furthermore, in Rodríguez et al. (2003) representation theorems establishing the connection between syntactic and semantic notions of implicative consequence relation, as discussed in Sect. 2.2 of Chap. 2, have been proved in the framework of a complete BL-algebra.

Thus, to a great extent the authors (Rodríguez et al. 2003) are successful to introduce a notion which meets both the ends of their target, viz. (i) extending to fuzzy sets of premises, considering a complete BL-algebra as the meta-level algebraic structure such that Chakraborty's graded consequence relation (Chakraborty 1995) can be obtained as a special case of the notion for implicative consequence relation, and (ii) retaining the both-way connection between consequence operator and consequence relation in the context of fuzzy sets of premises.

It is to be noted here that in introducing the notion of implicative consequence relation the authors need to use two kinds of inclusion relation between fuzzy sets; one is the standard $\subseteq$ in the crisp sense, and the other, the $\sqsubseteq_*$, in the graded sense. One can notice that $X \subseteq Y$ iff $X \sqsubseteq_* Y = 1$. Though, $\subseteq$ can be obtained from $\sqsubseteq_*$, one cannot ignore that two, one crisp and one graded, inclusions between fuzzy sets are required.

So, the summary of the development made so far for establishing a similar connection between consequence operator and consequence relation in fuzzy context, as available in classical case, is as follows. Chakraborty's (1995) notion of graded consequence is applicable for ordinary sets of premises only. Pavelka (1979) and Castro et al. (1994) proposed a setup where fuzzy sets of premises are considered but their versions for cut condition do not coincide with that of Chakraborty's version when restricted to crisp sets of premises. In Rodríguez et al. (2003), the authors proposed an axiomatization for consequence relation in fuzzy context addressing both the above-mentioned points. Their axiomatization uses two notions of inclusion between fuzzy sets of premises, namely, an ordinary inclusion $\subseteq$ in (I_c2) and a graded inclusion $\sqsubseteq_*$ in (I_c3). Formally, there is no problem in axiomatizing the notion of implicative consequence using two notions of inclusion. But our concern is to explore if using two inclusions for two natural properties, viz. monotonicity and cut, of a consequence relation in graded context yields some other insight. In this regard, in the following section, we shall present a study on revisiting the notion of implicative consequence relation from a different perspective so that without bringing in two different notions of inclusion the axiom (I_c3) can be interpreted in a different way not affecting the sense of cut. Besides, this new approach generalizes both the notions of graded consequence relation and implicative consequence relation (Cf. Theorem 5.23 and Proposition 5.3).

5.4 Implicative Consequence Relation in the Light of Consistency-Generating Relation

In connection with the problem of using two notions of inclusion in defining implicative consequence relation in Rodríguez et al. (2003), as discussed in Sect. 5.3, we propose a setup (Dutta and Chakraborty 2016) where only the crisp notion of inclusion ($\subseteq$) between fuzzy sets has been considered. The purpose for incorporating $\sqsubseteq_*$ in (I_c3) has been addressed from a different perspective. This in turn gives a different understanding of the condition for cut. We introduce a set of meta-theoretic notions, e.g. *CE*, *CCE*, *MCE*, in order to understand the meta-linguistic sentence $X \sqsubseteq_* \tilde{C}(Y)$ present in (I_c3). Additionally, these notions help us to understand the notion of consistency in a new way. It is to be noted that the construction of these notions in a meta-language which has to have a negation apart from the usual ones, viz. conjunction, implication and universal quantifier taken in the contexts of graded consequence relation (Chakraborty 1995) and implicative consequence relation (Rodríguez et al. 2003). The corresponding algebraic structure is supposed to be a complete BL-algebra with an operator $\neg_m$ for the above-mentioned negation satisfying the following conditions.
—$\neg_m$ preserves values of the classical negation and
—if $a \leq b$, then $\neg_m b \leq \neg_m a$.

Definition 5.13 (Dutta and Chakraborty 2016) *CE*, called consistency-generating relation, is defined to be a fuzzy binary relation on $\mathscr{F}(F)$ assuming values from a BL-algebra L, and characterized by the following axioms:

(CE1) If $X_1 \subseteq X_2$, then $CE(Y, X_2) \leq CE(Y, X_1)$ for all Y.
(CE2) $CE(F, X) = 0$ for any X.
(CE3) If $CE(Y, X) > 0$ for some X, then for all $Z \subseteq Y$, $CE(Y, Z) > 0$.

Before proceeding further we here introduce some notational changes. The implicative consequence relation I_c, from now onwards, would be denoted as $\mid\!\sim$, i.e. instead of $I_c(X, \alpha)$, to keep the uniformity of presentation with the notion of graded consequence relation, we would present the above as $gr(X \mid\!\sim \alpha)$, and the adjoint pair $(*, \rightarrow)$ of the BL-algebra also will be denoted as $(*_m, \rightarrow_m)$ reinforcing their meta-linguistic status.

So, let us start with a fuzzy relation $\mid\!\sim$ over $\mathscr{F}(F) \times F$. At this stage, we are not imposing any condition on $\mid\!\sim$.

Definition 5.14 In the setup of a complete BL-algebra, with an additional operator $\neg_m$, we define $S_X = \{Y \in \mathscr{F}(F) : \inf_\alpha(gr(X \mid\!\sim \alpha) \rightarrow_m Y(\alpha)) *_m \neg_m \inf_\alpha Y_{\mid\sim}(\alpha) \leq CE(Y, X)\}$, where $Y_{\mid\sim} \in \mathscr{F}(F)$ is defined as $Y_{\mid\sim}(x) = gr(Y \mid\!\sim x)$ for all $x \in F$. For any $X \in \mathscr{F}(F)$, CCE is a function from S_X to L defined by $CCE(Y) = CE(Y, X)$.

To indicate the reference fuzzy set X, we shall write $CCE(Y, X)$ for $CCE(Y)$.

Note 5.5 Through CE our attempt is to introduce a notion called 'consistently extended'. CCE additionally incorporates the idea that whatever follows from X belongs to Y, and it is not that everything follows from Y, i.e. Y is non-explosive. The word 'consistently extended' is used as a single term, it should not be mixed with the usual notions of 'consistency' and 'extension'.

Definition 5.15 Given a $\mid\sim$ satisfying (I_c1) and (I_c2), $CE_{syn}(Y, X)$ is a binary fuzzy relation on $\mathscr{F}(F)$, such that $CE_{syn}(Y, X) = \inf_x(X(x) \rightarrow Y_{\mid\sim}(x)) *_m \neg_m \inf_x Y_{\mid\sim}(x)$.

Note 5.6 At this point, we should note that though the notion of CCE only incorporates presence of a fuzzy relation $\mid\sim$, the definition of CE_{syn} presumes $\mid\sim$ to satisfy (I_c1) and (I_c2). (I_c3) is not required at this stage. So, in neither of Definitions 5.14, 5.15, presence of an implicative consequence relation is assumed.

Theorem 5.21 *CE_{syn} is a CE.*

Proof (CE1) Let $X_1 \subseteq X_2$. Then $X_1(x) \rightarrow_m Y_{\mid\sim}(x) \geq X_2(x) \rightarrow_m Y_{\mid\sim}(x)$.
$\inf_x(X_1(x) \rightarrow_m Y_{\mid\sim}(x)) *_m \neg_m \inf_x Y_{\mid\sim}(x) \geq \inf_x(X_2(x) \rightarrow_m Y_{\mid\sim}(x)) *_m \neg_m \inf_x Y_{\mid\sim}(x)$.
Hence, $CE_{syn}(Y, X_1) \geq CE_{syn}(Y, X_2)$.
(CE2) $CE_{syn}(F, X) = \inf_x(X_1(x) \rightarrow_m F_{\mid\sim}(x)) *_m \neg_m \inf_x F_{\mid\sim}(x)$.
$= 1 *_m \neg_m \inf_x F_{\mid\sim}(x) = 0$. [since by (I_c1), $F_{\mid\sim}(x) = 1$ for all $x \in F$]
(CE3) Let $CE_{syn}(Y, X) > 0$. That is, $\inf_x(X(x) \rightarrow_m Y_{\mid\sim}(x)) *_m \neg_m \inf_x Y_{\mid\sim}(x) > 0$.
That is, $\neg_m \inf_x Y_{\mid\sim}(x) > 0$.
Let $Z \subseteq Y$, then $Z(x) \leq Y(x) \leq Y_{\mid\sim}(x)$ [by (I_c1)]
That is, $Z(x) \rightarrow_m Y_{\mid\sim}(x) = 1$ for all $x \in F$.
Hence, $CE_{syn}(Y, Z) = \inf_x(Z(x) \rightarrow_m Y_{\mid\sim}(x)) *_m \neg_m \inf_x Y_{\mid\sim}(x) > 0$. □

Proposition 5.2 *For any $X, Y \in \mathscr{F}(F)$, $\inf_x(gr(X \mid\!\sim x) \rightarrow_m Y(x)) *_m \neg_m \inf_x Y_{\mid\sim}(x) \leq CE_{syn}(Y, X)$.*

Proof By (I_c1) we know $X(x) \leq gr(X \mid\!\sim x)$. Then, by properties of $\rightarrow_m$, we have $X(x) \rightarrow_m gr(Y \mid\!\sim x) \geq gr(X \mid\!\sim x) \rightarrow_m gr(Y \mid\!\sim x) \geq gr(X \mid\!\sim x) \rightarrow_m Y(x)$.
I.e. $\inf_x(X(x) \rightarrow_m gr(Y \mid\!\sim x)) *_m \neg_m \inf_x Y_{\mid\sim}(x)$
$\geq \inf_x(gr(X \mid\!\sim x) \rightarrow_m Y(x)) *_m \neg_m \inf_x Y_{\mid\sim}(x)$.
Hence, $CE_{syn}(Y, X) \geq \inf_x(gr(X \mid\!\sim x) \rightarrow_m Y(x)) *_m \neg_m \inf_x Y_{\mid\sim}(x)$. □

Note 5.7 So, as a corollary of Proposition 5.2, following the notion of CCE given in Definition 5.14, for any $X, Y \in \mathscr{F}(F)$, we can write $CCE_{syn}(Y, X) = CE_{syn}(Y, X)$. It is noticed that CCE_{syn} is a particular CE, and $CCE_{syn}(Y, X)$ specifies that every member of X follows from Y and Y is non-explosive. As CCE_{syn} is based on a $\mid\sim$ which has some properties of an implicative consequence relation, $CCE_{syn}(Y, X)$ may be regarded to be the abbreviation of 'X is consistently extended to Y, where Y is non-explosive and its syntactic consequence includes X'.

We now rewrite the axioms of the notion of implicative consequence relation I_c, in terms of a modified notion of the same.

Definition 5.16 I_c^m, a modified version of I_c, is a fuzzy relation $|\!\sim$ satisfying the following axioms:

(I_c^m1) $X(\alpha) \leq gr(X |\!\sim \alpha)$.
(I_c^m2) If $X \subseteq Y$, then $gr(X |\!\sim \alpha) \leq gr(Y |\!\sim \alpha)$.
(I_c^m3) $CCE_{syn}(Y, X) *_m gr(Y \cup X |\!\sim \alpha) \leq gr(Y |\!\sim \alpha)$.

The underlying meaning of (I_c^m3) may be given as follows. If X is consistently extended to a non-explosive Y such that X is included in the syntactic consequences of Y, then whatever follows from $Y \cup X$ follows from Y itself.

($\tilde{C}$3) can then be written as $CCE_{syn}(Y, X) *_m \tilde{C}(Y \cup X) \subseteq \tilde{C}(Y)$. The equivalence between I_c and $\tilde{C}$ naturally can be extended in this modified context too. Also, to be noted that the presence of $Y_{|\sim}$ in $CCE_{syn}(Y, X)$ of (I_c^m3) denotes the fuzzy set $Y_{|\sim}(x)$ $= gr(X |\!\sim x)$, whereas the same represents the fuzzy set $\tilde{C}(Y)$ when the axiom ($\tilde{C}$3) is considered.

Proposition 5.3 *Considering a complete BL-algebra as the meta-structure for* $|\!\sim$, *and computing* $\neg_m$ *by the drastic negation,* (I_c3) *and* (GC3) *are obtained as special cases of* (I_c^m3) *when* $Y_{|\sim} \neq F$.

Proof Let us consider $\neg_m$ to be the drastic negation Klir and Yuan (2006), i.e. $\neg_m(x)$ $= 0$, if $x = 1$, and 1, otherwise. As $Y_{|\sim} \neq F$, $Y_{|\sim}(x) \neq 1$ for some x. I.e. $\inf_x Y_{|\sim}(x) \neq 1$.
Hence, $\neg \inf_x Y_{|\sim}(x) = 1$. So, $CCE_{syn}(Y, X) = \inf_x(X(x) \rightarrow_m Y_{|\sim}(x)) *_m \neg_m \inf_x Y_{|\sim}(x)$
$= \inf_x(X(x) \rightarrow_m Y_{|\sim}(x))$.
Now for fuzzy sets X, Y, $CCE_{syn}(Y, X) = \inf_x(X(x) \rightarrow_m Y_{|\sim}(x)) = [X \sqsubseteq_{*_m} \tilde{C}(Y)]$.
That is, (I_c3) can be obtained as a special case (I_c^m3) when $Y_{|\sim} \neq F$.
On the other hand, when X, Y are ordinary sets, we have $CCE_{syn}(Y, X) = \inf_x (X(x) \rightarrow_m Y_{|\sim}(x)) = \inf_{x \in X}(1 \rightarrow_m Y_{|\sim}(x)) = \inf_{x \in X} Y_{|\sim}(x) = \inf_{x \in X} gr(Y |\!\sim x)$.
Hence, we obtain (GC3) as special case of (I_c^m3) when $Y_{|\sim} \neq F$. □

Note 5.8 For $Y_{|\sim} = F$, $CCE_{syn}(Y, X) = 0$, so the inequality (I_c^m3) becomes immediate. Besides, as for $Y_{|\sim} = F$, $gr(Y |\!\sim \alpha) = 1$, the inequalities presented in (I_c3) and (GC3) become immediate. So, we need not bother about the case for $Y_{|\sim} = F$.

While defining CCE_{syn} we have based on a $|\!\sim$ satisfying (I_c1) and (I_c2), and have showed that CCE_{syn} is a special kind of CCE. Below, we introduce another notion called MCE.

Definition 5.17 For $X, Y \in \mathscr{F}(F)$, the notion of *maximal consistent extension* is defined by $MCE(Y, X) = \inf_x(X(x) \rightarrow_m Y(x)) *_m \inf_x(Y(x) \vee Y(\neg x))$, for $Y \neq F$, $= 0$, for $Y = F$.

In Definition 5.17, we need to introduce an object-level negation $\neg$. This may not be the same as the meta-linguistic connective 'not', i.e. $\neg_m$.

Theorem 5.22 *MCE is a CE.*

Proof (CE1) Let $X_1 \subseteq X_2$. Then $X_1(x) \rightarrow_m Y(x) \geq X_2(x) \rightarrow_m Y(x)$.
$\inf_x(X_1(x) \rightarrow_m Y(x)) *_m \inf_x(Y(x) \vee Y(\neg x))$
$\geq \inf_x(X_2(x) \rightarrow_m Y(x)) *_m \inf_x(Y(x) \vee Y(\neg x))$.
I.e. $MCE(Y, X_1) \geq MCE(Y, X_2)$.
(CE2) $MCE(F, X) = 0$ by definition.
(CE3) Let $MCE(Y, X) > 0$. I.e. $\inf_x(X(x) \rightarrow_m Y(x)) *_m \inf_x(Y(x) \vee Y(\neg x)) > 0$.
So, $\inf_x(Y(x) \vee Y(\neg x)) > 0$.
Also, for $Z \subseteq Y$, $Z(x) \rightarrow_m Y(x) = 1$, for all $x \in F$.

So, $MCE(Y, Z) = \inf_x(Z(x) \rightarrow_m Y(x)) *_m \inf_x(Y(x) \vee Y(\neg x)) > 0$. □

The development made so far does not involve any semantics. To prove the completeness theorem for classical logic, a standard way (Rodríguez et al. 2003; Enderton 1972) is to view a collection of maximal consistent sets of formulae as a collection of valuation functions over formulae. Maximal consistent sets are consistent sets which cannot be extended consistently, which means for each formula α, either α or $\neg\alpha$ is a member of the set. Now we are at the stage of defining the semantic counterpart of the modified implicative consequence relation using the notion of *MCE*.

Definition 5.18 Given a collection of fuzzy sets $\{T_i\}_{i \in I}$ over formulae such that for each $i \in I$, $T_i \neq F$, $gr(X \approx_{I_c^m} \alpha) = \inf_{i \in I}(MCE(T_i, X) \rightarrow_m T_i(\alpha))$.

Theorem 5.23 *The semantic notions of implicative consequence relation, defined in Rodríguez et al. (2003), and the semantic graded consequence relation, defined in Chakraborty (1995), are special cases of* $\approx_{I_c^m}$, *when a complete BL-algebra is considered as the meta-structure for* $\mid\!\sim$, *and* $\neg$ *is computed by the drastic negation.*

Proof Let $\neg$ of the language be computed by the drastic negation.
Now if X is considered to be a fuzzy set of formulae, then $MCE(T_i, X) = \inf_x(X(x) \rightarrow_m T_i(x)) *_m \inf_x(T_i(x) \vee T_i(\neg x))$.
$= \inf_x(X(x) \rightarrow_m T_i(x))$ [since $T_i(x) \vee T_i(\neg x) = 1$].
Hence, $gr(X \approx_{I_c^m} \alpha) = \inf_{i \in I}[\inf_x(X(x) \rightarrow_m T_i(x)) \rightarrow_m T_i(\alpha)]$
$= \inf_{i \in I}([X \sqsubseteq_* T_i] \rightarrow_m T_i(\alpha))$.
So, we obtain the notion of semantic implicative consequence relation due to Rodríguez et al. (2003).
Now, let us consider X to be an ordinary set of formulae. Then, $MCE(T_i, X) = \inf_x(X(x) \rightarrow_m T_i(x)) *_m \inf_x(T_i(x) \vee T_i(\neg x))$.
$= \inf_{x \in X}(1 \rightarrow_m T_i(x)) = \inf_{x \in X} T_i(x)$.
Hence, $gr(X \approx_{I_c^m} \alpha) = \inf_{i \in I}[\inf_x(X(x) \rightarrow_m T_i(x)) \rightarrow_m T_i(\alpha)]$
$= \inf_{i \in I}(T_i(x) \rightarrow_m T_i(\alpha))$.
Thus, the notion of semantic graded consequence relation due to Chakraborty (1995) is obtained. □

So, in this section, we have introduced a notion of *consistency-generating relation* by a fuzzy relation *CE* between fuzzy sets. We have then proposed two different such consistency-generating relations, namely, CCE_{syn} and MCE. CCE_{syn} is a notion based on $\mid\!\sim$ with overlap and monotonicity in fuzzy context, and is used to present the classical condition for cut in fuzzy context. On the other hand, MCE is introduced independent of $\mid\!\sim$, and used to define the notion of semantic consequence in fuzzy context. It is shown that $(I_c^m 3)$ and $\approx_{I_c^m}$, thus defined, generalize the respective notions in Rodríguez et al. (2003) and Chakraborty (1995) with respect to a specific algebraic structure.

5.4.1 From the Notion of Consistency-Generating Relation to the Notion of Consistency

Now we shall further develop the notion of consistency with the help of the notions of consistency-generating relation, in particular, CCE_{syn} and MCE. The idea behind the notions *CE*, *CCE* and *MCE* perhaps can be better explained drawing an analogy from classical scenario. Through the notion of *CE* our attempt is to give the idea of 'consistently extended' a formal form. Let us start with a set of formulae X in classical context. Let X be extended to $Y_1, Y_2, \ldots Y_n, \ldots$ in a step-by-step manner just by adding formulas to X with the conditions that for each i, $Y_i \subseteq Y_{i+1}$ and $Y_i \neq F$. There could be several such chains rooted at X. Let us assume that the notion of 'consistently extended' is characterized by some specific means, i.e. axioms, following which there could be a chain such that for each Y_i, Y_{i-1} is *consistently extended* to Y_i, and for some j, Y_j contains all consequences of X. That is, Y_j is such an extension of X which is closed under all consequences of X. Now the Y_j containing all consequence of X might not reach the maximality criterion, i.e. it may not be the case that for each α, either α or $\neg\alpha$ belongs to Y_j. To reach at such an extension of X, Y_j may need to be extended further. X is said to be consistent if there is a consistently extended chain rooted at X such that for some i, Y_i contains all consequences of X, and Y_i is extended maximally.

The important point here is that the relation *CE* (consistently extended) is primitive and consistency is defined in terms of *CE*. In the fuzzy setup, the above idea is captured through the notions of *CE*, *CCE*, CCE_{syn}, *MCE*, etc., and the notion of consistency is obtained thereby as a derived notion. *CE* imposes the basic conditions for building a chain of fuzzy sets; *CCE*, CCE_{syn} and *MCE* add some more demands for building a chain which is already built satisfying properties of *CE*. X is said to be consistent if there is a chain rooted at it and that contains a fuzzy set entitled to be ascribed both the properties of CCE_{syn} and *MCE*.

Before proceeding for the definition of consistency we introduce one more notation, viz. $CCE_{syn}^{(X)}$, defined in the following way.
For any $X \in \mathscr{F}(F)$, we write $CCE_{syn}^{(X)} = \{Y \in \mathscr{F}(F) : CE_{syn}(Y, X) > 0\}$.

Definition 5.19 For any $X \in \mathscr{F}(F)$, $Cons(X) = 0$, if $X_{\mid\sim} = F$,
$= \sup_{Y \in CCE_{syn}^{(X_{\mid\sim})}} [MCE(Y, X_{\mid\sim})]$, otherwise.

Lemma 5.1 *If $X_1 \subseteq X_2$, then $Y \in CCE_{syn}^{(X_2)}$ implies $Y \in CCE_{syn}^{(X_1)}$.*

Proof Let $Y \in CCE_{syn}^{(X_2)}$ and $X_1 \subseteq X_2$. Then $CCE_{syn}(Y, X_2) = CE_{syn}(Y, X_2) > 0$.
As CCE_{syn} satisfies CE asioms, $0 < CE_{syn}(Y, X_2) \leq CE_{syn}(Y, X_1)$.
Hence, $Y \in CCE_{syn}^{(X_1)}$. That is $CCE_{syn}^{(X_2)} \subseteq CCE_{syn}^{(X_1)}$. □

Theorem 5.24 *If $X_1 \subseteq X_2$, then $Cons(X_2) \leq Cons(X_1)$.*

Proof Let $X_1 \subseteq X_2$. Now if $Cons(X_2) = 0$, we are done. So, let $Cons(X_2) \neq 0$.
Hence, $(X_2)_{\mid\sim} \neq F$, and as $X_1 \subseteq X_2$, $(X_1)_{\mid\sim} \neq F$ [since $(X_1)_{\mid\sim} \subseteq (X_2)_{\mid\sim}$.]
Therefore, $Cons(X_2) = \sup_{Y \in CCE_{syn}^{((X_2)_{\mid\sim})}} [MCE(Y, (X_2)_{\mid\sim})]$.
$\leq \sup_{Y \in CCE_{syn}^{((X_1)_{\mid\sim})}} [MCE(Y, (X_2)_{\mid\sim})]$ [by Lemma 5.1]
$\leq \sup_{Y \in CCE_{syn}^{((X_1)_{\mid\sim})}} [MCE(Y, (X_1)_{\mid\sim})]$ [by Theorem 5.22]
$= Cons(X_1)$. □

Lemma 5.2 *If $X_{\mid\sim} = Z_{\mid\sim}$, then $Cons(X) = Cons(Z)$.*

Proof If $X_{\mid\sim} = Z_{\mid\sim} = F$, then $Cons(X) = Cons(Z) = 0$.
Let $X_{\mid\sim} = Z_{\mid\sim} \neq F$. Then $Cons(X) = \sup_{Y \in CCE_{syn}^{(X_{\mid\sim})}} [MCE(Y, X_{\mid\sim})]$.
Now $Y \in CCE_{syn}^{(X_{\mid\sim})}$ means $CE_{syn}(Y, X_{\mid\sim}) > 0$.
So, $CE_{syn}(Y, X_{\mid\sim}) = \inf_x (X_{\mid\sim}(x) \rightarrow_m Y_{\mid\sim}(x)) *_m \neg_m \inf_x Y_{\mid\sim}(x)$.
$= \inf_x (Z_{\mid\sim}(x) \rightarrow_m Y_{\mid\sim}(x)) *_m \neg_m \inf_x Y_{\mid\sim}(x)$ [since $X_{\mid\sim} = Z_{\mid\sim}$].
$= CE_{syn}(Y, Z_{\mid\sim}) > 0$.
Hence, $Y \in CCE_{syn}^{(X_{\mid\sim})}$ implies $Y \in CCE_{syn}^{(Z_{\mid\sim})}$.
Similarly, the other direction can be proved, i.e. we have $CCE_{syn}^{(X_{\mid\sim})} = CCE_{syn}^{(Z_{\mid\sim})}$.
Also, $MCE(Y, X_{\mid\sim}) = \inf_x (X_{\mid\sim}(x) \rightarrow_m Y(x)) *_m \inf_x (Y(x) \vee Y(\neg x))$
$= \inf_x (Z_{\mid\sim}(x) \rightarrow_m Y(x)) *_m \inf_x (Y(x) \vee Y(\neg x))$
$= MCE(Y, Z_{\mid\sim})$.
Hence, $Cons(X) = \sup_{Y \in CCE_{syn}^{(X_{\mid\sim})}} [MCE(Y, X_{\mid\sim})]$.
$= \sup_{Y \in CCE_{syn}^{(Z_{\mid\sim})}} [MCE(Y, Z_{\mid\sim})]$.
$= Cons(Z)$. □

Theorem 5.25 *If $gr(X \mid\sim \alpha) = 1$, then $Cons(X) = Cons(X \cup \{\alpha\})$.*

Proof Let $gr(X \mid\sim \alpha) = 1$. Now two cases arise: (i) $X_{\mid\sim} = F$ and (ii) $X_{\mid\sim} \neq F$.
(i) If $X_{\mid\sim} = F$, $(X \cup \{\alpha\})_{\mid\sim} = F$, and hence $Cons(X) = Cons(X \cup \{\alpha\}) = 0$.
(ii) Let $X_{\mid\sim} \neq F$. Then $Cons(X) = \sup_{Y \in CCE_{syn}^{(X_{\mid\sim})}} [MCE(Y, X_{\mid\sim})]$.
Now as $gr(X \mid\sim \alpha) = 1$, i.e. $(X_{\mid\sim})(\alpha) = 1$, there could be two cases.
(a) $X(\alpha) = 1$. So, $X = X \cup \{\alpha\}$, and hence $Cons(X) = Cons(X \cup \{\alpha\})$.
(b) $X(\alpha) \neq 1$. Then as $(X \cup \{\alpha\})(\beta) = X(\beta) \leq X_{\mid\sim}(\beta)$ for $\beta \neq \alpha$, and $(X \cup \{\alpha\})(\alpha) = (X_{\mid\sim})(\alpha)$, we can conclude that $X \cup \{\alpha\} \subseteq X_{\mid\sim}$.
Hence, by (I_c3) $\inf_x[(X \cup \{\alpha\})(x) \rightarrow_m X_{\mid\sim}(x)] *_m gr(X \cup \{\alpha\} \mid\sim \beta) \leq gr(X \mid\sim \beta)$.
I.e. $1 *_m gr(X \cup \{\alpha\} \mid\sim \beta) \leq gr(X \mid\sim \beta)$ [since $X \cup \{\alpha\} \subseteq X_{\mid\sim}$].
Hence, $(X \cup \{\alpha\})_{\mid\sim} \subseteq X_{\mid\sim}$. Then by (I_c2), $(X \cup \{\alpha\})_{\mid\sim} = X_{\mid\sim}$. Hence, by Lemma 5.2, $Cons(X) = Cons(X \cup \{\alpha\})$. □

The above study shows that the construction of the meta-theoretic concepts needs a well-defined meta-language, and a definite algebraic structure interpreting its entities. So, study of a logic cannot be all about concentrating on its object language and its interpretation.

References

Biacino, L., Gerla, G.: Closure operators for fuzzy subsets. In: EUFIT 93, First European Congress on Fuzzy and Intelligent Technologies, pp. 1441–1447. Aachen (1993)

Biacino, L., Gerla, G.: Closure operators in fuzzy set theory. In: Bezdek, J.C., Dubois, D., Prade, H. (Eds.) Fuzzy Sets. Approximate Reasoning and Information Systems, pp. 243–278. Kluwer Academic Publishers, Dordrecht (1999)

Biacino, L., Gerla, G.: An extension principle for closure operators. J. Math. Anal. Appl. **198**, 1–24 (1996)

Castro, J.L., Trillas, E., Cubillo, S.: On consequence in approximate reasoning. J. Appl. Non-Class. Logics **4**(1), 91–103 (1994)

Chakraborty, M.K.: Use of fuzzy set theory in introducing graded consequence in multiple valued logic. In: Gupta, M.M., Yamakawa, T. (Eds.), Fuzzy Logic in Knowledge-Based Systems, Decision and Control, pp. 247–257. Elsevier Science Publishers, (B.V) North Holland (1988)

Chakraborty, M.K.: Graded consequence: further studies. J. Appl. Non-Class. Logics **5**, 227–237 (1995)

Dubois, D., Prade, H.: Operations in a fuzzy-valued logic. Inform. Control **43**(2), 224–240 (1979)

Dutta, S., Chakraborty, M.K.: Graded consequence with fuzzy set of premises. Fundam. Inform. **133**, 1–18 (2014)

Dutta, S., Chakraborty, M.K.: The role of metalanguage in graded logical approaches. Fuzzy Sets Syst. **298**, 238–250 (2016)

Enderton, H.B., A Mathematical Introduction to Logic. Harcourt Academic Press (1972)

Esteva, F., Godo, L., Noguera, C.: First order t-norm based fuzzy logics with truth-constants: distinguished semantics and completeness properties. Ann. Pure Appl. Logic **161**(2): Elsevier, 185–202 (2009)

Esteva, F., Godo, L., Noguera, C.: On completeness results for predicate Łukasiewicz, Product, Gödel and Nilpotent minimum logics expanded with truth constants. Mathware Soft Comput. XIV **3**, 233–246 (2007)

Gaines, B.R.: Foundations of fuzzy reasoning. In: Gupta, M.M., Saridis, G.N., Gaines, B.R. (Eds.) Fuzzy Automata and Decision Processes, pp. 19–75. Elsevier, North Holland Inc., New York (1977)

Gentzen, G.: Investigations into logical deductions. In: Szabo, M.E. (Ed.), Collected papers of G. Gentzen, pp. 68-131. North Holland Publications, Amsterdam (1969)

Gerla, G.: Fuzzy Logic : Mathematical Tools for Approximate Reasoning. Kluwer Academic Publishers (2001)

Gerla, G.: An extension principle for fuzzy logics. Math. Logic Quart. **40**, 357–380 (1994)

Gerla, G.: Graded consequence relations and closure operators. J. Appl. Non-class. Logic **6**, 369–379 (1996)

Giles, R.: Łukasiewicz logic and fuzzy set theory. Int. J. Man-Mach. Stud. **8**, 3–27 (1976)

Gottwald, S.: Fuzzy Sets and Fuzzy Logic. Vieweg, Wiesbaden (1993)

Gottwald, S.: An approach to handle partially sound rules of inference. In: Bouchon-Meunier, B., Yager, R.R., Zadeh, L.A. (Eds.) Advances in Intelligent Computing, IPMU'94, Selected papers, Lecture Notes Computer Sci., Vol. 945, etc., pp. 380–388. Springer: Berlin (1995)

Hájek, P.: Metamathematics of Fuzzy Logic. Kluwer Academic Publishers, Dordrecht (1998)

Klir, G.J., Yuan, B..: Fuzzy Sets And Fuzzy Logic: Theory and Applications. Prentice Hall of India Private Limited, New Delhi (2006)

Novak, V.: On syntactico-semantical completeness of first order fuzzy logic, parts I and II. Kybernetica **2**, 6(1, 2), 47–154 (1990)

Pavelka, J.: On fuzzy logic I, II, III Zeitscher for Math. Logik und Grundlagen d. Math. **25**, 45–52, 119–134, 447–464 (1979)

Rodríguez, R.O., Esteva, F., Garcia, P., Godo, L.: On implicative closure operators in approximate reasoning. Int. J. Approx. Reason. **33**, 159–184 (2003)

Surma, S.J.: The growth of logic out of the foundational research in mathematics. In: Agazzi, E. (ed.) Modern Logic—A Survey, pp. 15–33. D. Reidel Publishing Co., Dordrecht (1981)

Tarski, A.: Methodology of Deductive Sciences. Logic, Semantics, Metamathematics, pp. 60–109 (1956)

Chapter 6
Graded Consequence and Consequence in Different Approaches to Fuzzy Logics

Abstract This chapter presents a comparative analysis of the notion of graded consequence with other approaches to consequence in fuzzy logics. We shall first present different systems of fuzzy logics with respect to their proposed notions of consequence and then analyse in order to see how faithful those approaches are in incorporating many-valuedness in the notion of consequence.

6.1 Fuzzy Logic vis-á-vis Graded Consequence

What brings many-valued logic, fuzzy logic and theory of graded consequence (GCT) on a common platform is that each considers bivalence inadequate and embraces multivalence for interpreting well-formed formulae. Traditionally, different systems of many-valued logic admitted values other than truth and falsehood from different motivations. The second and the third decades of the last century saw a sprouting of systems of many-valued logic. However, treatment of vagueness was far from the original motive behind any of these many-valued systems (see Chap. 1). Generally, the set of truth values admitted in different many-valued logics is some subset of the real interval [0, 1]. For Zadeh's proposed system of fuzzy logic based on fuzzy set theory Zadeh (1965), the unit interval [0, 1] was a natural choice for interpreting vague sentences which occur either as premises or as conclusions of inferences in most cases of human reasoning. A further development of fuzzy logic in a more formal way was carried out in Esteva et al. (2009), Hájek (1998), Pavelka (1979), Novak (1990), Gottwald (2019). A general set-up for generating a system of logics with a notion of graded consequence, as has been discussed in previous chapters, is developed by Chakraborty et al. to associate a strength of derivability of a conclusion from a set of premises. Where the approach of GCT differs from other many-valued and fuzzy logical approaches is that it admits degrees of truth not only of predications at the object level, but of predications at the meta-level, and in principle at a level higher than that also.

Before taking up the issue proper, let us draw attention of the readers to the following aspect of logic—we mean classical logic. In the Hilbert type axiom system, the syntactic consequence relation $\vdash$ is defined in terms of a finite sequence of well-

M. K. Chakraborty and S. Dutta, *Theory of Graded Consequence*, Logic in Asia: Studia Logica Library, https://doi.org/10.1007/978-981-13-8896-5_6

formed formulae, called a derivation. $X \vdash \alpha$ holds if and only if there is a sequence $\alpha_1, \alpha_2, \ldots, \alpha_n$ of wffs such that $\alpha_n = \alpha$, and any α_i is either an axiom or is from the premise set X or is obtained by some rule of inference, usually MP.

Because of soundness theorem if the premises in X get the value 1 (true) in some interpretation the conclusion also gets the value 1. This is because of two factors; the axioms are always true and the rules are always valid. Since α has been derived from X, assumed to be true (1), by valid rules of inference the truth value transmitted in α via the derivation is 1. If the rule used had been invalid, just because the premises were true one could not claim that the conclusion would be true as well. Let us consider, for example, the following derivation (in the set-up of classical logic).

1.	$\alpha \supset \beta$	premise
2.	β	premise
3.	α	rule of inference 1 (an arbitrary inference rule)
4.	$\alpha \supset (\delta \supset \alpha)$	axiom
5.	$\delta \supset \alpha$	rule of inference 2 (MP)

Let in some interpretation both $\alpha \supset \beta$ and β get the value 1. Now just because of this, one cannot claim that $\delta \supset \alpha$ gets 1 as well. In fact, two rules have been used of which one is invalid. So, from this particular derivation that $\delta \supset \alpha$ is also 1 cannot be inferred, though it may in fact be 1 in the said interpretation. The (truth) value transmitted to the conclusion depends not only on the truth values of the premises but also on the validity/invalidity (or the strength) of the derivation process. In case of the above derivation, the value transmitted at the successive stages is as follows:

1.	1	
2.	1	
3.	$1 \times 1 \times 0$	[since the strength of the rule is 0]
4.	$1 \times 1 \times 0 \times 1$	[though step 4 is a tautology, the sequence $\alpha \supset \beta, \beta, \alpha, \alpha \supset (\delta \supset \alpha)$ cannot be considered as a derivation of $\alpha \supset (\delta \supset \alpha)$]
5.	$1 \times 1 \times 0 \times 1 \times 1$	[the value transmitted through the given sequence is 0]

To be noted that there are two values involved, one the step value (1 or 0) associated to each formula at each step n, and the value transmitted to the n-th wff because of considering the sequence 1, 2, ..., n of steps as a derivation of the n-th formula. It should be clear from step 4. The operation product ($\times$) has been taken to compute the meta-linguistic conjunction (which could as well be the operation minimum).

The strength of a rule is determined from the semantic considerations: the semantics is designed beforehand, axioms are chosen from within the tautologies, and the rules are chosen in such a way that truth-preserving property is satisfied. Though syntactically one is free to choose any set of wffs as axioms and any relation between a set of formulae and a formula as rule, to make a good/acceptable logic the above restrictions are to be observed. Once the validity of the rule is presumed, meaning thereby the strength of the rule is 1 in the classical scenario, the value transmitted to the inference by a derivation becomes solely dependent on the values of the premises, which can be calculated by some operation (e.g. product) on the values of the premises. We will see, in the case of rules of inference in fuzzy logic, the value of the conclusion has been calculated in terms of the value of the premises only. For example, if the two premises get values 'a' and 'b', the conclusion gets '$a.b$' (cf. Sect. 6.1.2). If a and b are both 1, the conclusion is 1 also. But we have seen before that it should not be so. If one applies a wrong rule, i.e. of strength 0, the value of the conclusion cannot be *determinately* claimed to be 1.

Existing approaches towards reasoning with imprecise concepts allow the object level to be many-valued, while maintaining meta-level statements to be of yes/no type. Motivations behind incorporating many-valuedness at the meta-logical notions have been described in the Introduction (cf. Chap. 1).

In this chapter, our aim would be to show that in many-valued and fuzzy logics, 'consequence' is either a crisp notion or it has been assigned a value which does not reflect the 'truth value' of the meta-linguistic sentence defining the concept of deduction. GCT makes a point of difference here.

The content of this chapter is distributed in the following way. In Sect. 6.1.1, we shall recall from Chap. 5 Pavelka's approach, the first generalization of Tarskian consequence operator in fuzzy context, and present the notion of proof as presented by Pavelka. Then in Sect. 6.1.2 we will briefly pass through the notion of inference rule as viewed by fuzzy logicians. Section 6.1.3 is meant to pinpoint the problem that lies in the notion of many-valued rule of inference, with special focus on Pavelka's approach. We shall throw light on the issue that though in fuzzy logics a notion of 'degree of consequence' is brought in, there are a number of difficulties in the formation of some meta-linguistic notions. These difficulties arise from the perspective of maintaining a distinction between 'use and mention' of a symbol in a logical discourse (Church 1956). In Sect. 6.2, we will present a rewriting of the theory of graded consequence following the principle of use and mention of a symbol. The last two sections, Sects. 6.3 and 6.4, are to revisit different approaches towards defining the notion of consequence in fuzzy logics and present how crispness is incorporated in disguise of many-valuedness of the proposed notions.

6.1.1 Notion of Proof in Pavelka's Fuzzy Logic

Pavelka's fuzzy consequence operator has been defined in Sect. 5.1.1. We here present the notion of proof.

Pavelka in (1979), introduced the idea of fuzzy syntax on an abstract set F incorporating two key concepts, namely fuzzy set of axioms $\mathscr{A}$ and fuzzy or many-valued rules of inference. A many-valued rule of inference r consists of two components $\langle r', r'' \rangle$ where the first (grammatical) component r' operates on formulae and the second (evaluation) component r'' operates on truth values; in Pavelka's own words Pavelka (1979) the second component prescribes how the '*...truth value of the conclusion is to be computed from the truth values of the premises*'. A fuzzy set X of F, over a complete lattice L, is said to be closed with respect to an n-ary rule $r = \langle r', r'' \rangle$ on F provided for all $(\alpha_1, \alpha_2, \ldots, \alpha_n)$ belonging to the domain of r', the grammatical component of r, $X(r'(\alpha_1, \alpha_2, \ldots, \alpha_n)) \geq r''(X(\alpha_1), X(\alpha_2), \ldots, X(\alpha_n))$ holds. One can notice that $X(r'(\alpha_1, \alpha_2, \ldots, \alpha_n))$ represents the belongingness degree of the derived formula $r'(\alpha_1, \alpha_2, \ldots, \alpha_n)$ to the fuzzy set X, and $r''(X(\alpha_1), X(\alpha_2), \ldots, X(\alpha_n))$ represents the truth value computed by the semantic counterpart r'' of r on the basis of belongingness degrees of the formulas in the premise of the rule r. Here we need to point out that the value which is attached to the conclusion is nothing but $r''(X(\alpha_1), X(\alpha_2), \ldots, X(\alpha_n))$. So, it is clear from Pavelka's own explanation that the value which is being attached to the concluding formula is not the value of the rule; rather it is the truth value of the conclusion as different instances of r may have different values.

The notion of proof, as introduced by Pavelka Pavelka (1979), is discussed below. We use Pavelka's definition of verbatim. Given a fuzzy set $\mathscr{A}$ of formulae, interpreted as axioms and a set $\mathscr{R}$ of rules of inference, an $\mathscr{R}$-proof is defined as a finite non-empty string $\omega = \langle \omega_1, \omega_2, \ldots, \omega_n \rangle$ over the set $F \cup (F \times \{0\}) \cup (F \times R \times N^+)$. Unlike classical proof or derivation here each ω_i does not represent the i-th formula used in the proof; rather the formula involved in the i-th step is denoted as $\ulcorner\omega_i$, and ω_i represents the justification by which the formula is regarded as one of the step of the proof. That is, for each ω_i $(i = 1, 2, \ldots, n)$ either ω_i is (x) (i.e. considered as a premise) or $(x, 0)$ (i.e. considered as an axiom) or $(x, r, \langle i_1, i_2, \ldots, i_n \rangle)$ (i.e. consdered as an application of the rule r), where $x = \ulcorner\omega_i$, the formula under consideration at the i-th term of ω.

If $\omega_i = (x, r, \langle i_1, i_2, \ldots, i_n \rangle)$, $(i = 1, 2, \ldots, n)$ then $x = r'\ (\ulcorner\omega_{i_1}, \ulcorner\omega_{i_2}, \ldots, \ulcorner\omega_{i_n})$ where r' is the grammatical component of the rule r, as mentioned above. For an $\mathscr{R}$-proof $\omega = \langle \omega_1, \omega_2, \ldots, \omega_n \rangle$ of $\ulcorner\omega_n$ from a fuzzy subset of formulae X, there is a corresponding function $\widehat{\omega} : L^F \mapsto L$ satisfying the following conditions.
(i) If length of ω is 1, then either $\omega = (x)$ or $(x, 0)$. Then accordingly $\widehat{\omega}(X) = Xx$, i.e. the membership degree of x in the fuzzy subset X when $\omega = (x)$ and $\mathscr{A}x$, i.e. the membership degree of x in the fuzzy set of axioms $\mathscr{A}$ when $\omega = (x, 0)$. (ii) If $\omega = \langle \omega_1, \omega_2, \ldots, \omega_n \rangle$ then

$$
\begin{aligned}
\widehat{\omega}(X) &= Xx && \text{if } \omega_n = (x) \\
&= \mathscr{A}x && \text{if } \omega_n = (x, 0) \\
&= r''(\widehat{\omega}_{i_1}(X), \widehat{\omega}_{i_2}(X), \ldots, \widehat{\omega}_{i_n}(X)) && \text{if } \omega_n = (x, r, \langle i_1, i_2, \ldots, i_n \rangle), i_1, i_2, \ldots, i_n < n
\end{aligned}
$$

So, it can be noticed that the value of $\langle \omega_1, \omega_2, \ldots, \omega_n \rangle$, a proof of $\ulcorner\omega_n$ from X, only takes care of the value of the last step of the derivation. This does not seem to be the value of the sentence '$\langle \omega_1 \rangle$ is a proof of $\ulcorner\omega_1$ from X and $\langle \omega_1, \omega_2 \rangle$ is a proof of $\ulcorner\omega_2$ from X and ...and $\langle \omega_1, \omega_2, \ldots, \omega_n \rangle$ is a proof of $\ulcorner\omega_n$ from X'.

Like any definition by recursion, in Pavelka's definition of proof also value of each step is computed with the help of the value of a segment of the proof preceding that particular step. But 'proof' as a whole should refer to the complete chain of steps and the same applies to the value of proof too.

6.1.2 The Expressed Meaning for Consequence: Goguen to Hájek

In designing a many-valued/fuzzy rule of inference, Pavelka clearly derives his motivation from Goguen (1968), in which the Modus Ponens rule is intended to capture the following:

"*If you know P is true at least to the degree a and $P \supset Q$ at least to the degree b then conclude that Q is true at least to the degree $a.b$*" ('.' is the multiplication operator in the unit interval [0, 1])

An immediate formalization of the above idea of fuzzy rule of inference is

$$\frac{(P, a) \quad (P \supset Q, b)}{(Q, a.b)}$$

That is, a rule of inference can be viewed as a subset of P($F \times$ [0, 1])$\times$ ($F \times$ [0, 1]). This is a crisp relation.

Pavelka demonstrated many-valued rule modus ponens as follows:

$$\left(\begin{array}{c|c} \mathrm{P} & \mathrm{a} \\ \mathrm{P} \supset \mathrm{Q} & \mathrm{b} \\ \hline \mathrm{Q} & \mathrm{a} * \mathrm{b} \end{array}\right).$$

This is a replica of the form proposed by Goguen. ['$*$' is the multiplication operator in a complete residuated lattice.]

The rule of inference r with components $\langle r', r'' \rangle$ is an abstraction of the intuitive idea of Goguen, further refined. Here r' is the rule MP and r'' is the product operation in Goguen's case whereas $*$ in Pavelka's case.

Following the same tradition Hájek proposed a fuzzy logic system Hájek (1998), called Rational Pavelka Logic (RPL). Identifying the pair (P, a) with the formula $\overline{a} \supset P$, the rule

$$\frac{(P, a) \quad (P \supset Q, b)}{(Q, a * b)}$$

can be obtained as a derived rule in RPL, where $\overline{a}$ is the wff denoting the truth value a. This again corroborates the idea, initiated by Goguen. We shall discuss RPL in Sect. 6.3.

Though Pavelka's work may be considered to be seminal in the area of fuzzy logic, there are several problematic philosophical issues connected with this approach.

First, can these rules of inference be called many-valued rules in the true sense of the term? What does the value $a * b$ represent? It is interpreted as the threshold below which the value of the conclusion Q cannot fall if the values of P and $P \supset Q$

be greater than or equal to a and b, respectively. So, it is related with the formula Q, not with the derivation.

Second, one might think that mathematically the crisp relation $r(\{(P, a), (P \supset Q, b)\}, (Q, a * b))$ may be presented as a fuzzy relation $r(\{(P, a), (P \supset Q, b)\}, Q)$ with the relatedness grade $a * b$. But doing so would lead to violation of the principle of use and mention (Church 1956), according to which

> *an object or symbol or word (finite string of symbols) cannot be both used and mentioned at the same level.* (Church 1956)

In order to do a proper scrutiny, we will follow Alonzo Church on the use–mention dichotomy in the following subsection.

Third, in Pavelka's logic whether the distinction of levels is carried out or is this distinction blurred? Could it be recast maintaining level distinction? Since there shall be two levels above the lowest one, the object level, two types of quotation have to be used. In Sect. 6.1.3, this tedious exercise will be performed in order to show that if one wants to view the rule of inference and/or the notion of proof, proposed in Pavelka (1979), as a many-valued/fuzzy rule then that will violate Church's principle of use and mention. In other words, the exercises of rewriting Pavelka's system in the framework of level distinction (cf. Sect. 6.1.3) is taken up to establish the claim that Pavelka's notion of proof or consequence is not a many-valued notion.

It is true that while building a theory sometimes one needs to compromise between the ease of presentation and being formally right. Despite all discussion pointing out what should make a presentation of logic formally right, we cannot but acknowledge that Pavelka's papers (1979) did a pioneering work to establish fuzzy logic within the domain of logic proper, and it has no mathematical ambiguity. But we notice that in the history of mathematics and logics, researchers have come across similar situation before. For example, the approach of the theory of types, systematized and developed in Principia Mathematica (Whitehead and Russell, 1910–13), is successful in eliminating the known paradoxes, but it is clumsy in practice and has certain other drawbacks as well. Our effort, in this context, is similar; we want to only pose the question to all the researchers whether we should or should not care about 'use and mention' of a symbol in building a logic.

6.1.3 Pavelka's Notion of Proof Reframed Maintaining Distinction of Levels of Logic

Let us recall from Introduction (Sect. 1.1) that usually three levels of languages are used in a logical discourse—object level, meta-level and metameta-level. In this section, the three levels will be distinctively written (and depicted) in the case of Pavelka logic. It is to be noted that there is a distinction between 'use' and 'mention' of a word. For example, of the sentences 'Man is a rational animal' and 'Man has three letters' in the first, the word 'Man' is used whereas in the second the word is mentioned. It is appropriate to write ' "Man" has three letters' in the second case.

In natural language most often sentences are not written with so much care; readers get the meaning from the context. Let us quote from Church in the context of logical languages.

> *Following the convenient and natural phraseology of Quine, we may distinguish between use and mention of a word or symbol. . . . As a precaution against univocation, we shall hereafter avoid the practice - which might otherwise sometimes be convenient - of borrowing formulas of the object language for use in the syntax language (or other meta-language) with the same meaning that they have in the object language* [Church (1956), pp. 61–63].

It might be mentioned that the practice, mentioned by Church, though convenient sometimes for human languages, is absolutely unusable in machine languages. To write an account of three layers of languages, we have to avoid univocation and have to follow the usual practice of putting a word within quotation marks or in different typeset when it is to be mentioned. Next, we quote below four more pieces from Church (1956) to lay bare the principles that would be followed in the sequel.

The quotations are scattered at different parts of the Introduction of the book; we put them in a sequential order keeping the present purpose in mind:

1. *The semantical rule must in the first instance be stated in a presupposed and therefore unformalized meta-language, here taken to be ordinary English. Subsequently, for their more exact study, we may formalize the meta-language (using a presupposed metameta-language and following the method already described for formalizing the object language) and restate the semantical rules in this formalized language (This leads to the subject of semantics) .... As a condition of rigour, we require that the proof of a theorem (of the object language) shall make no reference to or use of any interpretation, ...* [Church (1956), p. 55].
2. *The study of the purely formal part of a formalized language in abstraction from the interpretation, i.e. of the logistic system, is called ... logical syntax. The meta-language used in order to study the logistic system in this way is called the syntax language* [Church (1956), p. 58].
3. *... the reader must always understand that syntactical discussions are carried out in a syntax language whose formalization is ultimately contemplated, and distinctions based upon such formalization may be relevant to the discussion. In such informal development of syntax, we shall think of the syntax language as being a different language from the object language* [Church (1956), p. 59].
4. *After setting up the logistic system as described, we still do not have a formalized language until an interpretation is provided. This will require a more extensive meta-language than the restricted portion of English used in setting up the logistic system. However, it will proceed not by translations of the well-formed formulas into English phrases but rather by semantical rules which, in general, use rather than mention English phrases ... and which shall prescribe for every well-formed formula either how it denotes ... or else how it has values ...* [Church (1956), p. 54].

Following (1) let us make an attempt Dutta and Chakraborty (2014) to formalize the meta-language of Pavelka's system. As mentioned in (2) the meta-language basically consists of the *syntax language* and the *semantical meta-language*. The former is the language specifying the formation rules of the formulae and the formation rules of the derivation chains (proofs); the latter is the language describing the semantical rules. We have called this language as level-1 language (see Sect. 1.1). The language containing the object language, the *logical syntax* (cf. (3)), and the semantics of the object language (cf. (4)) together is called level-0 language (see Sect. 1.1).

We present the *Level-0* of Pavelka's system as follows:

(i) All formulae $\alpha, \beta, \gamma, \ldots$
(ii) All fuzzy sets of formulae $\mathscr{A}, X, Y, Z, \ldots, T_1, T_2, T_3, \ldots$.
($\mathscr{A}$ for axiom set, X, Y, Z, etc., for premises and T_i's for defining the semantics).
(iii) All finite sequences of formulae $\langle\alpha_1, \alpha_2, \ldots, \alpha_n\rangle, \langle\beta_1, \beta_2, \ldots, \beta_m\rangle, \ldots$.
(iv) All finite sequences over $F \times L$, i.e. $\langle(\alpha_1 a_1), (\alpha_2 a_2), \ldots, (\alpha_n a_n)\rangle, \ldots$.
(v) A particular set r' containing all sequences of formulae of the form $\langle\alpha, \alpha \supset \beta, \beta\rangle$, i.e. $r' = \{\langle\alpha, \alpha \supset \beta, \beta\rangle, \langle\gamma, \gamma \supset \delta, \delta\rangle, \ldots\}$.
(vi) All two length sequences over natural numbers, i.e. $\langle 1, 2\rangle, \langle 2, 3\rangle, \ldots$.
(vii) All elements $\omega_1, \ldots, \omega_n, \ldots$ where each ω_i is an entity of the following kinds:
$(\alpha), (\beta), (\gamma), \ldots$.
$(\alpha, 0), (\beta, 0), (\gamma, 0), \ldots$.
$(\alpha, r', \langle n_1, n_2\rangle), (\beta, r', \langle s_1, s_2\rangle), (\gamma, r', \langle m_1, m_2\rangle), \ldots$ where $n_1, n_2, s_1, s_2, m_1, m_2$
are finite natural numbers.
(viii) $(\mathrm{F}) = \{(\alpha), (\beta), (\gamma), \ldots\}$.
(ix) $(\mathrm{F}, 0) = \{(\alpha, 0), (\beta, 0), (\gamma, 0), \ldots\}$.
(x) $(\mathrm{F} \times r' \times N^+) = \{(\alpha, r', \langle n_1, n_2\rangle), (\beta, r', \langle s_1, s_2\rangle), (\gamma, r', \langle m_1, m_2\rangle), \ldots\}$.
(xi) All finite sequences of the form $\langle\omega_1, \omega_2, \ldots, \omega_n\rangle, \ldots$.
(xii) All finite sequences of the form $\langle(\omega_1\ a_1), (\omega_2\ a_2), \ldots, (\omega_n a_n)\rangle, \ldots$.
(xiii) All values $a_1, a_2, \ldots, a_n, \ldots$

The fuzzy set of formulae $\mathscr{A}$ is taken as a fuzzy set of axioms, and for convenience, we have considered only the presence of a rule r', which actually represents the grammatical component of the rule Modus Ponens, at the level-0 of Pavelka logic. Among all the above listed entities item (i) corresponds to the object language, and the rest belongs to the logical syntax and semantics of the object language.

The *Level-1* consists of the following:

- Names of each level-0 entity, i.e.
'α', 'β', 'γ', ..., '$\mathscr{A}$', 'X', 'Y', 'Z', ...,
'$\langle\alpha_1, \alpha_2, \ldots, \alpha_n\rangle$', '$\langle\beta_1, \beta_2, \ldots, \beta_m\rangle$', ..., '$\langle(\alpha_1\ a_1), (\alpha_2\ a_2), \ldots, (\alpha_n\ a_n)\rangle$', ..., '$r'$'
'ω_1', 'ω_2', ..., 'ω_n', ..., '(F)', '(F, 0)', '(F$\times r' \times N^+$)', '$\langle\omega_1, \omega_2, \ldots, \omega_n\rangle$', ...,
'$\langle(\omega_1 a_1), (\omega_2 a_2), \ldots, (\omega_n a_n)\rangle$', ...'$a_1$', '$a_2$', ..., '$a_n$', ...

These are the constant symbols of the level-1 language.

- Variables: There are three types of variables one of each kind.
  x (ranges over all formulae of level-0).
  T (ranges over T_1, T_2, T_3,...of level-0).
  $\langle(\,),(\,)\rangle$ (ranges over all 2-length sequences over $F \times L$).
- Function symbol: $\ulcorner$.
- Predicate symbol: $\in$, r (both binary).
- Propositional connectives: $\rightarrow_1$, $\rightarrow_2$, &, $\vee$
  (two implications, one conjunction and one disjunction).
- Quantifiers: $\forall$, $\exists$.
- Other symbols: (,), $\Big[\;,\;\Big]$.

Terms and well-formed formulae are constructed as follows:

- Terms: (1) Constant symbols and variables are terms.
  (2) $\ulcorner$'ω_1', $\ulcorner$'ω_2', $\ulcorner$'ω_3', ... are terms.
  (3)$\ulcorner$'$\langle$ (ω_1 a_1), (ω_2 a_2), ..., (ω_n a_n) $\rangle$', ... – expressions of this kind are also terms.
- Wffs: Following formulae represent each type of basic wff of the level-1 language.
  'α'$\in$ 'X', ..., 'α'$\in T$, ..., $x \in$ 'X', ... ,$x \in T$, ..., 'α' $\in$ '$\mathscr{A}$'.
  $\ulcorner$'ω_1' $\in$ 'X', ..., $\ulcorner$'ω_1' $\in \mathscr{A}$, 'ω_1' $\in$ '(F)', ..., 'ω_1'$\in$ '(F$\times\{0\}$)', ...,
  'ω_1'$\in$ '(F$\times r' \times N^+$)',..., r('$\langle(\alpha_1, a_1), (\alpha_2, a_2)\rangle$', '$\alpha_3$'),..., $r(\langle(\,),(\,)\rangle$, $\ulcorner$'ω_1') ....
- Following expressions are examples of some compound wffs formed by conjoining some of the above-mentioned basic wffs by connectives.
  - 'X' $\subseteq$ 'Y' $\equiv \forall\, x$ ($x \in$ 'X' $\rightarrow_2 x \in$ 'Y') [read as 'X' is included in 'Y'].
  - 'X' $\models$ 'α' $\equiv \forall\, T$ ('X' $\subseteq T \rightarrow_1$ 'α' $\in T$). [read as 'α' is a semantic consequence of 'X'].
  - Taut('α') $\equiv \forall T$ ('α' $\in T$) [read as 'α' is a tautology].

The notion of proof (in fact a derivation which Pavelka termed as proof) is to be defined by a level-1 sentence. Following Pavelka, we can write a sequence '$\langle\omega_1, \omega_2, \ldots, \omega_n\rangle$' as a proof of $\ulcorner$'ω_n' from a premise 'X' if and only if

$\Big[$ (if 'ω_1'is in '(F)' then $\ulcorner$'ω_1' is in 'X') and (if 'ω_1' is in '(F$\times\{0\}$)' then $\ulcorner$'ω_1' is an axiom) $\Big]$ and $\Big[$ (if 'ω_2' is in '(F)' then $\ulcorner$'ω_2' is in 'X') and (if 'ω_2' is in '(F$\times\{0\}$)' then $\ulcorner$'ω_2' is an axiom) $\Big]$ and $\Big[$ (if 'ω_3' is in '(F)' then $\ulcorner$'ω_3' is in 'X') and (if 'ω_3' is in '(F$\times\{0\}$)' then $\ulcorner$'ω_3' is an axiom) and (if 'ω_3' is in '(F$\times r' \times N^+$)' then $\ulcorner$'ω_3' is a result of a rule r) $\Big]$ and ... and $\Big[$ (if 'ω_n' is in '(F)' then $\ulcorner$'ω_n' is in 'X') and (if 'ω_n' is in '(F$\times\{0\}$)' then $\ulcorner$'ω_n' is an axiom) and (if 'ω_n' is in '(F$\times r' \times N^+$)' then $\ulcorner$'ω_n' is a result of a rule r) $\Big]$.

The above defining condition turns out to be the following wff of level-1:

$$\Big[(\text{`}\omega_1\text{'} \in \text{`}(F)\text{'} \rightarrow_1 \ulcorner\text{`}\omega_1\text{'} \in \text{`}X\text{'}) \,\&\, (\text{`}\omega_1\text{'} \in \text{`}(F \times \{0\})\text{'} \rightarrow_1 \ulcorner\text{`}\omega_1\text{'} \in \mathscr{A})\Big] \,\&$$

$$\Big[(\text{`}\omega_2\text{'} \in \text{`}(F)\text{'} \rightarrow_1 \ulcorner\text{`}\omega_2\text{'} \in \text{`}X\text{'}) \,\&\, (\text{`}\omega_2\text{'} \in \text{`}(F \times \{0\})\text{'} \rightarrow_1 \ulcorner\text{`}\omega_2\text{'} \in \mathscr{A})\Big] \,\&$$

$$\Big[(\text{`}\omega_3\text{'} \in \text{`}(F)\text{'} \rightarrow_1 \ulcorner\text{`}\omega_3\text{'} \in \text{`}X\text{'}) \,\&\, (\text{`}\omega_3\text{'} \in \text{`}(F \times \{0\})\text{'} \rightarrow_1 \ulcorner\text{`}\omega_3\text{'} \in \mathscr{A}) \,\&$$

$$(\text{`}\omega_3\text{'} \in \text{'}(F \times r' \times N^+)\text{'} \rightarrow_1 Res_r(\ulcorner\text{`}\omega_3\text{'}))\Big] \,\& \ldots \&$$

$$\Big[(\text{`}\omega_n\text{'} \in \text{`}(F)\text{'} \rightarrow_1 \ulcorner\text{`}\omega_n\text{'} \in \text{`}X\text{'}) \,\&\, (\text{`}\omega_n\text{'} \in \text{`}(F \times \{0\})\text{'} \rightarrow_1 \ulcorner\text{`}\omega_n\text{'} \in \mathscr{A})$$

$$\& \, (\text{`}\omega_n\text{'} \in \text{'}(F \times r' \times N^+)\text{'} \rightarrow_1 Res_r(\ulcorner\text{`}\omega_n\text{'}))\Big]. \qquad \ldots(1)$$

$Res_r\ulcorner$'ω_i' stands for $\ulcorner$'ω_i' is a result of some rule r, $(2 < i \leq n)$, and it is defined formally below. First of all the value of the sentence (1) is to be calculated by considering values of all its conjuncts. But Pavelka has taken only the last component, i.e. the sentence under the last $\Big[\;\Big]$ of (1), and the value of it has been tagged with $\ulcorner$'ω_n', claiming it as the *value of* '$\langle \omega_1, \omega_2, \ldots, \omega_n \rangle$' *is a proof of* $\ulcorner$'ω_n' from 'X'.

Now the *difficulty concerning the principle of use and mention* is as follows. We take the value taken by Pavelka, viz. the value of the last conjunct in (1):

(i) Let us concentrate on $Res_r(\ulcorner\text{`}\omega_i\text{'}).Res_r(\ulcorner\text{`}\omega_i\text{'})$ is actually the abbreviation of $\exists\langle (), ()\rangle\Big(\langle (), ()\rangle \subseteq \ulcorner\text{`}\langle(\omega_1\underline{a_1}), (\omega_2\underline{a_2}), \ldots, (\omega_{i-1}\underline{a_{i-1}})\rangle \& r(\langle (), ()\rangle, \text{`}\omega_i^{\prime}\text{'})\Big)$, which can be read as, there is a subsequence $\langle ( ), ( )\rangle$ such that $\langle ( ), ( )\rangle$ is included in $\ulcorner$'$\langle (\omega_1\, \underline{a_1}), (\omega_2\, \underline{a_2}), \ldots, (\omega_{i-1}\, \underline{a_{i-1}})\rangle$' and $\langle ( ), ( )\rangle$ is related with $\ulcorner$'ω_i' by the rule r.

Here $\underline{a_k}$, $k = 1, 2, \ldots, i-1$, refers to the truth value of the level-1 sentence $\Big[(\text{`}\omega_k\text{'} \in \text{`}(F)\text{'} \rightarrow_1 \ulcorner\text{`}\omega_k\text{'} \in \text{`}X\text{'}) \,\&\, (\text{`}\omega_k\text{'} \in \text{`}(F \times \{0\})\text{'} \rightarrow_1 \ulcorner\text{`}\omega_k\text{'} \in \mathscr{A}) \,\& (\text{`}\omega_k\text{'} \in \text{`}(F \times r' \times N^+)\text{'} \rightarrow_1 Res_r(\ulcorner\text{`}\omega_k\text{'}))\Big]$ that is the last conjuncts of '$\langle \omega_1, \omega_2, \ldots, \omega_k\rangle$' is a proof of $\ulcorner$'ω_k' from a premise 'X'.

(ii) Now, we arrive at the most crucial formal difficulty. The attention of readers is being drawn to the point that $\underline{a_k}$ is the name of the value of the following level-1 sentence $\Big[(\text{`}\omega_k\text{'} \in \text{`}(F)\text{'} \rightarrow_1 \ulcorner\text{`}\omega_k\text{'} \in \text{`}X\text{'}) \,\&\, (\text{`}\omega_k\text{'} \in \text{`}(F \times \{0\})\text{'} \rightarrow_1 \ulcorner\text{`}\omega_k\text{'} \in \mathscr{A}) \,\& (\text{`}\omega_k\text{'} \in \text{`}(F \times r' \times N^+)\text{'} \rightarrow_1 Res_r(\ulcorner\text{`}\omega_k\text{'}))\Big]$, which is being used at the same level. The name of the value of a level-1 sentence should be an entity of level-2, but here that has been used at level-1. But as a name, it is also mentioned at level-1.

Thus, we see that the notion of proof, as defined in Pavelka, does not take care of the prescription, viz. '*As a condition of rigour, we require that the proof of a theorem (of the object language) shall make no reference to or use of any interpretation*', mentioned in quotation (1). So, the main question is about the well-formedness of the wff $Res_r(\ulcorner`\omega_i')$, and hence the notion of proof, as presented in the above analysis.

In Sect. 6.2, we will see that the above philosophical difficulty does not arise while considering similar treatment for GCT, and to make this difference more visible in Sect. 6.3, we shall present an Example 6.2.

6.2 Rewriting the Theory of Graded Consequence Maintaining Level Distinction

Below we present the theory of graded consequence with all its distinguished levels of logic activities.
Level-0 consists of :

- Formulae α, β, γ ...with or without superscripts and subscripts.
- All ordinary sets of formulae A_X, X, Y, Z,...
- A collection of fuzzy sets of formulae $\widetilde{A}_X, T_1, T_2, T_3$,...
- All finite sequences of formulae $\langle\alpha_1, \alpha_2 ..., \alpha_n\rangle, \langle\beta_1, \beta_2 ..., \beta_m\rangle, \ldots$
- Some special sets of finite sequences of formulae of the same form, called the syntactic components of the rules of inference,

$S_i^1 = \{\langle\alpha_1^1, \alpha_2^1, \ldots \alpha_i^1, \alpha^1\rangle, \langle\beta_1^1, \beta_2^1, \ldots \beta_i^1, \beta^1\rangle, \ldots\}$

$\vdots$

$S_m^j = \{\langle\alpha_1^j, \alpha_2^j, \ldots \alpha_m^j, \alpha^j\rangle, \langle\beta_1^j, \beta_2^j, \ldots \beta_m^j, \beta^j\rangle, \ldots\}$.

Note 6.1 1. A_X is a special set of formulae consisting of the support of the special fuzzy set $\widetilde{A}_X$, called logical axioms.
2. T_1, T_2, T_3, ...are fuzzy subsets of formulae, identified with the truth valuation functions such that truth value of a formula under a valuation function is exactly the same as the membership degree of that formula to the respective fuzzy subset. These are the relevant states of affair.
3. $\widetilde{A}_X$ is a special fuzzy subset of formulae, identified with $A_X \cap (\cap_i T_i)$, i.e. $\widetilde{A}_X(\alpha)$ $= \cap_i T_i(\alpha)$ if α is a member of A_X. $= 0$ otherwise.
4. Each of $S_i^1, \ldots S_m^j$ contains finite sequences of formulae of the same form. What we intend to mean by the phrase 'finite sequences of formulae of the same form' is enunciated by an example: If we consider S_2^1 as the rule Modus Ponens then S_2^1 would be the set $\{\langle\alpha, \alpha \supset \beta, \beta\rangle, \langle\gamma, \gamma \supset \delta, \delta\rangle \ldots\}$, i.e. all instances of the rule. The subscript 2 denotes the arity of the rule, and the superscript denotes the enumeration number of the rule with arity 2.

Level-1 consists of:

- Names of some of the level-0 entities, viz.
  'α', 'β', 'γ' ...'X', 'Y', 'Z',..., '$\widetilde{A}_X$','$\langle\alpha_1, \alpha_2 ..., \alpha_n\rangle$', '$\langle\alpha_1, \alpha_2 ..., \alpha_m\rangle$', ..., '$S_i^1$', ...'$S_m^j$'.
  All these are constant symbols of level-1 language whose interpretations are the respective level-0 objects.
- Variables

  1. The formula variable x; $x_1^1, x_2^1, ..., x_i^1, x^1; ...; x_1^j, x_2^j, ..., x_m^j, x^j$ (range over all formulae of level-0).
  2. The crisp set variable V (ranges over all ordinary sets of formulae of level-0).
  3. The fuzzy set variable T (ranges over T_1, T_2, T_3, ...of level-0).
  4. The sequence variable $\langle\ \ \rangle_{SV}$ (ranges over all finite sequences of formulae of level-0).
  5. $\langle x_1^1, x_2^1, \ldots, x_i^1, x^1\rangle$ (ranges over S_i^1).

     ⋮

     $\langle x_1^j, x_2^j, \ldots, x_m^j, x^j\rangle$ (ranges over S_m^j).

Note 6.2 1. $\langle x_1^1, x_2^1, \ldots, x_i^1, x^1\rangle$, ...$\langle x_1^j, x_2^j, \ldots, x_m^j, x^j\rangle$ are used as a kind of sequence variables which range over some particular types of sequence of formulae, and are formed by the formula variables $x_1^1, x_2^1, ..., x_i^1, x^1; ...; x_1^j, x_2^j, ..., x_m^j, x^j$.

2. This is, hence a many-sorted language, variables ranging over different sets. It is not required to take infinitely many variables of any sort. Depending on different rules of inference, different sets of finitely many formula variables are needed to form a sequence of variables ranging over each of such rules.

- Function Symbols: dom, ran.
- Binary Predicate Symbols: $\in$, R, $\vdash$
- Propositional Connectives: $\rightarrow$, &, $\vee$.
- Quantifiers: $\forall$, $\exists$.
- Other Symbols: (,).
- Terms: Constant symbols and variables are terms. Also, the expressions of the following kinds are considered as terms.
  dom($\langle x_1^1, x_2^1, \ldots, x_i^1, x^1\rangle$), ..., dom($\langle x_1^j, x_2^j, \ldots, x_m^j, x^j\rangle$).
  ran($\langle x_1^1, x_2^1, \ldots, x_i^1, x^1\rangle$), ..., ran($\langle x_1^j, x_2^j, \ldots, x_m^j, x^j\rangle$).
  [dom($\langle x_1^1, x_2^1, \ldots, x_i^1, x^1\rangle$) denotes the set consisting of the subsequences formed by the first i component of each sequence of S_i^1, and ran($\langle x_1^1, x_2^1, \ldots, x_i^1, x^1\rangle$) denotes the set consisting of those formulae which form the last component of each sequence present in S_i^1 as it is already mentioned that the sequence variable $\langle x_1^1, x_2^1, \ldots, x_i^1, x^1\rangle$ ranges over the set S_i^1].

- Some basic wffs involving the predicate $\in$:
  'α' $\in$ 'X', ..., 'α'$\in T$, ..., $x \in$ 'X', ..., $x \in T$, ...,'α'$\in$ '$\widetilde{A}_X$', ...$x \in \langle\ \ \rangle_{SV}$, ... $x \in$ '$\langle\alpha_1, \alpha_2, \ldots, \alpha_n\rangle$', ..., $x \in \mathrm{dom}(\langle x_1^1, x_2^1, \ldots, x_i^1, x^1\rangle)$, ..., $\mathrm{ran}(\langle x_1^1, x_2^1, \ldots, x_i^1, x^1\rangle) \in T$ ....
- A few examples of compound wffs involving the predicate $\in$:

1. 'X' $\subseteq$ 'Y' $\equiv \forall x(x \in$ 'X' $\rightarrow x \in$ 'Y')
[$'X' \subseteq$ 'Y' (read as 'X' is included in 'Y') is an abbreviation of the level-1 formula $\forall\, x\ (x \in$ 'X' $\rightarrow x \in$ 'Y') which means that for all x if x is a member of 'X' then x is a member of 'Y'].
2. 'X' $\subseteq T \equiv \forall x(x \in$ 'X' $\rightarrow x \in T)$.
3. $\mathrm{dom}(\langle x_1^1, x_2^1, \ldots, x_i^1, x^1\rangle) \subseteq T \equiv \forall x(x \in \mathrm{dom}(\langle x_1^1, x_2^1, \ldots, x_i^1, x^1\rangle) \rightarrow x \in T)$.
4. $\langle\ \ \rangle_{SV} \subseteq$ '$\langle\alpha_1, \alpha_2, \ldots, \alpha_n\rangle$'$\equiv \forall x(x \in \langle\ \ \rangle_{SV} rightarrow x \in$ '$\langle\alpha_1, \alpha_2, \ldots, \alpha_n\rangle$').
5. 'X' $\models$ 'α' $\equiv \forall T($'X' $\subseteq T \rightarrow$ 'α' $\in)$.
   ['X' $\models$ 'α' (read as 'α' is a semantic consequence of 'X') is an abbreviation of the level-1 formula $\forall T($'X' $\subseteq T \rightarrow$ 'α' $\in T)$ which means that for all T, if 'X' is included in T then 'α' is a member of T]
6. Taut ('α') $\equiv \forall T($'α' $\in T)$.
   [Taut ('α') (read as 'α' is a tautology) is an abbreviation of the level-1 formula $\forall T($'$\alpha' \in T)$ which means that for all T, 'α' is a member of T]
7. Cons ('X') $\equiv \exists T($'X'$\subseteq T)$.
   [Cons ('X') (read as 'X' is consistent) is an abbreviation of the level-1 formula $\exists T($'X' $\subseteq T)$ which means that there is T such that 'X' is included in T].

- Some examples of basic wffs with the predicate $\vdash$:
  'X' $\vdash$ 'α', 'X' $\vdash x$.
- Compound wffs involving $\vdash$:
  Incons('X') $\equiv \forall\, x$ ('X' $\vdash x$).

Before introducing wffs, involving R, let us see, how rules of inferences are introduced in classical situation. Let us consider the classical rule, 'from α and $\alpha \supset \beta$ infer β'. (α, $\alpha \supset \beta$, β are object-level formulae and $\supset$ is object-level connective 'if-then'.) This corresponds to the syntactic part of the rule which is an instruction of how to act having some special form of formulae at hand. But to realize the justification of the instruction, one appeals to its meaning, i.e. in the present case, the justification of the above instruction is 'whenever α is true and $\alpha \supset \beta$ is true β is also true'. So in classical case, corresponding to a rule we need two meta-level sentences one pertaining to the syntactic part, which is an ordinary relation, and the other pertaining to the semantic part. In the theory of graded consequence, rules of inference are introduced to be fuzzy relations. Now to justify these fuzzy rules of inference, we need following level-1 formulae:

$$R_i^1 \equiv \forall \langle x_1^1, x_2^1, \ldots, x_i^1, x^1\rangle(\forall T(\mathrm{dom}(\langle x_1^1, x_2^1, \ldots, x_i^1, x^1\rangle) \subseteq T \rightarrow \mathrm{ran}(\langle x_1^1, x_2^1, \ldots, x_i^1, x^1\rangle) \in T).$$

$$\vdots$$

$R_m^j \equiv \forall \langle x_1^j, x_2^j, \ldots, x_m^j, x^j\rangle(\forall T(\text{dom}(\langle x_1^j, x_2^j, \ldots, x_m^j, x^j\rangle) \subseteq T \rightarrow \text{ran}(\langle x_1^j, x_2^j, \ldots, x_m^j, x^j\rangle) \in T)$.

These are to take care of the semantic part of a rule. For instance, R_i^1 represents the level-1 formula that *if for all interpretation* T, $dom(\langle x_1^1, x_2^1, \ldots, x_i^1, x^1\rangle)$ *is true under* T, *then* $ran(\langle x_1^1, x_2^1, \ldots, x_i^1, x^1\rangle)$ *is also true under* T, and R_i^1 is a many-valued sentence basically representing the degree of validity of the rule, i.e. each instance of sequences from S_i^1. The syntactic counterpart is given by the following wffs.

– Atomic wffs involving the predicate R:
$R(\langle\ \rangle_{SV}, `\alpha')$, $R(`\langle\alpha_1, \alpha_2, \ldots, \alpha_n\rangle', `\alpha')$.
[$R(`\langle\alpha_1, \alpha_2, \ldots, \alpha_n\rangle', `\alpha')$ is interpreted as 'α' is related to '$\langle\alpha_1, \alpha_2, \ldots, \alpha_n\rangle$' by a rule of inference.]
In classical logic, let us take as an instance rule Modus Ponens, viz. *from* α *and* $\alpha \supset \beta$ *infer* β. This syntactic part of the rule is considered to be valid because of the semantic counterpart, i.e. *whenever* α *is true and* $\alpha \supset \beta$ *is true* β *is true* gets the truth-value *true*. This principle may be accepted for any other rule. In accordance with the above understanding R ('$\langle\alpha_1, \alpha_2, \ldots, \alpha_n\rangle$', '$\alpha$') is interpreted by a fuzzy relation which assigns $|\ R_i^j\ |$, the truth value of the level-1 formula R_i^j to the level-1 formula R ('$\langle\alpha_1, \alpha_2, \ldots, \alpha_n\rangle$', '$\alpha$') if ('$\langle\alpha_1, \alpha_2, \ldots, \alpha_n\rangle$', '$\alpha$') $\in S_i^j$ and 0 otherwise.

– Compound wffs involving the predicate R:

1. $Res_R(`\alpha_i') \equiv \exists\langle\ \rangle_{SV}\ (\langle\ \rangle_{SV} \subseteq `\langle\alpha_1, \alpha_2, \ldots, \alpha_{i-1}\rangle'\ \&\ R\ (\langle\ \rangle_{SV}, `\alpha_i'))$.
[$Res_R(`\alpha_i')$ is read as 'α_i' is a result of a rule of inference applied on a subset of '$\langle\alpha_1, \alpha_2, \ldots, \alpha_{i-1}\rangle$'. The idea behind the right-hand side formula is, there is some sequence included in the sequence of formulae '$\langle\alpha_1, \alpha_2, \ldots, \alpha_{i-1}\rangle$' such that '$\alpha_i$' is related to that sequence by a rule of inference.]
2. '$\langle\alpha_1, \alpha_2, \ldots, \alpha_n\rangle$'$D$('$X$', '$\alpha_n$') $\equiv$ ('α_1' $\in$ 'X' $\vee$ 'α_1' $\in$ '$\widetilde{A}_X$') & ('α_2' $\in$ 'X' $\vee$ 'α_2' $\in$ '$\widetilde{A}_X$' $\vee$ Res_R('α_2')) & ... & ('α_n' $\in$ 'X' $\vee$ 'α_n' $\in$ '$\widetilde{A}_X$' $\vee$ Res_R('α_n')).
['$\langle\alpha_1, \alpha_2, \ldots, \alpha_n\rangle$'$D$('$X$', '$\alpha_n$') is read as '$\langle\alpha_1, \alpha_2, \ldots, \alpha_n\rangle$' is a derivation '$\alpha_n$' from '$X$'. The intended meaning of the formula is (either 'α_1' is a member of the premise 'X' or 'α_1' is an axiom) and (either 'α_2' is a member of 'X' or 'α_2' is an axiom or 'α_2' is a result of a rule of inference from a subset of '$\langle\ \alpha_1\ \rangle$') and ... and (either '$\alpha_n$' is a member of 'X' or 'α_n' is an axiom or 'α_n' is a result of a rule of inference from a subset of '$\langle\ \alpha_1, \alpha_2, \ldots, \alpha_{n-1}\ \rangle$').]
3. 'X' $\vdash_{AX}$ 'α' $\equiv \exists\langle\ \rangle_{SV}\ (\langle\ \rangle_{SV}\ D$ ('X', 'α')).
['X' $\vdash_{AX}$ 'α'(read as 'α' is a syntactic consequence of 'X') is an abbreviation of the level-1 formula $\exists\langle\ \rangle_{SV}\ (\langle\ \rangle_{SV}\ D$ ('X', 'α')) which means that there is a sequence of formulae which is a derivation of 'α' from 'X'].

Interpretation of the level-1 language:

Truth set-structure: Lattice meet being the simplest monoidal composition among all other possible varieties over a lattice structure, here for simplicity the value set for the level-1 language is chosen to be a complete pseudo-Boolean algebra, i.e. Heyting algebra, $(L, \wedge, \vee, \rightarrow_m, 0, 1)$, which is a particular kind of complete residuated lattice. Operators $\wedge$ and $\rightarrow_m$, that form an adjoint pair, are used to compute the level-1 connectives & and $\rightarrow$, respectively. The lattice join is used to compute the disjunction ($\vee$) present in the level-1 language, and the operators 'inf'and 'sup' are used to take care of the quantifiers '$\forall$' and '$\exists$', respectively.

Determination of the truth values of the basic wffs:

1. Formulae of the type 'α' $\in$ 'X', would get the value 1 if interpretation of 'α', i.e. α is a member of the set referred to by 'X' and 0 otherwise.
2. Let us consider the wffs of the type 'α'$\in T$. As the wff contains a variable, viz. T this wff does not have a truth value as such but after being bounded in presence of quantifiers, i.e. $\exists$('α' $\in T$) and $\forall$('α' $\in T$), the values are $sup_i T_i(\alpha)$ and $inf_i T_i(\alpha)$, respectively.
3. Similar interpretation is applicable to the wffs of the type $x \in$ 'X', $x \in T$.
4. The sentences of the type 'X' $\vdash$ 'α' would be interpreted by a fuzzy relation between the set of sets of formulae and the set of formulae. In fact, there can be many such predicate symbols $\vdash_1, \vdash_2, \ldots$, at level-1, and their respective interpretations, i.e. several fuzzy relations at the corresponding algebraic structure of level-1.
5. Value of the wffs of the type R ('$\langle \alpha_1, \alpha_2, \ldots, \alpha_n \rangle$', '$\alpha$') would be determined by the fuzzy relation, defined earlier in this section. That is, the value of the level-1 formula R ('$\langle \alpha_1, \alpha_2, \ldots, \alpha_n \rangle$', '$\alpha$') is $\mid R_i^j \mid$, if ('$\langle \alpha_1, \alpha_2, \ldots, \alpha_n \rangle$', '$\alpha$') $\in S_i^j$ and 0 otherwise.

So after the determination of the truth values of the basic wffs it is quite immediate to see how the compound wffs, some of which are our well-known meta-logical concepts, are being offered a many-valued reading.

Example 6.1 To elucidate, let us consider the sentence 'X' $\models$ 'α' which stands for $\forall\, T$ ('X' $\subseteq T \rightarrow$ 'α' $\in T$), i.e. $\forall\, T$ ($\forall x$ ($x \in$ 'X' $\rightarrow x \in T$)$\rightarrow$ 'α' $\in T$). Given a collection of fuzzy subsets $\{T_i\}_{i \in I}$, T will range over the collection, and '$\forall$' and '$\rightarrow$' will be computed by 'inf' and '$\rightarrow_m$', respectively. Hence L being a set consisting of truth values other than true (1) and false (0), the truth value of the sentence would be $inf_i\ (inf_{x \in X}(1 \rightarrow_m T(x))) \rightarrow_m T(\alpha)) = inf_i\ (inf_{x \in X}\ T(x) \rightarrow_m T(\alpha))$. This may be a value other than 0 and 1.

Note 6.3 One point we would like to mention here. For simplicity, supported by the work of Shoesmith and Smiley, at level-0 we have taken one collection of fuzzy subsets, viz. $T_1, T_2, T_3, \ldots$ instead of all fuzzy subsets over formulae, as is done in the classical fuzzy logics. But it is possible to incorporate several sets of fuzzy subsets of formulae at level-0. In that case, we would take several collections of fuzzy subsets of formulae, i.e. $\{T_i\}_{i \in I}$, $\{T_j\}_{j \in J}$, $\ldots$ at level-0 and define the notion of semantic consequence, consistency, tautologihood, etc., with respect to each of the collections of fuzzy subsets of formulae.

Level-2 consists of the following.
Alphabet: Constant symbols: 'w', (where w is a wff of level-1)
'$\vdash$', '$\models$', '$\vdash_{AX}$' (names of $\vdash$, $\models$, and $\vdash_{AX}$, respectively)
Binary predicate Symbol: $\gg$ (a symbol for metameta-level *implies*)

Note 6.4 It is to be noted that $\models$ and $\vdash_{AX}$ are defined symbols at level-1. We need to give names to these symbols too to talk about them at level-2. Analogy may be taken from classical logic. With $\neg$ and $\supset$ as primitive, $\wedge$ and $\vee$ are defined, and then we talk about conjunction ('$\wedge$') and disjunction ('$\vee$').

Terms: Constant symbols are terms.
Basic wffs: $t_1 \gg t_2$, where t_1, t_2 are names of level-1 formulae.

Interpretation of level-2 language:

1. Interpretation of each constant symbol of level-2 language is the corresponding level-1 entity. That is, the interpretation of 'w' is the level-1 formula w, and that of '$\vdash$', '$\models$', '$\vdash_{AX}$' are $\vdash$, $\models$, $\vdash_{AX}$, respectively.
2. Interpretation of the predicate symbol $\gg$ is such that the wff $t_1 \gg t_2$ gets the value 1 (true) if the value of the level-1 formula referred to by t_1 is less or equal to the value of the level-1 formula referred to by t_2 and 0 otherwise.

Now in the meta-language of level-1 language, i.e. at level-2 we talk about the behaviour of the notion of consequence, a level-1 concept, by the following description.

Definition: A fuzzy relation '$\vdash$' is a graded consequence relation if '$\vdash$' satisfies,
' "α" $\in$ "X" ' $\gg$ ' "X" $\vdash$ "α" ',
' "X" $\subseteq$ "Y" & "X" $\vdash$ "α" ' $\gg$ ' "Y" $\vdash$ "α" ', and
' $\forall x$ ($x \in$ "Z" $\rightarrow$ "X" $\vdash x$) & "$X \cup Z$" $\vdash$ "α" ' $\gg$ ' "X" $\vdash$ "α" '
for all wffs α and sets X, Y, Z of wffs.

This definition constitutes the formal representation of the Gentzen type generalization of consequence relation presented in Definition 2.2. Also, it is to be noted that the expressions under the quote are meta-level (or level-1) sentences.

Note 6.5 1. As we mentioned earlier in Note 6.3, if we have various sets of fuzzy sets of formulae at level-0 as sets of valuations like $\{T_i\}_{i\in I}$, $\{T_j\}_{j\in J}$, ..., then at level-1 there would be several notions of semantic consequence like $\models_{`\{T_i\}'_{i\in I}}$, $\models_{`\{T_j\}'_{j\in J}}$, ... Now to talk about all these notions of semantic consequence we need to have the name of each such notions, i.e. '$\models_{``\{T_i\}''_{i\in I}}$', '$\models_{``\{T_j\}''_{j\in J}}$', ... and a variable $\models_{\{T_v\}_{v\in V}}$ to range over all such $\models_{`\{T_i\}'_{i\in I}}$, $\models_{`\{T_j\}'_{j\in J}}$, ... at level-2. With this language background of level-2, we are able to write down the following level-2 assertions:

Representation Theorem:
(i) For any $\models_{\{T_v\}_{v\in V}}$, $\models_{\{T_v\}_{v\in V}}$ is a graded consequence relation.
(ii) If '$\vdash$' is a graded consequence relation, then there exists $\models_{\{T_v\}_{v\in V}}$ such that $\models_{\{T_v\}_{v\in V}}$ coincides with '$\vdash$'.

These theorems are presented in Theorem 2.1. Here are their formal presentations only in the current framework.
2. If we admit to have many-valuedness for the notions of soundness and completeness too, then we can accommodate sentences like Sound (‘ $\vdash$’, ‘ $\models_{\text{“}\{T_i\}''_{i\in I}}$)’ and Complete (‘ $\vdash$’, ‘ $\models_{\text{“}\{T_i\}''_{i\in I}}$’) by introducing Sound and Complete as two basic predicate symbols of level-2 and associate the values of the level-1 well-formed formulas $\forall X \forall x((X \vdash x) \to (X \models_{\text{‘}\{T_i\}'_{i\in I}} x))$ and $\forall X \forall x((X \models_{\text{‘}\{T_i\}'_{i\in I}} x) \to (X \vdash x))$, respectively, to the sentences.

6.3 A Brief Revisit to Hájek's Logic RPL From the Perspective of Distinction of Levels

In Sect. 6.1.3, we have seen the formal difficulties in writing the notion of proof due to Pavelka in a framework of distinguished languages for different levels of logic activities. Apart from the notion of proof, there are a few more points to be mentioned. We shall come back to this issue; before that let us first have a brief look at the Hájek's system of Rational Pavelka Logic (RPL).

As found in Hájek (1998), the ‘provability degree of α from X’, is defined as follows. Provability degree of α from $X = sup \ \{a : X \vdash (\alpha, a)\ \}$.
Hájek presented RPL as an extension of Łukasiewicz logic. Language of RPL is extended by the names of the rational truth values in [0, 1] and then two axioms, viz.
$\overline{a *_Ł b} \equiv \overline{a} \&_Ł \ \overline{b}$, $\overline{a \Rightarrow_Ł b} \equiv \overline{a} \supset_Ł \overline{b}$ (where $*_Ł$ and $\Rightarrow_Ł$are Łukasiewicz's t-norm and its corresponding residuum) are added with the Łukasiewicz logic axioms. Introducing names of rational truth constants Hájek used (α, a) as an abbreviation of the formula $\overline{a} \supset_Ł \alpha$ and then with above-mentioned set of axioms and rule modus ponens Hájek proved

$$\frac{\begin{array}{l}\overline{a} \supset_Ł \alpha \\ \overline{b} \supset_Ł (\alpha \supset_Ł \beta)\end{array}}{\overline{a *_Ł b} \supset_Ł \beta}$$

as a derived rule of RPL. This is the same as Pavelka's many-valued rule MP. Now in RPL, $X \vdash (\alpha, a)$, i.e. $X \vdash \overline{a} \supset_Ł \alpha$ is a wff of level-1. But as discussed earlier, from a crisp set X of formulae, following a crisp set of axioms and crisp rules of inferences a formula of the form (α, a) can be derived or cannot be derived, i.e. the notion of derivation is two-valued. The definition of ‘provability degree of α from X’ suggests to compute the supremum of all those a for which (α, a) is derived from X. This leads to a number of problems.

1. ‘(α, a) is derived from X’ is a wff of level-1. To compute the ‘provability degree of α from X’ one needs to extract out those values a for which (α, a), i.e. $\overline{a} \supset_Ł \alpha$ is derivable from X, and put them together for computing the supremum. But the question is, where at which level this process of extracting out of the value a is taking place? Provability is a notion of level-1. So, keeping that in mind,

if we assume that extracting out a from $X \vdash \overline{a} \supset_Ł \alpha$ is a level-1 activity then dequoting $\overline{a}$ means *using* the value a which is *referred to* by $\overline{a}$ at the same level, takes place. So, computing '*sup*' over all those a for which $X \vdash \overline{a} \supset_Ł \alpha$ holds is an external device which cannot be placed at any level.

2. It should be natural to think that the provability degree of a formula α from a set of formulae X would be the truth value of the statement 'α is provable from X'. The value should not be a mere number, counted by some external device, depending on the notion '(α, a) is derived from X'.
3. Usually, 'α' is provable from 'X' should be understood by the sentence, viz. *there is a derivation of* 'α' *from* 'X' or some variant form of it. '*sup*' is usually used as an operator for existential quantifier 'there is'. But the way 'sup' is used in the definition of provability degree shows that neither it is used to compute the truth value of the sentence *there is a derivation of* 'α' *from* 'X' nor for computing the truth value of the sentence *there is a derivation of* '(α, a)' *from* 'X'.

So, Hájek's proposed RPL also cannot address many-valuedness at meta-logical concepts genuinely.

Before winding up this section, let us look back to the Pavelka's notion of proof, as a many-valued inference, through an example, and present the difference from the context of GCT as well.

Example 6.2 (1) **A derivation in the context of Pavelka's logic**: We will now consider one example explaining both the difficulties regarding the notion of proof mentioned in Sect. 6.1.3. Let X be the fuzzy subset such that $X(\alpha) = 0.7$, $X(\alpha \supset \beta) = 0.3$, $X(\beta \supset \gamma) = 0.9$ and 0 for the rest of the formulae. Let us consider the proof ω of γ from X as proposed by Pavelka. Let $\omega = \langle\omega_1, \omega_2, \omega_3, \omega_4, \omega_5\rangle$, where $\omega_1 = (\alpha)$, $\omega_2 = (\alpha \supset \beta)$, $\omega_3 = (\beta, r, \langle 1, 2\rangle)$, $\omega_4 = (\beta \supset \gamma)$, $\omega_5 = (\gamma, r, \langle 3, 4\rangle)$.
Hence $\widehat{\langle\omega_1\rangle}(X) = 0.7$, $\langle\omega_1, \widehat{\omega_2\rangle}(X) = 0.3$, $\langle\omega_1, \omega_2, \widehat{\omega_3\rangle}(X) = 0.7 * .3$, $\langle\omega_1, \omega_2, \omega_3, \widehat{\omega_4\rangle}(X) = 0.9$, $\langle\omega_1, \omega_2, \omega_3, \omega_4, \widehat{\omega_5\rangle}(X) = 0.7 * 0.3 * 0.9$.

(i) As given by the expression (1) of Sect. 6.1.3, in order to calculate the value of the sentence '$\langle\omega_1, \omega_2, \omega_3, \omega_4, \omega_5\rangle$ is a proof of γ from X, if we consider value of each step of the proof $\langle\omega_1, \omega_2, \omega_3, \omega_4, \omega_5\rangle$, i.e. the value of each conjunct under $\Big[\ \Big]$, then the value should be $.7 * .3 * (.7 * .3) * .9 * (.7 * .3 * .9)$, but instead of that Pavelka considered the value of $\Big[\ \Big]$ under the last conjunct of the entire expression for proof presented in (1), and $.7 * .3 * .9$ is attached as the value of the proof of γ from X.

(ii) As proposed by Pavelka, the value of $\langle\omega_1, \widehat{\omega_2, \omega_3\rangle}(X)$ would be

$$\Big[(`\omega_3\text{'} \in `(F)\text{'} \to_1 \ulcorner`\omega_3\text{'} \in `X\text{'})\&(`\omega_3\text{'} \in `(F \times \{0\})\text{'} \to_1 \ulcorner`\omega_3\text{'} \in \mathscr{A})\&$$

$$(`\omega_3\text{'} \in `(F \times r' \times N^+)\text{'} \to_1 Res_r(\ulcorner`\omega_3\text{'}))\Big].$$

But as $\omega_3 = (\beta, r, \langle 1, 2\rangle) \in (F \times r' \times N^+)$, the value of the first two conjuncts of the above expression would be 1. Hence the value of the whole expression would be value($Res_r(\ulcorner`\omega_3') )$. Now $Res_r(\ulcorner`\omega_3') \equiv$
$\exists\langle(), ()\rangle\big(\langle(), ()\rangle \subseteq \ulcorner`\langle(\omega_1\widehat{\langle\omega_1\rangle}(X)), (\omega_2\widehat{\langle\omega_1, \omega_2\rangle}(X))\rangle \& r(\langle(), ()\rangle, \ulcorner`\omega_3'))$
$\equiv \exists\langle(), ()\rangle(\langle(), ()\rangle \subseteq \ulcorner`\langle(\omega_1, `.7'), (\omega_2, `.3')\rangle' \& r(\langle(), ()\rangle, \ulcorner`\omega_3'))$.
So, we need to refer to $\widehat{\langle\omega_1\rangle}(X)$, *the value of* $\langle\omega_1\rangle$ and $\widehat{\langle\omega_1, \omega_2\rangle}(X)$, *the value of* $\langle\omega_1, \omega_2\rangle$ at level-1 where the value of the proof itself is an entity.

(2) **A derivation in the context of GCT**: Let us now consider the same sequence of derivation in the context of GCT. So, we have a set of premises $X = \{\alpha, \alpha \supset \beta, \beta \supset \gamma\}$. We start with an axiomatic base $(\mathscr{A}, \mathscr{R})$ where $\mathscr{R} = \{R_{MP}\}$, and a semantic base $\{T_i\}_{i \in I}$ as well. With respect to the semantic base we can calculate the value of the level-1 sentence $R_{MP} = \forall_T[\text{dom}(\langle x_1, x_2, x_3\rangle) \subseteq T \rightarrow \text{ran}(\langle x_1, x_2, x_3\rangle) \in T]$. Let us denote the value of R_{MP} be $|R_{MP}|$.

Now the derivation of γ from X is given by the sequence $\langle\alpha, \alpha \supset \beta, \beta, \beta \supset \gamma, \gamma\rangle$, where the respective step values of the derivation are 1 (as α is a premise), 1 (as $\alpha \supset \beta$ is a premise), $|R_{MP}|$ (as β is obtained by applying R_{MP}), 1 (as $\beta \supset \gamma$ is a premise), and $|R_{MP}|$ (as γ is obtained by R_{MP}). So, the value of the derivation '$\langle\alpha, \alpha \supset \beta, \beta, \beta \supset \gamma, \gamma\rangle$'$D$('$X$', '$\gamma$') is $1 \wedge 1 \wedge |R_{MP}| \wedge 1 \wedge |R_{MP}| = |R_{MP}|$.

So, here we can notice the following advantages. (i) The derivation sequence does not need to refer to the values of its sub-derivations. (ii) The value of a step obtained by R_{MP} does not vary from one instance of the application of the rule to the other instance; it is fixed, based on a semantic justification, for each instance of the rule. (iii) The value of the derivation takes into account the value of each step in the derivation.

Note 6.6 1. As is clear from the Sects. 6.1.3 and 6.2, according to the methodology each level consists of two parts, namely, a language and its corresponding algebra. To talk about a sentence or value of a sentence, present at some level, one has to go to the next higher level. That is an entity of a level cannot be both used and mentioned at the same level. But, in designing $Res_r(\ulcorner`\omega_i')$ the above-mentioned principle has to be violated if one desires to accommodate Pavelka's notion of proof with the assumption that r is a fuzzy rule (i.e. a fuzzy relation). The proof is a notion that has been 'used' at level-1 of Pavelka logic and so the value of a proof cannot be referred to, i.e. 'mentioned' at that level. Besides this, while writing a proof, as the value of a sub-proof is being incorporated within the body of the rule, it is also 'used' at the same level. So, maintaining of level distinction properly does not allow us to see the meta-logical notion, namely, 'α' is a syntactic consequence of 'X' as a many-valued notion.
That is we can say that $(\beta, a * b)$ follows from (α, a) and $(\alpha \supset \beta, b)$. But if we want to say that from (α, a) and $(\alpha \supset \beta, b)$, β follows and the strength of this following is to the extent $a * b$ then we would make a mistake in the categories of 'using' and 'mentioning'.

2. Hájek has presented a comparatively simpler version of Pavelka logic and named it as Rational Pavelka Logic. We have tried to analyse Hájek's presentation from the same angle and faced a similar problem there too.
3. Moreover, in Pavelka (1979), assumptions regarding meta-level truth structure had not been made explicit. But some hints of thoughts, proposing a complete lattice to be a structure for meta-level language, can be found there. But to see the value attached with the notion of semantic consequence in Pavelka's case as a value of the *sentence* determining *the notion of semantic consequence* we need to have two implications, as is clear from the syntax of level-1 language of Pavelka logic presented in Sect. 6.1.3. In this regard, the level-1 connective $\rightarrow_1$ needs to be interpreted as $a \Rightarrow_1 b = sup\ \{z : a \wedge z \leq b\ \}$ which is available in a complete lattice. So, as the notion of semantic consequence, i.e. 'X' $\models$ 'α' is to be defined by both the connectives $\rightarrow_1$, $\rightarrow_2$, (i.e. two implications are needed at the same level) one can be critical about the notion of semantic consequence itself.

6.4 Łukasiewicz Fuzzy Propositional Logic

In Sect. 4.3.1, we have seen how the notion of semantic consequence, based on designated set, of many-valued logics can be obtained as a particular case of GCT. Below we present another approach, where a series of entailment relations are defined starting from that of the many-valued notion of semantic consequence.

In Łukasiewicz fuzzy propositional logic Bergmann (2008), concepts like Fuzzy$_L$ entailment, degree entailment, n-degree entailment and fuzzy consequence are introduced. We present the definitions as follows Bergmann (2008).

Definition 6.1 A formula α is a Fuzzy$_L$ entailment of a set of formulae Γ if for every fuzzy truth-value assignment, which is a fuzzy set from the set of all formulae to the set [0, 1], if every member of Γ gets the value 1 then α gets the value 1.

It is quite clear that the notion of Fuzzy$_L$ degree entailment is actually the same as the notion of $\models_{ML}$ where the designated set is {1}, and we know $\models_{ML}$ is a two-valued concept.

Definition 6.2 α is said to be a degree entailment of Γ if for any fuzzy truth-value assignment, the infimum of the values assumed by all members of Γ is less or equal to the value obtained by α.

This seems close to another way of defining semantic consequence ($\models_{\leq}$) in the context of many-valued logics when the lattice meet is taken to be the operator for the object language conjunction. In contrast to $\models_{ML}$ based on the designated set of truth values, in the literature of many-valued logics Bou et al. (2009), $X \models_{\leq} \alpha$ holds if and only if for every truth assignments the infimum of the truth values of the premise set X under a truth assignment is less or equal to the truth value of α under the same truth assignment. In the literature, the former notion is known as *truth-preserving consequence*, and the latter is known as *degree-preserving consequence*.

An argument is said to be degree valid if its premises degree entails (following Definition 6.2) its conclusion. So, degree entailment and degree valid both are 'yes/no concepts'.

For an argument which is not degree valid, a notion like approximate degree of validity has been introduced. Given an argument having Γ as the set of premises and α as the conclusion, and a fuzzy truth-value assignment V, a function say d_V, called downward distance, has been defined as follows.

Definition 6.3 $d_V(\Gamma, \alpha) = \inf_{\gamma \in \Gamma} V(\gamma) - V(\alpha)$, if $\inf_{\gamma \in \Gamma} V(\gamma) - V(\alpha) > 0$
$= 0$, otherwise.

Now the argument is said to be n-degree valid if $n = 1 - \sup_V d_V(\Gamma, \alpha)$. So, here an attempt to attach a value to the meta-linguistic notion 'an argument is valid' is found. But this again proposes a method of computation which does not have any connection with the defining criterion of the concept 'an argument is valid'. Also, the presence of the operation '$-$' implies that the definition only applies to [0, 1].

In summing up we can say, that in all the above approaches towards the notion of consequence, be it a proof or derivation or rule of inference, either the notion is crisp or some grade is attached not paying attention to the underlying defining criterion for the concept. The whole theory of graded consequence is meticulously designed keeping these issues in mind.

References

Bergmann, M.: An Introduction to Many-valued and Fuzzy Logic: Semantics, Algebras and Derivation Systems. Cambridge University Press (2008)

Bou, F., Esteva, F., Font, J.M., Gil, A., Godo, L., Torrens, A., Verdú, V.: Logics preserving degrees of truth from varieties of residuated lattices. J. Logic Comput. **19**(6), 1031–1069 (2009)

Church, A.: Introduction to Mathematical Logic, vol. 1. Princeton University Press, N.J. (1956)

Dutta, S., Chakraborty, M.K.: Grade in metalogical notions: a comparative study of fuzzy logics. Mathware Soft Comput. Mag. **21**(2), 20–32 (2014)

Esteva, F., Godo, L., Noguera, C.: First order t-norm based fuzzy logics with truth-constants: distinguished semantics and completeness properties. Ann. Pure and Appl. Logic **161**(2): Elsevier, 185–202 (2009)

Goguen, J.A.: The logic of inexact concept. Synthese **19**, 325–373 (1968)

Gottwald, S.: An approach to handle partially sound rules of inference. In: Bouchon-Meunier, B., Yager, R.R., Zadeh, L.A. (Eds.) Advances in Intelligent Computing, IPMU'94, Selected papers, Lecture Notes Computer Science, Vol. 945, pp. 380–388. Springer: Berlin etc. (1995)

Hájek, P.: Metamathematics of Fuzzy Logic. Kluwer Academic Publishers, Dordrecht (1998)

Novak. V.: On syntactico-semantical completeness of first order fuzzy logic, parts I and II. Kybernetica **2**, 6(1, 2), 47–154 (1990)

Pavelka, J.: On fuzzy logic I, II, III Zeitscher for Math. Logik und Grundlagen d. Math. **25**, 4552, 119134, 447464 (1979)

Zadeh, L.A.: Fuzzy sets. Inf. Control **8**, 338–353 (1965)

Chapter 7
Graded Consequence in Decision-Making: A Few Applications

Abstract In this chapter, we shall present some aspects of decision-making based on the basic idea of the theory of graded consequence. In the first case, we shall extend the theory in the context of interval semantics and present how different ways of aggregating information collected from different sources/experts/agents incorporates different attitudes of a decision-maker. In the second case, we shall propose an extension of GCT in a distributed network of decision-making among agents (sources of information) and decision-maker, each having their own local logics in the sense of Barwise and Seligman. The third will be on dealing with Sorites like paradox, which is usually considered to be one testing criterion for any theory of reasoning with imprecise information.

7.1 GCT with Interval Semantics: Different Approaches Towards Aggregating Information

From different aspects of the theory GCT, it is already explained that apart from the philosophical ground for distinguishing different levels of a logic discourse, there is a practical ground too; and that is about viewing the initial semantic context $\{T_i\}_{i \in I}$ as different sources of information, which mathematically are functions assigning values to object-level sentences. The meta-level logic is the logic of a decision-maker who in order to make a decision collects information from the sources, and aggregates them for determining whether a formula (semantically) follows from a set of formulae. In the framework of classical logic in order to come to a decision, which is of yes/no type, we do aggregate information gathered from all possible valuation functions. As discussed in Sect. 2.1 of Chap. 2, logicians (Shoesmith and Smiley 1978) even shifted from that consideration and proposed the notion of (semantic) consequence relative to a collection $\{T_i\}_{i \in I}$, which is not necessarily the whole (cf. expressions (II) and (A) of Sect. 2.1). In the context of GCT, this relativization of the notion of consequence with respect to a collection of $\{T_i\}_{i \in I}$ has been further generalized considering T_i as fuzzy sets (cf. expression (B) of Sect. 2.1). In order to have a faithful representation of the quantifier $\forall$, present in the statement $\forall_{T_i \in \{T_i\}_{i \in I}}[X \subseteq T_i \to \alpha \in T_i]$, in many-valued context, the

M. K. Chakraborty and S. Dutta, *Theory of Graded Consequence*, Logic in Asia: Studia Logica Library, https://doi.org/10.1007/978-981-13-8896-5_7

lattice meet 'inf' comes into play, and we aggregate the views of T_i's by computing $\inf_{i \in I}[\inf_{x \in X} T_i(x) \rightarrow_m T_i(\alpha)]$. But the use of infimum in aggregating views of a set of agents has some disadvantages. In case of infimum, even if most of the agents consider that 'if X is true then α is true' to be of higher value, one single opinion with lower value gets the privilege. Making a balance between the theoretical expectation to a logical consequence and the above-mentioned practical lacuna is quite challenging. In this respect, we present some trade-off moving from single-valued semantics to interval-valued semantics in the following sequel.

There are plenty of instances where it is impossible to claim *precisely* that *an imprecise concept applies to an object to a specific degree*. As a result, when an imprecise concept is quantized by a single value, the inherent impreciseness of the concept is somewhat lost. Assigning an interval-value, to some extent, manages to retain the uncertainty *of understanding an imprecise concept* as it only attaches a set of possible interpretations to the concept. In this regard let us quote a few lines from Cornelis et al. (2004). *IVFS theory emerged from the observation that in a lot of cases no objective procedure is available to select the crisp membership degrees of elements in a fuzzy set. It was suggested to alleviate that problem by allowing to specify only an interval ...to which the actual membership degree is assumed to belong*. Thus, interval mathematics and its application in the context of imprecise reasoning are quite significant. So, developing GCT in the context of interval-valued semantics is meaningful both from the angle of theory building and applications.

In this section, we shall present three different attitudes of decision-making based on GCT. The information coming from different sources T_i's, as well as the attitude (conservative, liberal, moderate) of the decision-maker plays a role in the process of decision-making and in the final conclusion. Keeping this practical motivation in mind we here propose three different notions for deriving conclusion, two of which satisfy the graded consequence axioms (Chakraborty 1995), and the third is a close variant of that. In each of these cases, the object language formulae are interpreted by closed sub-intervals of [0, 1], but the notion of consequence is made single-valued. This value assignment is done taking either the left-hand end point or the right-hand end point or some value in between from the final interval that is computed as an outcome. It is not completely unrealistic to think that experts, i.e. T_i's are entitled to assign a range of values, but the decision-maker is constrained to conclude a single value, and such a practice of precisification in final result prevails in the literature of fuzzy set theory, especially in the area of application of the theory. Before entering into the main concern of this section, we first present a brief overview of the interval mathematics (Alcalde et al. 2005; Bedregal and Takahashi 2006; Bedregal et al. 2010; Bedregal and Santiago 2013; Cornelis et al. 2004; Deschrijver and Cornelis 2007; Deschrijver and Kerre 2003; Li and Li 2012; Gasse et al. 2006) below.

7.1.1 Interval Mathematics: Some Basic Notions

Assigning a specific grade to an imprecise sentence often pushes us into a situation where from a range of possible values we are to choose a single one for the sake of the mathematical ease of computation. Lifting the whole mathematics of fuzzy set theory in the context of interval-valued fuzzy set theory, researchers (Alcalde et al. 2005; Bedregal and Takahashi 2006; Bedregal et al. 2010; Bedregal and Santiago 2013; Cornelis et al. 2004; Deschrijver and Cornelis 2007; Deschrijver and Kerre 2003; Li and Li 2012; Gasse et al. 2006) to a great extent could manage to resolve this problem. In this section we present some part of the development made in Bedregal and Takahashi (2006), Bedregal et al. (2010), Bedregal and Santiago (2013), Cornelis et al. (2004), Deschrijver and Cornelis (2007), Deschrijver and Kerre (2003), Gasse et al. (2006) depending upon the need of this paper.

Let us consider $\mathscr{U} = \{[a, b]/0 \leq a \leq b \leq 1\}$ along with two order relations $\leq_I$ and $\subseteq$, defined by: $[x_1, x_2] \leq_I [y_1, y_2]$ iff $x_1 \leq y_1$ and $x_2 \leq y_2$ and $[x_1, x_2] \subseteq [y_1, y_2]$ iff $y_1 \leq x_1 \leq x_2 \leq y_2$.

$(\mathscr{U}, \leq_I)$ forms a complete lattice, and $(\mathscr{U}, \subseteq)$ forms a poset. Let $\bigwedge$ be the lattice meet corresponding to the order relation $\leq_I$.

Definition 7.1 (*Bedregal and Takahashi* 2006) An interval t-norm is a mapping $T : \mathscr{U} \times \mathscr{U} \mapsto \mathscr{U}$ such that T is commutative, associative, monotonic with respect to $\leq_I$ and $\subseteq$, and $[1, 1]$ is the identity element with respect to T.

Definition 7.2 (*Gasse et al.* 2006) Let T be an interval t-norm. T is called t-representable if there exist t-norms t_1, t_2 on $[0, 1]$ such that $T([x_1, x_2], [y_1, y_2]) = [t_1(x_1, y_1), t_2(x_2, y_2)]$.

Definition 7.3 (*Gasse et al.* 2006): For any t-norm $*$ on $[0, 1]$ and $a \in [0, 1]$, $T_{*,a}$ is defined below:
$T_{*,a}([x_1, x_2], [y_1, y_2]) = [x_1 * y_1, max((x_2 * y_2) * a, x_1 * y_2, x_2 * y_1)]$.

In Deschrijver and Kerre (2003) it has been shown that for any t-norm $*$ on $[0, 1]$ and any $a \in [0, 1]$, $T_{*,a}$ is an interval t-norm. Moreover, for $a = 1$, $T_{*,a}$ becomes a t-representable t-norm (Gasse et al. 2006); i.e. $T_{*,1}([x_1, x_2], [y_1, y_2]) = [x_1 * y_1, x_2 * y_2]$. For the purpose of this paper, we shall consider such an $T_{*,1}$ and denote this interval t-norm based on $*$ as $*_I$.

Definition 7.4 Given $\rightarrow$, the residuum of $*$ on $[0, 1]$, and for $a = 1$, the operation $\rightarrow_I$: $\mathscr{U} \times \mathscr{U} \mapsto \mathscr{U}$ is defined by:

$$\begin{aligned}[x_1, x_2] \rightarrow_I [y_1, y_2] &= [min\{x_1 \rightarrow y_1, x_2 \rightarrow y_2\}, min\{(x_2 * 1) \rightarrow y_2, x_1 \rightarrow y_2\}].\\ &= [min\{x_1 \rightarrow y_1, x_2 \rightarrow y_2\}, min\{x_2 \rightarrow y_2, x_1 \rightarrow y_2\}].\end{aligned}$$

In Gasse et al. (2006) it is shown that $\rightarrow_I$ is an interval fuzzy implication with the adjoint pair $(*_I, \rightarrow_I)$ on $\mathscr{U}$. For $I_1, I_2, I' \in \mathscr{U}$, the following properties of $(*_I, \rightarrow_I)$ are of particular interest here:

(i) If $I_1 \leq_I I_2$ then $I_2 \rightarrow_I I \leq_I I_1 \rightarrow_I I$.

(ii) If $I_1 \leq_I I_2$ then $I \rightarrow_I I_1 \leq_I I \rightarrow_I I_2$.
(iii) $I \rightarrow_I I' \geq_I I'$.
(iv) $[1, 1] \rightarrow_I I = I$.
(v) If $I \leq_I I'$ then $I \rightarrow_I I' = [1, 1]$.
(vi) $(I_1 \rightarrow_I I_2) *_I ((I_1 \bigwedge I_2) \rightarrow_I I) \leq_I (I_1 \rightarrow_I I)$.
(vii) $\bigwedge_k (I' \rightarrow_I I_k) = (I' \rightarrow_I \bigwedge_k I_k)$.
(viii) $I_1 *_I I_2 \leq_I I_3$ iff $I_2 \leq_I I_1 \rightarrow_I I_3$.

In the sequel below we propose (Dutta et al. 2015) a few different definitions for the semantic notion of graded consequence. These definitions incorporate different decision-making attitudes from the practical perspectives. Moreover, when the semantics for the object language formulae is restricted to the degenerate intervals, each of the notions yields the original notion of graded consequence (Chakraborty 1995; Chakraborty and Basu 1997; Chakraborty and Dutta 2010).

7.1.2 Graded Consequence: Form (Σ)

Definition 7.5 Given a collection of interval-valued fuzzy sets, say $\{T_i\}_{i \in I}$, the grade of $X \approx\!\!\!\!\models \alpha$, i.e. $gr(X \approx\!\!\!\!\models \alpha) = l([\bigwedge_{i \in I}\{\bigwedge_{x \in X} T_i(x) \rightarrow_I T_i(\alpha)\}])$, ...($\Sigma$)
where $l([.])$ represents the left-hand end point of an interval; that is, $l([x_1, x_2]) = x_1$.

The similarity and the differences between the notions of $gr(X \approx\!\!\!\!\models \alpha)$, given in Definition 2.3 and form (Σ), are quite visible from their respective forms. According to (Σ), to find out the degree to which α follows from X one has to first find out the truth interval assignment to the formulae by a set of experts $\{T_i\}_{i \in I}$. Then, the left-hand end point of the least interval-value assigned to $\bigwedge_{x \in X} T_i(x) \rightarrow_I T_i(\alpha)$ (if every member of X is true then α is true) needs to be computed. In order to stick to a single value for the notion of derivation, in this case, the left-hand end point of the resultant interval is taken.

One might think that value for (Σ) would be the same if before computing $\rightarrow_I$ with the component intervals, the left-hand end point of the concerned intervals is taken out, and the corresponding implication operation for single-valued case is applied. In order to show that (Σ) is not the same as $\inf_i [l(\wedge_{x \in X} T_i(x)) \rightarrow l(T_i(\alpha)]$, let us consider $l([.3, .7] \rightarrow_I [.2, .3])$, where $\rightarrow_I$ is defined in terms of the ordinary Łukasiewicz implication following Definition 7.4. Then it can be checked that $l([.3, .7] \rightarrow_I [.2, .3]) = .6$, and $l([.3, .7]) \rightarrow_Ł l([.2, .3]) = .9$.

Lemma 7.1 $\inf_i l(\{[x_i, y_i]\}_{i \in K}) = l(\bigwedge_i \{[x_i, y_i]\}_{i \in K})$.

Lemma 7.2 $\bigwedge_i I_i *_I \bigwedge_i I'_i \leq_I \bigwedge_i (I_i *_I I'_i)$, *where* I_i, I'_i *are intervals.*

Lemma 7.3 $\bigwedge_i \bigwedge_j I_{ij} = \bigwedge_j \bigwedge_i I_{ij}$.

Lemma 7.4 *If* $*_I$ *is a t-representable t-norm with respect to an ordinary t-norm* $*$ *then* $I_1 *_I I_2 \leq_I I_3$ *implies* $l(I_1) * l(I_2) \leq l(I_3)$.

Theorem 7.1 (Representation theorems with respect to (Σ))

(1) For any $\{T_i\}_{i\in I}$, $\approx$ in the sense of (Σ) is a graded consequence relation.
(2) For any graded consequence relation $\mid\!\sim$, there is a collection of interval-valued fuzzy sets such that $\approx$ generated in the sense of (Σ) coincides with $\mid\!\sim$.

Proof (1) (GC1) For $\alpha \in X$, $\bigwedge_{x\in X} T_i(x) \leq_I T_i(\alpha)$. By (v) we have $gr(X \approx \alpha) = 1$.
(GC2) For $X \subseteq Y$, $\bigwedge_{x\in Y} T_i(x) \leq_I \bigwedge_{x\in X} T_i(x)$. Hence by (i) GC2 is immediate.
(GC3) $\inf_{\beta\in Z} gr(X \approx \beta) * gr(X \cup Z \approx \alpha)$
$= \inf_{\beta\in Z} l[\bigwedge_{i\in I}\{\bigwedge_{x\in X} T_i(x) \rightarrow_I T_i(\beta)\}] * l[\bigwedge_{i\in I}\{\bigwedge_{x\in X\cup Z} T_i(x) \rightarrow_I T_i(\alpha)\}]$
$= l[\bigwedge_{\beta\in Z}\{\bigwedge_{i\in I}(\bigwedge_{x\in X} T_i(x) \rightarrow_I T_i(\beta))\}] * l[\bigwedge_{i\in I}\{\bigwedge_{x\in X\cup Z} T_i(x) \rightarrow_I T_i(\alpha)\}]$
(Lemma 7.1)
$= l[\bigwedge_{i\in I}\{\bigwedge_{\beta\in Z}(\bigwedge_{x\in X} T_i(x) \rightarrow_I T_i(\beta))\}] * l[\bigwedge_{i\in I}\{\bigwedge_{x\in X\cup Z} T_i(x) \rightarrow_I T_i(\alpha)\}]$
(Lemma 7.3)
$= l[\bigwedge_{i\in I}(\bigwedge_{x\in X} T_i(x) \rightarrow_I \bigwedge_{\beta\in Z} T_i(\beta))\}] * l[\bigwedge_{i\in I}\{\bigwedge_{x\in X\cup Z} T_i(x) \rightarrow_I T_i(\alpha)\}]$ (by (vii))
…(a)
Now for each T_i, $\{\bigwedge_{x\in X} T_i(x) \rightarrow_I \bigwedge_{\beta\in Z} T_i(\beta)\} *_I \{\bigwedge_{x\in X\cup Z} T_i(x) \rightarrow_I T_i(\alpha)\}$
$= \{\bigwedge_{x\in X} T_i(x) \rightarrow_I \bigwedge_{\beta\in Z} T_i(\beta)\} *_I \{[(\bigwedge_{x\in X} T_i(x)) \bigwedge (\bigwedge_{x\in Z} T_i(x))] \rightarrow_I T_i(\alpha)\}$
$\leq_I \{\bigwedge_{x\in X} T_i(x) \rightarrow_I T_i(\alpha)\}$. (by (vi)).
Therefore, $\bigwedge_{i\in I}[\{\bigwedge_{x\in X} T_i(x) \rightarrow_I \bigwedge_{\beta\in Z} T_i(\beta)\} *_I \{\bigwedge_{x\in X\cup Z} T_i(x) \rightarrow_I T_i(\alpha)\}]$
$\leq_I \bigwedge_{i\in I}[\bigwedge_{x\in X} T_i(x) \rightarrow_I T_i(\alpha)]$. …(b)
Also, $\bigwedge_{i\in I}[\{\bigwedge_{x\in X} T_i(x) \rightarrow_I \bigwedge_{\beta\in Z} T_i(\beta)\}] *_I \bigwedge_{i\in I}[\{\bigwedge_{x\in X\cup Z} T_i(x) \rightarrow_I T_i(\alpha)\}]$
$\leq_I \bigwedge_{i\in I}[\{\bigwedge_{x\in X} T_i(x) \rightarrow_I \bigwedge_{\beta\in Z} T_i(\beta)\} *_I \{\bigwedge_{x\in X\cup Z} T_i(x) \rightarrow_i T_i(\alpha)\}]$ (Lemma 7.2)
$\leq_I \bigwedge_{i\in I}[\bigwedge_{x\in X} T_i(x) \rightarrow_I T_i(\alpha)]$ (by (b))
$l[\bigwedge_{i\in I}(\bigwedge_{x\in X} T_i(x) \rightarrow_I \bigwedge_{\beta\in Z} T_i(\beta))\}] * l[\bigwedge_{i\in I}\{\bigwedge_{x\in X\cup Z} T_i(x) \rightarrow_I T_i(\alpha)\}]$
$\leq l[\bigwedge_{i\in I}\{\bigwedge_{x\in X} T_i(x) \rightarrow_I T_i(\alpha)\}]$ (Lemma 7.4 as $*_I$ is t-representable) …(c)
Hence from (a) and (c), we have $\inf_{\beta\in Z} gr(X \approx \beta) * gr(X \cup Z \approx \alpha) \leq gr(X \approx \alpha)$.
(2) Given $\mid\!\sim$, a graded consequence relation, let us consider the collection $\{T_X\}_{X\in P(F)}$ of interval-valued fuzzy sets over formulae such that $T_X(\alpha) = [gr(X \mid\!\sim \alpha)]_{BIR}$, where $[x]_{BIR}$ represents the best interval representation (Bedregal et al. 2010) of x, i.e. the interval $[x, x]$.
We want to prove $gr(X \mid\!\sim \alpha) = l[\bigwedge_{Y\in P(F)}\{\bigwedge_{x\in X} T_Y(x) \rightarrow_I T_Y(\alpha)\}]$.
By (GC3) and Lemma 7.1, we have $l(T_Y(\beta)) \geq l(T_{X\cup Y}(\beta)) * l(\wedge_{\alpha\in X} T_Y(\alpha))$.
As T_Y's are degenerate intervals by proposition 4.1 of Gasse et al. (2006) every true identity expressible in [0, 1] is expressible in $\mathscr{U}$. So, $T_{X\cup Y}(\beta) *_I \wedge_{\alpha\in X} T_Y(\alpha) \leq_I T_Y(\beta)$.
Then following the proof outline of (2) of Theorem 2.1, presented in Chap. 2, we can show, $T_X(\beta) \leq_I \wedge_{Y\in P(F)}[\wedge_{\alpha\in X} T_Y(\alpha) \rightarrow_I T_Y(\beta)]$ …(d)
$\wedge_{Y\in P(F)}[\wedge_{\alpha\in X} T_Y(\alpha) \rightarrow_I T_Y(\beta)] \leq_I \wedge_{X\subseteq Y\in P(F)}[\wedge_{\alpha\in X} T_Y(\alpha) \rightarrow_I T_Y(\beta)]$ …(e)
Now for $X \subseteq Y$, by (GC2), $T_X(\alpha) \leq_I T_Y(\alpha)$. Then by (GC1), we have $[1, 1] = \wedge_{\alpha\in X} T_X(\alpha) \leq_I \wedge_{\alpha\in X} T_Y(\alpha)$, i.e. $\wedge_{\alpha\in X} T_Y(\alpha) \rightarrow_I T_Y(\beta) = T_Y(\beta)$. Hence (e) becomes $\wedge_{Y\in P(F)}[\wedge_{\alpha\in X} T_Y(\alpha) \rightarrow_I T_Y(\beta)] \leq_I \wedge_{X\subseteq Y\in P(F)} T_Y(\beta) = T_X(\beta)$ …(f)

Combining (d) and (f), we have $gr(X \mid\sim \beta) = l[\wedge_{Y \in P(F)}\{\wedge_{\alpha \in X} T_Y(\alpha) \rightarrow_I T_Y(\beta)\}]$. □

Note 7.1 In this new context, the meta-level algebraic structure may be viewed as $\langle \mathscr{U}, *_I, \rightarrow_I, [0,0], [1,1], l \rangle$, a complete residuated lattice with $l : \mathscr{U} \mapsto [0,1]$. The structure is formed out of a complete residuated lattice $([0,1], *, \rightarrow, 0, 1)$, serving as base. Specifically, the adjoint pair $(*_I, \rightarrow_I)$ is defined in terms of the adjoint pair $(*, \rightarrow)$.

Definition (Σ) endorses the minimum truth-value assignment ensuring the limit; that is, no individual expert's assignment to $\bigwedge_{x \in X} T_i(x) \rightarrow_I T_i(\alpha)$ ('if every member of X is true then α is true'), should lie below the value aggregated for the notion of consequence. As an instance, let two experts assign [.1, .3] and [.7, .9] to $\bigwedge_{x \in X} T_i(x) \rightarrow_I T_i(\alpha)$. Then following ($\Sigma$) the value for '$\alpha$ follows from X' only considers the left-hand end point of the least interval, i.e. [.1, .3]. So, this value assignment does not take care of the consensus of all. It only emphasizes that the lower bound of everyone's point of agreement is .1, no matter whether someone really has marked a high grade.

7.1.3 Extension of ⊆ as a Lattice Order Relation on Intervals

Let us now explore a method of assigning the value to 'α follows from X' in such a way that it takes care of every individual's opinion. That is, we are looking for an interval which lies in the intersection of everyone's opinion. For this, we need a complete lattice structure with respect to the order relation $\subseteq$ on $\mathscr{U}$. According to the existing definition of $\subseteq$, given any degenerate interval say $[.a, .a]$ and an interval $[.b, .c]$ such that $.a < .b$, neither they are related nor they have any common lower bound with respect to $\subseteq$. So, in order to extend the present relation $\subseteq$ into a lattice order relation, we extend the definition of the existing partially ordered relation $\subseteq$ by the following definition.

Definition 7.6 $\subseteq_e$ is a binary relation on $\mathscr{U}$ defined as below:
$[x_1, x_2] \subseteq_e [y_1, y_2]$ if $y_1 \leq x_1 < x_2 \leq y_2$,
$[x_1, x_1] \subseteq_e [y_1, y_2]$ if $x_1 \leq y_2$.

Proposition 7.1 *If* $[x_1, x_2] \subseteq [y_1, y_2]$ *then* $[x_1, x_2] \subseteq_e [y_1, y_2]$.

Note 7.2 The converse of the above proposition does not hold. For the intervals [.2, .2] and [.3, .7], $[.2, .2] \subseteq_e [.3, .7]$, but $[.2, .2] \not\subseteq [.3, .7]$. Also to be noted that for two intervals $[x_1, x_2]$ and $[y_1, y_2]$, $[x_1, x_2] \subseteq_e [y_1, y_2]$ does not hold if $x_1 < x_2$ and $y_1 = y_2$. Let $[x_1, x_2] \subseteq_e [y_1, y_2]$ be such that $x_1 = x_2 < y_1 \leq y_2$ holds. This pair of intervals are called non-overlapping intervals under the relation $\subseteq_e$; other pairs of intervals under the relation $\subseteq_e$ are known as overlapping intervals under the relation $\subseteq_e$.

Proposition 7.2 *($\mathcal{U}$, $\subseteq_e$) forms a poset.*

Proposition 7.3 *($\mathcal{U}$, $\subseteq_e$) is a lattice where the greatest lower bound, say $\bigcap$, and the least upper bound, say $\bigcup$ are defined as follows:*

$$[x_1, x_2] \bigcap [y_1 y_2] = [\max(x_1, y_1), \min(x_2, y_2)] \textit{ if } \max(x_1, y_1) \leq \min(x_2, y_2)$$
$$= [\min(x_2, y_2), \min(x_2, y_2)], \textit{ otherwise.}$$
$$[x_1, x_2] \bigcup [y_1 y_2] = [\max(x_2, y_2), \max(x_2, y_2)], \textit{ if } x_1 = x_2,\ y_1 = y_2.$$
$$= [\max(x_1, y_1), \max(x_2, y_2)], \textit{ if } x_1 = x_2 < y_1 < y_2$$
$$= [\min(x_1, y_1), \max(x_2, y_2)], \textit{ otherwise.}$$
$$\textit{(i.e. either } x_1 = x_2,\ y_1 < y_2,\ y_1 \leq x_1, \textit{ or } x_1 < x_2,\ y_1 < y_2\textit{).}$$

Proposition 7.4 *($\mathcal{U}$, $\subseteq_e$) forms a complete lattice.*

Proof For any collection $\{[x_i, y_i]\}_{i \in K}$, $\bigcap_{i \in K}[x_i, y_i] = [\sup_i x_i, \inf_i y_i]$ if $\sup_i x_i \leq \inf_i y_i = [\inf_i y_i, \inf_i y_i]$, otherwise.
$\bigcup_{i \in K}[x_i, y_i] = [\inf_i x_i, \sup_i y_i]$, if $\inf_i x_i < \inf_i y_i \leq \sup_i y_i = [\sup_i x_i, \sup_i y_i]$, if $\inf_i x_i = \inf_i y_i \leq \sup_i y_i$.
Rest is straightforward as being a closed set, $[0, 1]$ contains infimum and supremum of x_i's and y_i's. □

7.1.4 Graded Consequence: Form (Σ')

Now let us define an alternative definition for semantic graded consequence relation which focuses on such a method of aggregation that selects the zone where the common consensus of the experts lie.

Definition 7.7 Given any collection of interval-valued fuzzy sets, say $\{T_i\}_{i \in I}$, $gr(X \approx\!\!\!| \alpha) = r([\bigcap_{i \in I}\{\bigwedge_{x \in X} T_i(x) \rightarrow_I T_i(\alpha)\}])$, where $r([x_1, x_2]) = x_2$. ...(Σ')

In this definition for graded semantic consequence the value of the sentence 'α follows from X' is the right-hand end point of the common interval-value assigned to the sentence 'whenever every member of X is true α is also true'. Thus, it takes care of every expert's opinion, and the method counts the maximum value of the interval where all the experts agree.

Lemma 7.5 $\inf_i r([x_i, y_i]_{i \in K}) = r(\bigcap_{i \in K}[x_i, y_i])$.

Lemma 7.6 $\bigcap_l \bigcap_k I_{lk} = \bigcap_k \bigcap_l I_{lk}$ *for each* $I_{lk} \in \mathcal{U}$.

Lemma 7.7 *If $*_I$ is a t-representable t-norm with respect to an ordinary t-norm $*$ then $I_1 *_I I_2 \leq_I I_3$ implies $r(I_1) * r(I_2) \leq r(I_3)$.*

Lemma 7.8 *If $\{[x_i^1, x_i^2]\}_{i \in K}$ and $\{[y_i^1, y_i^2]\}_{i \in K}$ are two collections of intervals such that $[x_i^1, x_i^2] \leq_I [y_i^1, y_i^2]$ for each $i \in K$, then $\bigcap_{i \in K}[x_i^1, x_i^2] \leq_I \bigcap_{i \in K}[y_i^1, y_i^2]$.*

Proof $x_i^1 \leq y_i^1$ and $x_i^2 \leq y_i^2$ for each $i \in K$.
Hence, $\sup\{x_i^1\}_{i\in K} \leq \sup\{y_i^1\}_{i\in K}$ and $\inf\{x_i^2\}_{i\in K} \leq \inf\{y_i^2\}_{i\in K}$. ...(1)
As $x_i^1 \leq x_i^2$ for $i \in K$, possibilities are (i) $\sup_i x_i^1 \leq \inf_i x_i^2$ and (ii) $\sup_i x_i^1 > \inf_i x_i^2$.
(i) $\sup_i x_i^1 \leq \inf_i x_i^2 \leq x_i^2 \leq y_i^2$ for each $i \in K$. So, $\sup_i x_i^1 \leq \inf_i x_i^2 \leq \inf_i y_i^2$.
Here again, two subcases arise. (ia) $\sup_i y_i^1 \leq \inf_i y_i^2$ and (ib) $\sup_i y_i^1 > \inf_i y_i^2$.
(ia) If $\sup_i y_i^1 \leq \inf_i y_i^2$, then $\bigcap_{i\in K}[y_i^1, y_i^2] = [\sup_i y_i^1, \inf_i y_i^2]$.
Hence inequalities of (1) ensure that $\bigcap_{i\in K}[x_i^1, x_i^2] \leq_I \bigcap_{i\in K}[y_i^1, y_i^2]$.
(ib) If $\sup_i y_i^1 > \inf_i y_i^2$, then $\bigcap_{i\in K}[y_i^1, y_i^2] = [\inf_i y_i^2, \inf_i y_i^2]$.
Again from (i) $\sup_i x_i^1 \leq \inf_i x_i^2 \leq \inf_i y_i^2$ implies $\bigcap_{i\in K}[x_i^1, x_i^2] \leq_I \bigcap_{i\in K}[y_i^1, y_i^2]$.
(ii) $\inf_i x_i^2 < \sup_i x_i^1$ implies $\inf_i x_i^2 < \sup_i x_i^1 \leq \sup_i y_i^1$.
So, $\bigcap_{i\in K}[x_i^1, x_i^2] = [\inf_i x_i^2, \inf_i x_i^2] \leq_I \bigcap_{i\in K}[y_i^1, y_i^2]$ since $\inf_i x_i^2 \leq \sup_i y_i^1$ and $\inf_i x_i^2 \leq \inf_i y_i^2$. □

Lemma 7.9 *For any collection* $\{I_j\}_{j\in J}$ *of intervals,* $\bigwedge_{j\in J} I_j \leq_I \bigcap_{j\in J} I_j$.

Lemma 7.10 *For any collection* $\{I_j\}_{j\in J}$ *of intervals,* $\bigcap_{j\in J} I_j \subseteq_e \bigwedge_{j\in J} I_j$.

Corollary 7.1 $r[\bigcap_{j\in J}\bigcap_{l\in L} I_{lj}] \leq r[\bigcap_{j\in J}\bigwedge_{l\in L} I_{lj}]$.

Theorem 7.2 (Representation theorems with respect to (Σ'))

(a) For any $\{T_i\}_{i\in I}$, $\approx$ *in the sense of* (Σ') *is a graded consequence relation.*
(b) For any graded consequence relation $\mid\sim$, *there is a collection of interval-valued fuzzy sets such that* $\approx$ *generated in the sense of* (Σ') *coincides with* $\mid\sim$.

Proof (a) (GC1) is proved as in Theorem 7.1. (GC2) is obtained using (i) and Lemma 7.8.
(GC3) $\inf_{\beta\in Z} gr(X \approx \beta) * gr(X \cup Z \approx \alpha)$
$= \inf_{\beta\in Z} r[\bigcap_{i\in I}\{\bigwedge_{x\in X} T_i(x) \rightarrow_I T_i(\beta)\}] * r[\bigcap_{i\in I}\{\bigwedge_{x\in X\cup Z} T_i(x) \rightarrow_I T_i(\alpha)\}]$
$= r[\bigcap_{\beta\in Z}\{\bigcap_{i\in I}(\bigwedge_{x\in X} T_i(x) \rightarrow_I T_i(\beta))\}] * r[\bigcap_{i\in I}\{\bigwedge_{x\in X\cup Z} T_i(x) \rightarrow_I T_i(\alpha)\}]$ (Lemma 7.5)
$= r[\bigcap_{i\in I}\{\bigcap_{\beta\in Z}(\bigwedge_{x\in X} T_i(x) \rightarrow_I T_i(\beta))\}] * r[\bigcap_{i\in I}\{\bigwedge_{x\in X\cup Z} T_i(x) \rightarrow_I T_i(\alpha)\}]$ (Lemma 7.6)
$\leq r[\bigcap_{i\in I}\{\bigwedge_{\beta\in Z}(\bigwedge_{x\in X} T_i(x) \rightarrow_I T_i(\beta))\}] * r[\bigcap_{i\in I}\{\bigwedge_{x\in X\cup Z} T_i(x) \rightarrow_I T_i(\alpha)\}]$ (Corollary 7.1)
$= r[\bigcap_{i\in I}(\bigwedge_{x\in X} T_i(x) \rightarrow_I \bigwedge_{\beta\in Z} T_i(\beta))\}] * r[\bigcap_{i\in I}\{\bigwedge_{x\in X\cup Z} T_i(x) \rightarrow_I T_i(\alpha)\}]$ (by (vii))
...(1)
Also we obtain, $r[\bigcap_{i\in I}[\{\bigwedge_{x\in X} T_i(x) \rightarrow_I \bigwedge_{\beta\in Z} T_i(\beta)\} *_I \{\bigwedge_{x\in X\cup Z} T_i(x) \rightarrow_I T_i(\alpha)\}]]$
$\leq r[\bigcap_{i\in I}\{\bigwedge_{x\in X} T_i(x) \rightarrow_I T_i(\alpha)\}]$ (by (vi) and Lemma 7.8)...(2)

Now $\bigwedge_{i\in I}(\bigwedge_{x\in X} T_i(x) \rightarrow_I \bigwedge_{\beta\in Z} T_i(\beta))\} *_I \bigwedge_{i\in I}\{\bigwedge_{x\in X\cup Z} T_i(x) \rightarrow_I T_i(\alpha)\}$
$\leq_I \bigwedge_{i\in I}[\{\bigwedge_{x\in X} T_i(x) \rightarrow_I \bigwedge_{\beta\in Z} T_i(\beta)\} *_I \{\bigwedge_{x\in X\cup Z} T_i(x) \rightarrow_I T_i(\alpha)\}]$ (Lemma 7.2)
$\leq_I \bigcap_{i\in I}[\{\bigwedge_{x\in X} T_i(x) \rightarrow_I \bigwedge_{\beta\in Z} T_i(\beta)\} *_I \{\bigwedge_{x\in X\cup Z} T_i(x) \rightarrow_I T_i(\alpha)\}]$.
(Lemma 7.9)
i.e. $r[\bigwedge_{i\in I}(\bigwedge_{x\in X} T_i(x) \rightarrow_I \bigwedge_{\beta\in Z} T_i(\beta))\}] * r[\bigwedge_{i\in I}\{\bigwedge_{x\in X\cup Z} T_i(x) \rightarrow_I T_i(\alpha)\}]$

$\leq r[\bigcap_{i\in I}[\{\bigwedge_{x\in X} T_i(x) \to_I \bigwedge_{\beta\in Z} T_i(\beta)\} *_I \{\bigwedge_{x\in X\cup Z} T_i(x) \to_I T_i(\alpha)\}]]$. (Lemma 7.7) ...(3)

$\bigcap_{i\in I}(\bigwedge_{x\in X} T_i(x) \to_I \bigwedge_{\beta\in Z} T_i(\beta))\} \subseteq_e \bigwedge_{i\in I}(\bigwedge_{x\in X} T_i(x) \to_I \bigwedge_{\beta\in Z} T_i(\beta))\}$. (Lemma 7.10)

So, $r[\bigcap_{i\in I}(\bigwedge_{x\in X} T_i(x) \to_I \bigwedge_{\beta\in Z} T_i(\beta))\}] \leq r[\bigwedge_{i\in I}(\bigwedge_{x\in X} T_i(x) \to_I \bigwedge_{\beta\in Z} T_i(\beta))\}]$.

Thus, $r[\bigcap_{i\in I}(\bigwedge_{x\in X} T_i(x) \to_I \bigwedge_{\beta\in Z} T_i(\beta))\}] * r[\bigcap_{i\in I}\{\bigwedge_{x\in X\cup Z} T_i(x) \to_I T_i(\alpha)\}]$

$\leq r[\bigwedge_{i\in I}(\bigwedge_{x\in X} T_i(x) \to_I \bigwedge_{\beta\in Z} T_i(\beta))\}] * r[\bigwedge_{i\in I}\{\bigwedge_{x\in X\cup Z} T_i(x) \to_I T_i(\alpha)\}]$

$\leq r[\bigcap_{i\in I}[\{\bigwedge_{x\in X} T_i(x) \to_I \bigwedge_{\beta\in Z} T_i(\beta)\} *_I \{\bigwedge_{x\in X\cup Z} T_i(x) \to_I T_i(\alpha)\}]]$ (by 3)

$\leq r[\bigcap_{i\in I}\{\bigwedge_{x\in X} T_i(x) \to_I T_i(\alpha)\}]$ (by 2).

Hence (GC3) is proved.

(b) Given a graded consequence relation $\mid\sim$, let $\{T_X\}_{X\in P(F)}$ be such that $T_X(\alpha) = [gr(X \mid\sim \alpha)]_{BIR}$. We want to show $gr(X \mid\sim \alpha) = r[\bigcap_{Y\in P(F)}\{\bigwedge_{x\in X} T_Y(x) \to_I T_Y(\alpha)\}]$.

Arguing as Theorem 7.1 we have, $T_X(\beta) \leq_I \wedge_{Y\in P(F)}[\wedge_{\alpha\in X} T_Y(\alpha) \to_I T_Y(\beta)]$

$\subseteq_e \cap_{Y\in P(F)}[\wedge_{\alpha\in X} T_Y(\alpha) \to_I T_Y(\beta)]$ (Lemma 7.9)

$r(T_X(\beta)) \leq r(\wedge_{Y\in P(F)}[\wedge_{\alpha\in X} T_Y(\alpha) \to_I T_Y(\beta)]) \leq r(\cap_{Y\in P(F)}[\wedge_{\alpha\in X} T_Y(\alpha) \to_I T_Y(\beta)])$ ...(1)

$\cap_{Y\in P(F)}[\wedge_{\alpha\in X} T_Y(\alpha) \to_I T_Y(\beta)] \subseteq_e \cap_{X\subseteq Y\in P(F)}[\wedge_{\alpha\in X} T_Y(\alpha) \to_I T_Y(\beta)]$ ...(2)

Following the proof of Theorem 7.1, for $X \subseteq Y$, $\wedge_{\alpha\in X} T_Y(\alpha) \to_I T_Y(\beta) = T_Y(\beta)$. From (2) and Lemma 7.10, $\cap_{Y\in P(F)}[\wedge_{\alpha\in X} T_Y(\alpha) \to_I T_Y(\beta)] \subseteq_e \wedge_{X\subseteq Y} T_Y(\beta) = T_X(\beta)$. Thus, $r(T_X(\beta)) = r(\cap_{Y\in P(F)}[\wedge_{\alpha\in X} T_Y(\alpha) \to_I T_Y(\beta)]) = gr(X \mid\sim \alpha)$. □

Here the same structure $\langle \mathscr{U}, *_I, \to_I, [0,0], [1,1], r\rangle$, as mentioned in Note 7.1, is taken; the only differences are: (i) $\mathscr{U}$ is endowed with both the lattice order relations $\leq_I$ and $\subseteq_e$, and (ii) a function $r : \mathscr{U} \mapsto [0,1]$, different from l, is considered here.

Let us present, below, a diagram to visualize the beauty and purpose of dealing with two lattice structures over the same domain. We consider a linear scale D and intervals over D. We consider $D = \{3, 5, 7, 9\}$ and $I_D = \{[a, b] : a \leq b$ and $a, b \in D\}$.

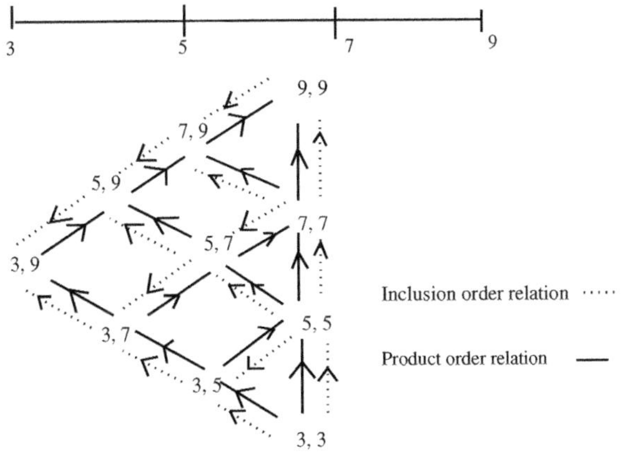

The diagram represents two lattice structures with respect to $\leq_I$ (product order relation) and $\subseteq_e$ (inclusion order relation) over I_D. Let $\{T_1, T_2, T_3\}$ be the set of experts who assign interval-value from I_D to every formula. Assuming the ordinary t-norm $\wedge$ on D, one can immediately obtain the corresponding residuum $\rightarrow$ on D, and hence $\wedge_{I_D}$, $\rightarrow_{I_D}$ on I_D can be constructed. Given a set of formulae X and a formula α, to compute $gr(X \approx\!\!| \alpha)$ in the sense of both (Σ) and (Σ'), first for each T_i $(i = 1, 2, 3)$, the value $\bigwedge_{x \in X} T_i(x) \rightarrow_{I_D} T_i(\alpha)$ needs to be computed. Let the respective values corresponding to T_1, T_2, T_3 be $[3, 5]$, $[5, 7]$, $[7, 7]$. Then, for (Σ), the least interval $[3, 5]$ will be selected and the left-hand end point 3 would be counted as the grade of $X \approx\!\!| \alpha$. Following (Σ'), $[5, 5]$ will be chosen as the interval included in all the intervals in the sense of $\subseteq_e$, and its right-hand end point 5 would be counted as the grade of $X \approx\!\!| \alpha$. The first method pulls down one of the experts high opinion, which is here 7, drastically to 3, whereas the second admits some room for adjustment between different opinions, and pulls down the value to 5. Thus, (Σ') provides a good way of respecting individual's opinion.

7.1.5 Graded Consequence: Form (Σ'')

Among the above two forms of graded consequence, (Σ) is based on a *conservative* attitude as it chooses the left-hand end point of the interval lying below each expert's (T_i) opinion for computing the value of $\wedge_{x \in X} T_i(x) \rightarrow_I T_i(\alpha)$. On the other hand, form (Σ') admits very *liberal* attitude as it takes the right-hand end point of the interval-value which lies at the zone of common consensus of the values for $\wedge_{x \in X} T_i(x) \rightarrow_I T_i(\alpha)$ taking care of every expert's opinion. Both of these reflect two extremities of decision-making attitude. Below we would look for an approach where considering each expert's opinion first an interval for $\wedge_{x \in X} T_i(x) \rightarrow_I T_i(\alpha)$ is assigned. Then, a number of times the assigned interval can be revised; the number is being stipulated by different constraints. Finally among these iterations for the values of $\wedge_{x \in X} T_i(x) \rightarrow_I T_i(\alpha)$ one would be chosen, and from all such revised interval values the common zone will be selected. This idea of iterative revision of an interval-value assignment is taken care of in the following series of definitions. Finally, Theorem 7.6 of this section throws light on the fact that the forms generated from iterative revisions (Σ_1'', Σ_2'') obtain a place between the two extreme attitudes of decision-making.

Definition 7.8 $\mathscr{I}_{[a,b]}$ is a collection of iterated revisions $[x^i, y^i]$'s of $[a, b]$, given by: $I_{[a,b]} = \{[x^i, y^i] : x^0 = a, y^0 = b, [x^i, y^i] \subseteq_e [x^{i-1}, y^{i-1}]$ and $[x^{i-1}, y^{i-1}]$ is non-degenerate$\}$. Let $\mathscr{I}^j_{[a,b]} (\subseteq \mathscr{I}_{[a,b]})$ be a set containing iterated revisions of $[a, b]$ up to j-th iterations.

Definition 7.9 $\mathscr{C}_{\mathscr{I}}$ is a choice function over $\{\mathscr{I}^j_{[a,b]} : j \geq 0, [a, b] \in \mathscr{U}\}$ such that (i) $\mathscr{C}_{\mathscr{I}}(\mathscr{I}^j_{[a,b]}) \in \mathscr{I}^j_{[a,b]}$ and (ii) $[a, b] \leq_I [c, d]$ implies $\mathscr{C}_{\mathscr{I}}(\mathscr{I}^j_{[a,b]}) \leq_I \mathscr{C}_{\mathscr{I}}(\mathscr{I}^j_{[c,d]})$.

Let us present one such case of iterative revision of intervals below.

Definition 7.10 For $\varepsilon > 0$, $\underline{C}_\varepsilon: \mathscr{U} \mapsto \mathscr{U}$ such that $\underline{C}_\varepsilon$ $([a, b]) = [a + \varepsilon, b]$ if $a + \varepsilon < b = [b, b]$, otherwise.
$\overline{C}_\varepsilon: \mathscr{U} \mapsto \mathscr{U}$ such that $\overline{C}_\varepsilon([a, b]) = [a, b - \varepsilon]$ if $a < b - \varepsilon = [a, a]$, otherwise.
$\mathscr{C}_\varepsilon: \mathscr{U} \mapsto \mathscr{U}$ such that $\mathscr{C}_\varepsilon([a, b]) = \underline{C}_\varepsilon([a, b]) \cap \overline{C}_\varepsilon([a, b])$.

Let us choose an arbitrarily fixed number n. Given $[a, b]$, fixing $\varepsilon \geq \frac{b-a}{n}$ and applying $\mathscr{C}_\varepsilon$ finite number of times on $[a, b]$, one instance of $I_{[a,b]}$, we call $\mathscr{I}^\varepsilon_{[a,b]}$, can be obtained in the following way.

Definition 7.11 $\mathscr{I}^\varepsilon_{[a,b]} = \{\mathscr{C}^i_\varepsilon([a, b]) : i \geq 0, \mathscr{C}^0_\varepsilon([a, b]) = [a, b], \mathscr{C}^{i-1}_\varepsilon([a, b])$ is non-degenerate$\}$.

Note 7.3 Taking $\varepsilon = \frac{1}{2}$ we have $\mathscr{I}^{\frac{1}{2}}_{[.3,.7]} = \{[.3, .7], [.3, .3]\}$, $\mathscr{I}^{\frac{1}{2}}_{[.1,.9]} = \{[.1, .9], [.4, .4]\}$. It is to be noted that as $\mathscr{I}^{\frac{1}{2}}_{[.3,.7]}$ contains only two iterations $\mathscr{I}^{\frac{1}{2}}_{[.3,.7]} = \mathscr{I}^{\frac{1}{2},j}_{[.3,.7]} = \{[.3, .7], [.3, .3]\}$ for any number of iterations $j \geq 2$, where $\mathscr{I}^{\frac{1}{2},j}_{[.3,.7]}$ contains intervals up to j-th iterations.

Definition 7.12 $\mathscr{C}_\mathscr{I}(\mathscr{I}^{\varepsilon,j}_{[a,b]}) = \cap\, \mathscr{I}^{\varepsilon,j}_{[a,b]}$, where j denotes the number of iterations.

Note 7.4 As $\mathscr{I}^\varepsilon$ is obtained by finitely many iterations, $\mathscr{C}_\mathscr{I}(\mathscr{I}^\varepsilon_{[a,b]}) = \cap\, \mathscr{I}^\varepsilon_{[a,b]}$, which is a degenerate interval. Clearly, $\mathscr{C}_\mathscr{I}$ (of Definition 7.12) satisfies condition (i) of the Definition 7.9 as $\mathscr{C}_\mathscr{I}(\mathscr{I}^{\varepsilon,j}_{[a,b]}) = \mathscr{C}^j_\varepsilon([a, b]) \in \mathscr{I}^\varepsilon_{[a,b]}$. To check that $\mathscr{C}_\mathscr{I}$ also satisfies condition (ii) of Definition 7.9, we need to prove a series of results below.

Proposition 7.5 *$\mathscr{C}_\varepsilon([a, b]) \subseteq_e [a, b]$.*

Proof Two cases arise. (i) $a + \varepsilon < b$, i.e. $a < b - \varepsilon$ (ii) otherwise.
For all these cases the result is straightforward from the definitions of $\subseteq_e$ and $\cap$. □

Theorem 7.3 *$[a, b] \leq_I [c, d]$ implies $\mathscr{C}_\varepsilon([a, b]) \leq_I \mathscr{C}_\varepsilon([c, d])$.*

Proof Let $[a, b] \leq_I [c, d]$, then $a \leq c, b \leq d$, and $a \leq b, c \leq d$.
Now (i) $\underline{C}_\varepsilon([a, b]) = [a + \varepsilon, b]$ or (ii) $\underline{C}_\varepsilon([a, b]) = [b, b]$.
(i) Let $\underline{C}_\varepsilon([a, b]) = [a + \varepsilon, b]$, i.e. $a + \varepsilon < b \leq d$. Also $a + \varepsilon \leq c + \varepsilon$.
Now either $c + \varepsilon < d$ or $d \leq c + \varepsilon$. Both the cases yield $\underline{C}_\varepsilon([a, b]) \leq_I \underline{C}_\varepsilon([c, d])$, as for $c + \varepsilon < d$, $\underline{C}_\varepsilon([c, d]) = [c + \varepsilon, d]$, i.e. $[a + \varepsilon, b] \leq_I [c + \varepsilon, d]$, and for $d \leq c + \varepsilon$, $\underline{C}_\varepsilon([c, d]) = [d, d]$, i.e. $[a + \varepsilon, b] \leq_I [d, d]$, since $a + \varepsilon < b \leq d$.
(ii) Let $\underline{C}_\varepsilon([a, b]) = [b, b]$, i.e. $b \leq a + \varepsilon \leq c + \varepsilon$ and also $b \leq d$.
Now either $c + \varepsilon < d$ or $d \leq c + \varepsilon$. If $c + \varepsilon < d$, $\underline{C}_\varepsilon([c, d]) = [c + \varepsilon, d]$. So, $[b, b] \leq_I [c + \varepsilon, d]$.
And if $d \leq c + \varepsilon$, then $\underline{C}_\varepsilon([c, d]) = [d, d]$, i.e. $[b, b] \leq_I [c + \varepsilon, d]$.
Hence $[a, b] \leq_I [c, d]$ implies $\underline{C}_\varepsilon([a, b]) \leq_I \underline{C}_\varepsilon([c, d])$.
Now we shall check the case for $\overline{C}_\varepsilon$. Let $[a, b] \leq_I [c, d]$.
There are two cases. (a) $\overline{C}_\varepsilon([a, b]) = [a, b - \varepsilon]$ (b) $\overline{C}_\varepsilon([a, b]) = [a, a]$.
(a) If $\overline{C}_\varepsilon([a, b]) = [a, b - \varepsilon]$, then $a < b - \varepsilon < b \leq d$ and $b - \varepsilon \leq d - \varepsilon$.
Now either $c < d - \varepsilon$ or $d - \varepsilon \leq c$.

If $c < d - \varepsilon$, $\overline{C}_\varepsilon([c, d]) = [c, d - \varepsilon]$, i.e. $[a, b - \varepsilon] \leq_I [c, d - \varepsilon]$ as $a \leq c, b \leq d$. If $d - \varepsilon \leq c$, $\overline{C}_\varepsilon([c, d]) = [c, c]$, So, $[a, b - \varepsilon] \leq_I [c, c]$ as $a \leq c, b - \varepsilon \leq d - \varepsilon \leq c$.

(b) Let $\overline{C}_\varepsilon([a, b]) = [a, a]$. Then $b - \varepsilon \leq a \leq b \leq d, a \leq c$.

Now either $c < d - \varepsilon$ or $d - \varepsilon \leq c$.

If $c < d - \varepsilon$, $\overline{C}_\varepsilon([c, d]) = [c, d - \varepsilon]$, i.e. $[a, a] \leq_I [c, d - \varepsilon]$ since $a \leq c < d - \varepsilon$. And if $d - \varepsilon \leq c$, $\overline{C}_\varepsilon([c, d]) = [c, c]$. So, $[a, a] \leq_I [c, c]$.

Therefore, $[a, b] \leq_I [c, d]$ implies $\overline{C}_\varepsilon([a, b]) \leq_I \overline{C}_\varepsilon([c, d])$.

Hence, $\mathscr{C}_\varepsilon([a, b]) = \underline{C}_\varepsilon([a, b]) \cap \overline{C}_\varepsilon([a, b]) \leq_I \underline{C}_\varepsilon([c, d]) \cap \overline{C}_\varepsilon([c, d]) = \mathscr{C}_\varepsilon([c, d])$. □

Corollary 7.2 $[a, b] \leq_I [c, d]$ *implies* $\mathscr{C}_\mathscr{I}(\mathscr{I}^{\varepsilon,j}_{[a,b]}) \leq_I \mathscr{C}_\mathscr{I}(\mathscr{I}^{\varepsilon,j}_{[c,d]})$.

Theorem 7.4 $[a, b] \leq_I [c, d]$ *implies* $\mathscr{C}_\mathscr{I}(\mathscr{I}^{\varepsilon}_{[a,b]}) \leq_I \mathscr{C}_\mathscr{I}(\mathscr{I}^{\varepsilon}_{[c,d]})$.

Proof Let $\mathscr{C}^i_\varepsilon([a, b])$ and $\mathscr{C}^j_\varepsilon([c, d])$ be degenerate. Then either (i) $i = j$ or (ii) $i \leq j$, or (iii) $j \leq i$.

(i) If $i = j$, by Theorem 7.3, $\mathscr{C}_\mathscr{I}(\mathscr{I}^{\varepsilon}_{[a,b]}) = \mathscr{C}^i_\varepsilon([a, b]) \leq_I \mathscr{C}^i_\varepsilon([c, d]) = \mathscr{C}_\mathscr{I}(\mathscr{I}^{\varepsilon}_{[c,d]})$.

(ii) By Theorem 7.3, $l(\mathscr{C}_\mathscr{I}(\mathscr{I}^{\varepsilon}_{[a,b]})) = l(\mathscr{C}^i_\varepsilon([a, b])) \leq l(\mathscr{C}^i_\varepsilon([c, d])) \leq l(\mathscr{C}^j_\varepsilon([c, d]))$ $= l(\mathscr{C}_\mathscr{I}(\mathscr{I}^{\varepsilon}_{[c,d]}))$.

That is, $\mathscr{C}_\mathscr{I}(\mathscr{I}^{\varepsilon}_{[a,b]}) \leq_I \mathscr{C}_\mathscr{I}(\mathscr{I}^{\varepsilon}_{[c,d]})$.

(iii) If $j \leq i$, $r(\mathscr{C}_\mathscr{I}(\mathscr{I}^{\varepsilon}_{[c,d]})) = r(\mathscr{C}^j_\varepsilon([c, d])) \geq r(\mathscr{C}^j_\varepsilon([a, b]))$. Hence, by Note 7.3 we have $r(\mathscr{C}^j_\varepsilon([a, b])) = r(\mathscr{C}^i_\varepsilon([a, b])) = r(\mathscr{C}_\mathscr{I}(\mathscr{I}_{[a,b]}))$, i.e. $\mathscr{C}_\mathscr{I}(\mathscr{I}^{\varepsilon}_{[a,b]}) \leq_I \mathscr{C}_\mathscr{I}(\mathscr{I}^{\varepsilon}_{[c,d]})$. □

Thus, we have shown the existence of a $\mathscr{C}_\mathscr{I}$ which satisfies conditions of Definition 7.9. Now we come back to the general context of the Definition 7.9 for $\mathscr{C}_\mathscr{I}$.

Note 7.5 $\mathscr{C}_\mathscr{I}(\mathscr{I}^{j}_{[a,b]})$, $\mathscr{C}_\mathscr{I}(\mathscr{I}_{[a,b]}) \subseteq_e [a, b]$, and $\mathscr{C}_\mathscr{I}(\mathscr{I}^{j}_{[a,b]})$, $\mathscr{C}_\mathscr{I}(\mathscr{I}_{[a,b]})$, $[a, b]$ are mutually overlapping pairs of intervals. For $\mathscr{I}^\varepsilon$, $\mathscr{C}_\mathscr{I}(\mathscr{I}^{\varepsilon}_{[a,b]}) \subseteq_e \mathscr{C}_\mathscr{I}(\mathscr{I}^{\varepsilon,j}_{[a,b]}) \subseteq_e [a, b]$.

We now propose different notions of $\approx$ based on the notion of iterative revision.

Definition 7.13 Given any collection of interval-valued fuzzy sets $\{T_i\}_{i \in I}$,

(Σ_1'') $gr(X \approx^{(\Sigma_1'')} \alpha) = r(\cap_{i \in I}\{\mathscr{C}_\mathscr{I}(\mathscr{I}^{j}_{[\wedge_{x \in X} T_i(x) \rightarrow_I T_i(\alpha)]})\})$ for arbitrarily fixed $j \geq 0$,

(Σ_2'') $gr(X \approx^{(\Sigma_2'')} \alpha) = r(\cap_{i \in I}\{\mathscr{C}_\mathscr{I}(\mathscr{I}^{\varepsilon}_{[\wedge_{x \in X} T_i(x) \rightarrow_I T_i(\alpha)]})\})$
$= l(\cap_{i \in I}\{\mathscr{C}_\mathscr{I}(\mathscr{I}^{\varepsilon}_{[\wedge_{x \in X} T_i(x) \rightarrow_I T_i(\alpha)]})\})$.

Theorem 7.5 *For any graded consequence relation* $\mid\sim$, *there is a collection of interval-valued fuzzy sets such that* $\approx$ *generated in the sense of* (Σ_1''), (Σ_2'') *coincide with* $\mid\sim$.

Proof Given $\mid\sim$, a graded consequence relation, let us consider $\{T_X\}_{X \in P(F)}$ such that $T_X(\alpha) = [gr(X \mid\sim \alpha)]_{BIR}$.

We want to prove $gr(X \mid\sim \alpha) = r[\bigcap_{Y \in P(F)}\{\mathscr{C}_\mathscr{I}(\mathscr{I}^{i}_{[\bigwedge_{x \in X} T_Y(x) \rightarrow_I T_Y(\alpha)]})\}]$ considering (Σ_1''), and $gr(X \mid\sim \alpha) = r[\bigcap_{Y \in P(F)}\{\mathscr{C}_\mathscr{I}(\mathscr{I}^{\varepsilon}_{[\bigwedge_{x \in X} T_Y(x) \rightarrow_I T_Y(\alpha)]})\}]$ considering

(Σ_2''). As for $\alpha \in F$ and $Y \subseteq F$, $T_Y(\alpha)$ is a degenerate interval, $\bigwedge_{x\in X} T_Y(x) \to_I T_Y(\alpha)$ is a degenerate interval. Hence by the construction of $\mathscr{I}$ and $\mathscr{I}^\varepsilon$, we have the following:
$\mathscr{C}_{\mathscr{I}}(\mathscr{I}^i_{[\bigwedge_{x\in X} T_Y(x)\to_I T_Y(\alpha)]}) = \bigwedge_{x\in X} T_Y(x) \to_I T_Y(\alpha) = \mathscr{C}_{\mathscr{I}}(\mathscr{I}^\varepsilon_{[\bigwedge_{x\in X} T_Y(x)\to_I T_Y(\alpha)]})$.
i.e. $r(\cap_{Y\in P(F)}\{\mathscr{C}_{\mathscr{I}}(\mathscr{I}^i_{[\wedge_{x\in X} T_i(x)\to_I T_i(\alpha)]})\}) = r(\cap_{Y\in P(F)}\{\mathscr{C}_{\mathscr{I}}(\mathscr{I}^\varepsilon_{[\wedge_{x\in X} T_i(x)\to_I T_i(\alpha)]})\})$
$= r[\bigcap_{Y\in P(F)}[\bigwedge_{x\in X} T_Y(x) \to_I T_Y(\alpha)]$.
Rest follows from the Theorem 7.2. □

Let us distinguish the notions of graded semantic consequence by superscribing $\approx\!\!|$ with their respective forms. So, we have $|\!\!\approx^{(\Sigma)}$, $|\!\!\approx^{(\Sigma')}$, and $|\!\!\approx^{(\Sigma_n'')}$, $n = 1, 2$.

Theorem 7.6 *$gr(X |\!\!\approx^{(\Sigma)} \alpha) \le gr(X |\!\!\approx^{(\Sigma_1'')} \alpha) \le gr(X |\!\!\approx^{(\Sigma')} \alpha)$ and $gr(X |\!\!\approx^{(\Sigma)} \alpha) \le gr(X |\!\!\approx^{(\Sigma_2'')} \alpha) \le gr(X |\!\!\approx^{(\Sigma')} \alpha)$.*

Proof By Lemma 7.9, $l(\wedge_{i\in I}[\wedge_{x\in X} T_i(x) \to_I T_i(\alpha)]) \le l(\cap_{i\in I}[\wedge_{x\in X} T_i(x) \to_I T_i(\alpha)])$
$\le r(\cap_{i\in I}[\wedge_{x\in X} T_i(x) \to_I T_i(\alpha)])$ …(i)
$\cap_{i\in I}\mathscr{C}_{\mathscr{I}}(\mathscr{I}_{[\wedge_{x\in X} T_i(x)\to_I T_i(\alpha)]})$, $\cap_{i\in I}\mathscr{C}_{\mathscr{I}}(\mathscr{I}^j_{[\wedge_{x\in X} T_i(x)\to_I T_i(\alpha)]}) \subseteq_e \cap_{i\in I}[\wedge_{x\in X} T_i(x) \to_I T_i(\alpha)]$
and they are overlapping. So, we have the following inequalities:
$l(\cap_{i\in I}[\wedge_{x\in X} T_i(x) \to_I T_i(\alpha)]) \le l(\cap_{i\in I}\mathscr{C}_{\mathscr{I}}(\mathscr{I}^j_{[\wedge_{x\in X} T_i(x)\to_I T_i(\alpha)]}))$
$\le r(\cap_{i\in I}\mathscr{C}_{\mathscr{I}}(\mathscr{I}^j_{[\wedge_{x\in X} T_i(x)\to_I T_i(\alpha)]}))$
$\le r(\cap_{i\in I}[\wedge_{x\in X} T_i(x) \to_I T_i(\alpha)])$ …(ii)
(i) and (ii) imply, $gr(X |\!\!\approx^{(\Sigma)} \alpha) \le gr(X |\!\!\approx^{(\Sigma_1'')} \alpha) \le gr(X |\!\!\approx^{(\Sigma')} \alpha)$.
Also following Note 7.5, $l(\cap_{i\in I}[\wedge_{x\in X} T_i(x) \to_I T_i(\alpha)]) \le r(\cap_{i\in I}\mathscr{C}_{\mathscr{I}}(\mathscr{I}^\varepsilon_{[\wedge_{x\in X} T_i(x)\to_I T_i(\alpha)]}))$
$\le r(\cap_{i\in I}[\wedge_{x\in X} T_i(x) \to_I T_i(\alpha)])$ …(iii).
Hence, combining (i) and (iii), $gr(X |\!\!\approx^{(\Sigma)} \alpha) \le gr(X |\!\!\approx^{(\Sigma_2'')} \alpha) \le gr(X |\!\!\approx^{(\Sigma')} \alpha)$. □

Note 7.6 If for (Σ_1'') $\mathscr{I}^{\varepsilon,j}$ (of Definition 7.12) is chosen instead of $\mathscr{I}^j$ then $gr(X |\!\!\approx^{(\Sigma)} \alpha) \le gr(X |\!\!\approx^{(\Sigma_2'')} \alpha) \le gr(X |\!\!\approx^{(\Sigma_1'')} \alpha) \le gr(X |\!\!\approx^{(\Sigma')} \alpha)$.

Theorem 7.7 *Given $\{T_i\}_{i\in I}$, $|\!\!\approx$ in the sense of (Σ_n'') $(n = 1, 2)$, satisfies (GC1), (GC2), and a variant of (GC3).*

Proof (GC1) is immediate. If $X \subseteq Y$, $\mathscr{C}_{\mathscr{I}}(\mathscr{I}^j_{[\wedge_{x\in X} T_i(x)\to_I T_i(\alpha)]}) \le_I \mathscr{C}_{\mathscr{I}}(\mathscr{I}^j_{[\wedge_{x\in Y} T_i(x)\to_I T_i(\alpha)]})$ and $\mathscr{C}_{\mathscr{I}}(\mathscr{I}^\varepsilon_{[\wedge_{x\in X} T_i(x)\to_I T_i(\alpha)]}) \le_I \mathscr{C}_{\mathscr{I}}(\mathscr{I}^\varepsilon_{[\wedge_{x\in Y} T_i(x)\to_I T_i(\alpha)]})$ are obtained by condition (ii) of Definition 7.9 and Theorem 7.4, respectively. Hence by Lemma 7.8, GC2 holds for (Σ_1''), (Σ_2'').
Now for each of (Σ_n''), $n = 1, 2$, we prove that a variant form of (GC3), i.e.
$\inf_{\beta\in Z} gr(X |\!\!\approx^{\Sigma_n''} \beta) * gr(X \cup Z |\!\!\approx^{\Sigma_n''} \alpha) \le gr(X |\!\!\approx^{\Sigma'} \alpha)$ holds.
$\inf_{\beta\in Z} gr(X |\!\!\approx^{\Sigma_2''} \beta) * gr(X \cup Z |\!\!\approx^{\Sigma_2''} \alpha)$
$= \inf_{\beta\in Z} [r\{\cap_{i\in I}\mathscr{C}_{\mathscr{I}}(\mathscr{I}^\varepsilon_{[\wedge_{x\in X} T_i(x)\to_I T_i(\beta)]})\}] * r[\cap_{i\in I}\mathscr{C}_{\mathscr{I}}(\mathscr{I}^\varepsilon_{[\wedge_{x\in X\cup Z} T_i(x)\to_I T_i(\alpha)]})\}]$.
$= r[\cap_{\beta\in Z}\{\cap_{i\in I}\mathscr{C}_{\mathscr{I}}(\mathscr{I}^\varepsilon_{[\wedge_{x\in X} T_i(x)\to_I T_i(\beta)]})\}] * r[\cap_{i\in I}\mathscr{C}_{\mathscr{I}}(\mathscr{I}^\varepsilon_{[\wedge_{x\in X\cup Z} T_i(x)\to_I T_i(\alpha)]})\}]$.
(Lemma 7.5)
$= r[\cap_{i\in I}\{\cap_{\beta\in Z}\mathscr{C}_{\mathscr{I}}(\mathscr{I}^\varepsilon_{[\wedge_{x\in X} T_i(x)\to_I T_i(\beta)]})\}] * r[\cap_{i\in I}\mathscr{C}_{\mathscr{I}}(\mathscr{I}^\varepsilon_{[\wedge_{x\in X\cup Z} T_i(x)\to_I T_i(\alpha)]})\}]$
(Lemma 7.6) …(i)
By Note 7.5, $\cap_{\beta\in Z}\mathscr{C}_{\mathscr{I}}(\mathscr{I}^\varepsilon_{[\wedge_{x\in X} T_i(x)\to_I T_i(\beta)]}) \subseteq_e \cap_{\beta\in Z}\{\wedge_{x\in X} T_i(x) \to_I T_i(\beta)\}$
$\subseteq_e \wedge_{\beta\in Z}\{\wedge_{x\in X} T_i(x) \to_I T_i(\beta)\}$ (Lemma 7.10).

Hence, $r(\cap_{i\in I}[\cap_{\beta\in Z}\mathscr{C}_{\mathscr{I}}(\mathscr{I}^{\varepsilon}_{[\wedge_{x\in X}T_i(x)\to_I T_i(\beta)]})]) \leq r(\cap_{i\in I}[\wedge_{\beta\in Z}\{\wedge_{x\in X}T_i(x)\to_I T_i(\beta)\}])$.
Similarly, $r(\cap_{i\in I}[\mathscr{C}_{\mathscr{I}}(\mathscr{I}^{\varepsilon}_{[\wedge_{x\in X\cup Z}T_i(x)\to_I T_i(\alpha)]})]) \leq r(\cap_{i\in I}[\{\wedge_{x\in X\cup Z}T_i(x)\to_I T_i(\alpha)\}])$.
(i) becomes, $\inf_{\beta\in Z} gr(X \approx^{\Sigma_2''} \beta) * gr(X\cup Z \approx^{\Sigma_2''} \alpha)$

$$\leq r(\cap_{i\in I}[\wedge_{\beta\in Z}\{\wedge_{x\in X}T_i(x)\to_I T_i(\beta)\}]) * r(\cap_{i\in I}[\{\wedge_{x\in X\cup Z}T_i(x)\to_I T_i(\alpha)\}])$$
$$= r(\cap_{i\in I}[\{\wedge_{x\in X}T_i(x)\to_I \wedge_{\beta\in Z}T_i(\beta)\}]) * r(\cap_{i\in I}[\{\wedge_{x\in X\cup Z}T_i(x)\to_I T_i(\alpha)\}]).$$

Now following the steps below the inequality (1) of the proof (a) of Theorem 7.2, we get: $\inf_{\beta\in Z} gr(X \approx^{\Sigma_2''} \beta) * gr(X\cup Z \approx^{\Sigma_2''} \alpha) \leq r[\cap_{i\in I}\{\wedge_{x\in X}T_i(x)\to_I T_i(\alpha)\}]$
$= gr(X \approx^{\Sigma'} \alpha)$.
Similar is the argument for $\inf_{\beta\in Z} gr(X \approx^{\Sigma_1''} \beta) * gr(X\cup Z \approx^{\Sigma_1''} \alpha) \leq gr(X \approx^{\Sigma'} \alpha)$.

□

From the development made in this section we observe that, in the context of interval semantics, GCT is simultaneously exploiting two different lattice structures ($\leq_I$, $\subseteq_e$) over the set of sub-intervals of [0, 1]. This adds an important dimension. Having endowed with both the notions, viz. *interval lying below a set of intervals* ($\leq_I$) and *interval lying in the intersection of a set of intervals* ($\subseteq_e$) we are able to address different attitudes of decision-making. Given a database based on the opinions of a set of experts, different notions of $\approx$ are introduced to address the following aspects of decision-making. (i) (Σ) proposes a set-up where the interval lying below all the experts' opinion would be counted. (ii) (Σ') proposes a set-up where the interval lying in the zone of common consensus would be counted. (iii) (Σ_1'') proposes a set-up where each expert can revise their initial interval of opinion finite number of times, and the number of such chances of revision is equal for each of the expert. Then the interval lying in the zone of common consensus is counted. (iv) (Σ_2'') proposes a set-up where the interval, taking care of expert's opinion, are revised (following a specific method, viz. $\mathscr{I}^{\varepsilon}$) till they reach a concrete value, and then the zone of common consensus is considered. We call the approach (i) as *conservative*, (ii) as liberal, and (iii), (iv) as moderate. Thus, this study provides a theoretical framework where a decision-maker having some of the above attitudes derives, with certain degree, a decision from a set of imprecise information.

7.2 GCT in the Context of Logic Infomorphism: A Case for Decision-Making in a Distributed Network

Development of GCT, as presented in the previous chapters, assumes a semantic base which is an arbitrary collection $\{T_i\}_{i\in I}$ of fuzzy sets assigning values (degree of truth or degree of belief) to the object language formulae. Each T_i may be counted as an expert whose opinion, i.e. values assigned to the object language formulae, forms the initial context or the database. Based on the collective data of $\{T_i\}_{i\in I}$ decisions are made. The decision-maker's concern is to decide whether a particular formula α is a consequence of a set X of formulae, and that is derived based on the opinions of the experts (that is, values assigned to the object language formulae by T_i's) as well as the reasoning base of the decision-maker (that is, the meta-linguistic algebraic structure). While assigning values to the basic sentences of the object language, experts can follow some logic suitable for the object language; the decision-maker may choose her own reasoning base, which is not necessarily the same as that of the experts. That is, reasoning bases for the object language and that of the meta-language need not be the same. So, in this context presence of two levels of reasoning, one for the object language sentences which are the direct concern of the experts/agents, and the meta-linguistic sentences which are generally about deriving the decisions, are accommodated. But in this framework, we have not yet looked into the reasoning base, which may be called local logic, of each individual agent/expert. In this regard, we would propose addition of a more concrete structure for each T_i in our work, based on the idea of logic infomorphism developed in Barwise and Seligman (1997) in a four-valued set-up (Vitória et al. 2009). Though this proposed extension (Dutta et al. 2019) will greatly depend on the work of Barwise and Seligman, there will be some additional consideration of a notion of belief as well, need for which is found explicitly in the intuitive discussion on decision-making in a distributed network (Barwise and Seligman 1997), but which did not appear in the formal development (Barwise and Seligman 1997).

Let us restate here the story which John Barwise and Jerry Seligman have used in Barwise and Seligman (1997) to exhibit that the way an individual reasons every day is part of a distributed logical framework where information flows among different components of the system. Decisions, taken at some components, use the shared information from the other components.

'*Judith, a keen but inexperienced mountaineer, embarked on an ascent of Mt. Ateb. She took with her a map, …, a flashlight. …Encouraged by the ease of the day's climb, she decided to take a different route down. …By 4 P.M Judith was hopelessly lost. …the loose stones betrayed her, and she tumbled a hundred feet before breaking her fall against a hardy uplands thorn. Clinging to the bush and wincing at the pain in her left leg, she took stock. It would soon be dark. …Suddenly she remembered the flashlight. …She began to flash out into the twilight. By a miracle, her signal was seen by another day hiker, …. Miranda quickly recognized the dots and dashes of the SOS and …phoned Mountain Rescue. Only twenty minutes later the searchlight from a helicopter scanned …, illuminating the frightened Judith, …*'

Let us analyse how in the story of Judith and Miranda information is carried from one to the other. The three important principles of information flow, as mentioned in Barwise and Seligman (1997) based on the theory of knowledge and flow of information by Dretske (1981), are (a) information flow results from regularities in the system, (b) to a person with prior knowledge k, r being F carries the information that s is G if the person could legitimately infer that s is G from r being F together with k, and (iii) knowledge is justified true belief and believing that p, carries the information that p. The factors behind the flow of information in the story, thus, can be sequentially arranged as follows.

The flashlight caught Miranda's attention. She *believes* that in dark mountain flashlight may indicate someone is in danger. Miranda *knows* that there is some *statistical regularity* established between flashes from the mountainside and climbers in distress. Her belief about someone is in trouble is supported by the *prior knowledge* of the statistical regularity. Thus, from the flashlight Miranda gains some information, using *legitimate sense of inference*, that someone might be in trouble.

Neither *believing* something, and hence having *justified true belief*, i.e. *knowing* something, nor *legitimately inferring* something, in an uncertain, incomplete information scenario like above, is a yes/no concept. Judith might have some uncertainty regarding *flashing out light catches someone's attention*. Miranda as well might not be completely certain about her belief that *light in mountain indicates someone is in distress*. These need to be reflected somewhere in the formal theory also. So, incorporation of a graded notion of inference might be advantageous. One more aspect is to incorporate a notion of *belief profile* of a reasoner. As described in the principles of information flow, inferring new information is generated from the relationship among prior knowledge, gathered information and belief of an agent. But while developing the formal theory, the component of belief remains missing in Barwise and Seligman (1997). Our proposal, in this section, is to incorporate a graded notion of inference based on the theory of graded consequence, and a notion of belief structure of an agent based on the theory developed in Keplicz-Dunin and Szałas (2013).

With this aim, this section is organized as follows. In Sect. 7.2.1, we shall first provide the basic relevant notions and results from Barwise and Seligman (1997). Section 7.2.3 comprises relating and extending the framework of GCT in the light of logic infomorphism and distributed network of decision-making, prescribed in Barwise and Seligman (1997). In order to achieve this task, we would take a middle way between two-valuedness and many-valuedness, and this would be obtained by using the notion of paraconsistent set (Vitória et al. 2009). The basic relevant details, in this regard, will be discussed in Sect. 7.2.2.

7.2.1 Barwise and Seligman's Logic for Distributed System

In Barwise and Seligman (1997), the formal counterpart of information available to different sources/agents, including their prior knowledge, is captured through the notion of classification; a classification specifies an agent's information and knowledge regarding which object satisfies which property or is of which type. That Miranda is aware of the statistical regularity between flashes from the mountainside and climbers in distress, mathematically can be incorporated in her classification of information and prior knowledge. The formal definition is as follows.

Definition 7.14 A classification $A = \langle Tok(A), Typ(A), \models_A \rangle$ consists of
(i) a set, $Tok(A)$, of objects to be classified, called tokens of A,
(ii) a set, $Typ(A)$, of properties used to classify the tokens, called the types of A, and
(iii) a binary relation, $\models_A$, between $Tok(A)$ and $Typ(A)$.

If $a \models_A \alpha$, then a is said to be of type α in A. That is, $\models_A$ basically specifies which token is of which type. Following the literature of rough sets (Pawlak 1982; Pawlak and Skowron 2007a, b, c), the notion of classification, presented in Barwise and Seligman (1997), can be viewed as a special kind of information system, which is a tuple $(U, \mathscr{A}, \{V_a\}_{a\in\mathscr{A}}, \{f_a\}_{a\in\mathscr{A}})$ consisting of, respectively, sets of objects, attributes, a set of values for each attribute, and a set of functions for each attribute specifying which object satisfies which attribute with what value. In the context of classification, U is basically $Tok(A)$, $\{(a, v) : a \in \mathscr{A}, v \in V_a\}$ is $Typ(A)$, and for $u \in U$, $f_a(u) = v$ can be associated with $u \models_A (a, v)$ for each $(a, v) \in Typ(A)$.

Now the notion of infomorphism, defined below, represents the relationship between classifications and provides a way of moving information back and forth between them.

Definition 7.15 Let $A = \langle Tok(A), Typ(A), \models_A\rangle$ and $B = \langle Tok(B), Typ(B), \models_B\rangle$ be two classifications. An infomorphism $f : A \rightleftarrows B$ from A to B is a contravariant pair of functions $f = (\hat{f}, \check{f})$ such that $\hat{f} : Typ(A) \mapsto Typ(B)$ and $\check{f} : Tok(B) \mapsto Tok(A)$ satisfying the following fundamental property of infomorphism. $\check{f}(b) \models_A \alpha$ iff $b \models_B \hat{f}(\alpha)$ for each $b \in Tok(B)$ and $\alpha \in Typ(A)$.

The notion of an *interpretation*, sometimes also called a *translation* of one language into another, is an example of infomorphism between classifications. There are two aspects of an interpretation; one is to do with tokens (structures), and the other is to do with types (sentences). An interpretation $\mathscr{I} : L_1 \rightleftarrows L_2$ of languages L_1 into L_2 does two things. At the level of types, it associates with every sentence α of L_1, a sentence $\mathscr{I}(\alpha)$ of L_2, its translation. At the level of tokens, it associates with every structure M for L_2, a structure $\mathscr{I}(M)$ for L_1. The relation that $\mathscr{I}(M) \models_{L_1} \alpha$ iff $M \models_{L_2} \mathscr{I}(\alpha)$ presents that what $\mathscr{I}(\alpha)$ says about the structure M is equivalent to what α says about the structure $\mathscr{I}(M)$.

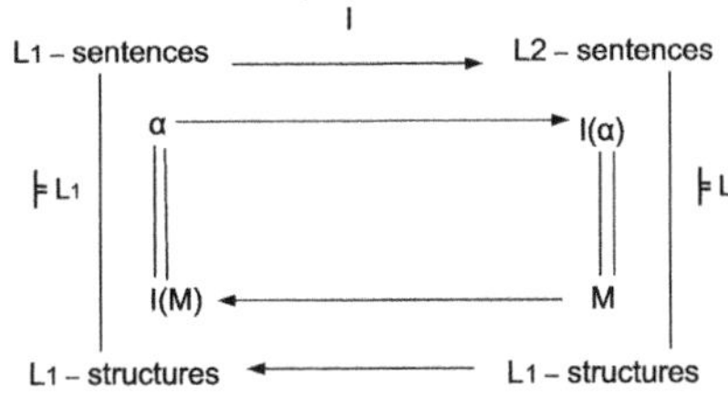

Definition 7.16 Given the infomorphisms $f : A \rightleftarrows B$ and $g : B \rightleftarrows C$, the composition $gf : A \rightleftarrows C$ of f and g is the infomorphism defined by $\hat{gf} = \hat{g}\hat{f}$ and $\check{gf} = \check{f}\check{g}$.

Given a classification of information, often it is found that some tokens are identical with respect to some types, and distinct with respect to the rest. The example, as given in Barwise and Seligman (1997), might render a better understanding in this regard.

My copy of today's edition of the local newspaper bears much in common with that of my next door neighbour. If mine has a picture of President Clinton on page 2, so does hers. If mine has three sections, so does hers. ...Mine has orange juice spilled on it, hers does not. Hers has the crossword puzzle solved, mine does not.

This aspect, in the theory of classification, is captured by the following notions of invariant and quotient classification.

Definition 7.17 Given a classification A, an invariant is a pair $I = (\Sigma, R)$ consisting of a set $\Sigma \subseteq Typ(A)$ of types of A and a binary relation R between tokens of A such that if aRb, then for each $\alpha \in \Sigma$, $a \models_A \alpha$ if and only if $b \models_A \alpha$.

In the above definition though R needs not to be an equivalence relation, in the further considerations R is considered to be the smallest equivalence relation containing the concerned relation.

Definition 7.18 Let $I = (\Sigma, R)$ be an invariant on the classification A. The quotient of A by I, denoted as A/I, is the classification with types Σ, whose tokens are the R-equivalence classes of tokens of A, and with $[a]_R \models_{A/I} \alpha$ if and only if $a \models_A \alpha$.

Definition 7.19 Given a classification A and an invariant $I = (\Sigma, R)$ on A, the canonical quotient infomorphism $\tau : A/I \rightleftarrows A$ is the inclusion function on types, and on tokens, it maps each token of A to its R-equivalence class.

Definition 7.20 Given an invariant $I = (\Sigma, R)$ on A, an infomorphism $f : B \rightleftarrows A$ respects I if (i) for each $\beta \in Typ(B)$, $\hat{f}(\beta) \in \Sigma$, and (ii) if $a_1 R a_2$, then $\check{f}(a_1) = \check{f}(a_2)$.

Given any invariant I, the canonical quotient infomorphism $\tau : A/I \rightleftarrows A$ clearly respects I. Moreover, it does so canonically in the following sense.

Proposition 7.6 *Let I be an invariant on A. Given any infomorphism $f : B \rightleftarrows A$ that respects I, there is a unique infomorphism $f' : B \rightleftarrows A/I$ such that the following diagram commutes.*

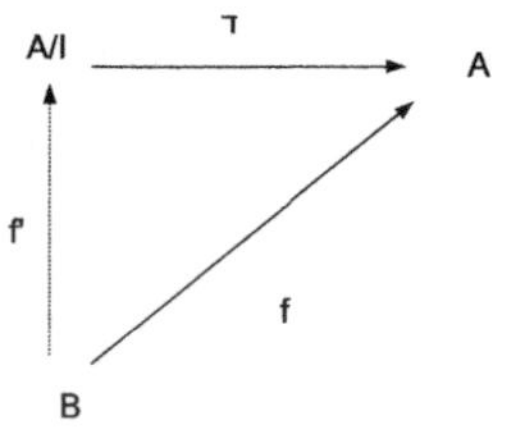

One can notice that the notion of invariance, as defined in Barwise and Seligman (1997), also corresponds to the notion of indiscernibility in the context of rough set literature. In an information system, given by the tuple $(U, A, \{V_a\}_{a\in A}, \{f_a\}_{a\in A})$, two objects x, y of U are said to be indiscernible if $f_a(x)$ and $f_a(y)$ receive the same value from V_a for any $a \in A$. Moreover, the notion of sequent, defined below, also has a counterpart in rough set literature. A sequent can be viewed as a decision rule, i.e. relation between two (finite) sets of attributes describing the available data of the information system.

As pointed out in Barwise and Seligman (1997), '*one way to think about information flow in a distributed system is in terms of a "theory" of the system, that is, a set of known laws that describes the system*'. Based on this general notion of classification of information, the notion of sequent or notion of consequence of a deductive logic is captured as follows.

Definition 7.21 Let $cl(A) = (Tok(A), Typ(A), \models_A)$ be a classification of A.
(i) For any $\Gamma, \Delta \subseteq Typ(A)$, $\langle \Gamma, \Delta \rangle$ is considered to be a sequent of $Typ(A)$.
(ii) A sequent $\langle \Gamma, \Delta \rangle$ is a partition of $\Sigma' \subseteq Typ(A)$ if $\Gamma \cup \Delta = \Sigma'$ and $\Gamma \cap \Delta = \phi$.
(iii) A binary relation $\vdash$ between subsets of $Typ(A)$ is called a consequence relation.
(iv) A theory $T = (\Sigma, \vdash)$ is a pair, where $\Sigma \subseteq Typ(A)$ and $\vdash$ is a consequence relation on Σ. A constraint of the theory T is a sequent $\langle \Gamma, \Delta \rangle$ such that $\Gamma \vdash \Delta$.
(v) A token a of $Tok(A)$ satisfies $\langle \Gamma, \Delta \rangle$ provided that if a is of type α for every $\alpha \in \Gamma$, then a is of type β for some $\beta \in \Delta$. A token not satisfying a sequent is called a counterexample to the sequent.
(vi) The theory $T(cl(A)) = (Typ(A), \vdash_A)$ generated by $cl(A)$ is the theory whose constraints are the set of sequents satisfied by every token of $Tok(A)$.
(vii) A theory whose constraints are satisfied by every token of the classification is called a complete theory.

Here it is to be noted that sequents are all possible pairs of sets of types, and some of them come under the consequence relation. Below we present such conditions in the definition of the *regular theory*, which basically presents the classical way of viewing a deductive logic.

Definition 7.22 A theory $T = (\Sigma, \vdash)$ is regular if it satisfies the following properties, viz. identity, weakening and global cut for all types α, and all set $\Gamma, \Gamma', \Delta, \Delta'$, Σ', Σ_0, Σ_1 of types.

Identity $\alpha \vdash \alpha$
Weakening If $\Gamma \vdash \Delta$, then $\Gamma, \Gamma' \vdash \Delta, \Delta'$.
Global cut If $\Gamma, \Sigma_0 \vdash \Delta, \Sigma_1$ for each partition $\langle \Sigma_0, \Sigma_1 \rangle$ of Σ', then $\Gamma \vdash \Delta$.

Proposition 7.7 *The theory $T(cl(A)) = (Typ(A), \vdash_A)$ generated by the classification $cl(A)$ of A is a regular theory.*

Proposition 7.8 *Any regular theory $T = (\Sigma, \vdash)$ satisfies the following condition.*
Finite cut: If $\Gamma, \alpha \vdash \Delta$ and $\Gamma \vdash \Delta, \alpha$, then $\Gamma \vdash \Delta$.

Definition 7.23 Given two theories $T_1 = (Typ(T_1), \vdash_{T_1})$ and $T_2 = (Typ(T_2), \vdash_{T_2})$, a (regular theory) interpretation $f : T_1 \mapsto T_2$ is a function from $Typ(T_1)$ to $Typ(T_2)$ such that for each $\Gamma, \Delta \subseteq Typ(T_1)$ if $\Gamma \vdash_{T_1} \Delta$, then $f(\Gamma) \vdash_{T_2} f(\Delta)$.

The notion of local logic puts the idea of a classification together with that of a regular theory. Also by introducing a notion of normal tokens, it models reasonable but unsound inferences.

Definition 7.24 A local logic $\mathscr{L} = (Tok(\mathscr{L}), Typ(\mathscr{L}), \models_{\mathscr{L}}, \vdash_{\mathscr{L}}, N_{\mathscr{L}})$ consists of
(i) a classification $cl(\mathscr{L}) = (Tok(\mathscr{L}), Typ(\mathscr{L}), \models_{\mathscr{L}})$,
(ii) a regular theory $Th(\mathscr{L}) = (Typ(\mathscr{L}), \vdash_{\mathscr{L}})$, and
(iii) a subset $N_{\mathscr{L}} \subseteq Tok(\mathscr{L})$, called the normal tokens of $\mathscr{L}$, which satisfies all the constraints of $Th(\mathscr{L})$.

Definition 7.25 A logic infomormhism $f : \mathscr{L}_1 \rightleftarrows \mathscr{L}_2$ consists of a contravariant pair $f = (\hat{f}, \check{f})$ of functions such that
(i) $f : cl(\mathscr{L}_1) \rightleftarrows cl(\mathscr{L}_2)$ is an infomorphism of classifications,
(ii) $\hat{f} : Th(\mathscr{L}_1) \mapsto Th(\mathscr{L}_2)$ is a theory interpretation, and
(iii) $\check{f}(N_{\mathscr{L}_2}) \subseteq N_{\mathscr{L}_1}$.

It can be observed that through these notions of classification, local logic and logic infomorphism, the target of the authors (Barwise and Seligman 1997) was to formalize, respectively, an individual's information base, logical reasoning base, and flow of information from one individual to another in the process of decision-making. But as mentioned before, the notion of local logic is developed bypassing the role of belief. Apart from that through the notion of local logic, it is endorsed that a sequent, which can be legitimately derived in the agent's theory of inference, is not necessarily supported by every object/situation (i.e. tokens) available to the agent. Only normal tokens satisfy every constraint of the theory of the agent. So, for a sequent $\langle \Gamma, \Delta \rangle$, the following possibilities appear natural:

- For any token a, a satisfies $\langle \Gamma, \Delta \rangle$.
- For any token a, a does not satisfy $\langle \Gamma, \Delta \rangle$.
- For a token a, a satisfies $\langle \Gamma, \Delta \rangle$, and for another token b, b does not satisfy $\langle \Gamma, \Delta \rangle$. And if we go beyond this two-valued set-up of Barwise and Seligman, then from information theoretic perspective at some point of time an agent may have incomplete information about some token with respect to some types. Thus, there might be neither any evidence (situation) in favour of a type nor any evidence denying the type. That is we can also have the following possibility.
- For any token a, there are neither evidences in favour of a satisfies $\langle \Gamma, \Delta \rangle$, nor in favour of a does not satisfy $\langle \Gamma, \Delta \rangle$.

Here we suddenly bring in a term 'evidence'. Appearance of the term 'evidence' also will be seen when we present the notion of paraconsistent sets (Vitória et al. 2009). Neither in Vitória et al. (2009) nor in the present context, any formal definition for 'evidence' is presented; but an intuition behind calling 'evidence' may be given. While defining local logic we have seen that the constraints of the theory of an agent are supported by some tokens. In intuitive term, these tokens are some objects or situations which satisfy the constraints, and thus a kind of *evidence* in favour of the constraints is established. So, one may identify a situation as an evidence in favour of a sequent if the situation satisfies the sequent. Moreover, when neither we can say a satisfies $\langle \Gamma, \Delta \rangle$ nor can say a does not satisfy $\langle \Gamma, \Delta \rangle$ this uncertainty is due to a lack of information regarding whether a sequent is satisfied. The notion of paraconsistent set, on which we shall base for further development, seems to accommodate this view. The formal definition of paraconsistent set is given in the next section.

So, in Sect. 7.2.3, our attempt would be to introduce a notion of graded, more specifically four-valued, satisfiability relation as well as a notion of graded inference in the framework of local logic; moreover, the theory of the local logic will be based on a notion of belief structure of an individual.

7.2.2 Paraconsistent Set

During the integration of information, one might need to deal with incomplete and conflicting information. Keeping this in mind, Vitória et al. (2009) developed a notion of paraconsistent set, which addresses set membership as a four-valued function. The value set L consists of the four values $\{u, f, t, i\}$, where the values, respectively, stand for 'unknown', 'false', 'true' and 'inconsistent' (or both true and false). The notion of paraconsistent set has been formalized as follows (Vitória et al. 2009).

Definition 7.26 Given a universe U, let $\neg U = \{\neg x : x \in U\}$. Intuitively, $x \in A$ represents the fact that there is an evidence that x is in A, and $\neg x \in A$ represents that there is an evidence that x is not in A. A paraconsistent set A on U is a subset of $U \cup \neg U$. In terms of the four-valued membership function A can be represented as follows.

A paraconsistent set is given by its membership function $A : U \mapsto L$ if we denote the degree of belongingness of x to A by $A(x)$, or can also be presented as a function from $U \times 2^{U \cup \neg U} \mapsto L$ if we refer to the same by $v(x \in A)$. The assignment of four values is given as follows.

$$\begin{aligned} A(x) = v(x \in A) &= t \text{ if } x \in A \text{ and } \neg x \notin A,\\ &= i \text{ if } x \in A \text{ and } \neg x \in A,\\ &= u \text{ if } x \notin A \text{ and } \neg x \notin A,\\ &= f \text{ if } x \notin A \text{ and } \neg x \in A, \end{aligned}$$

As $v(x \in A) = i$ indicates both $x \in A$ and $\neg x \in A$ from the context of information content $v(x \in A) = i$ has more information than $v(x \in A) = t$. In the same way, it can be observed that $v(x \in A) = f$ has more information than $v(x \in A) = u$, whereas $v(x \in A) = f$ and $v(x \in A) = t$ are having non-comparable information. So, from the context of information content, the natural ordering among these values turns out to be as follows.

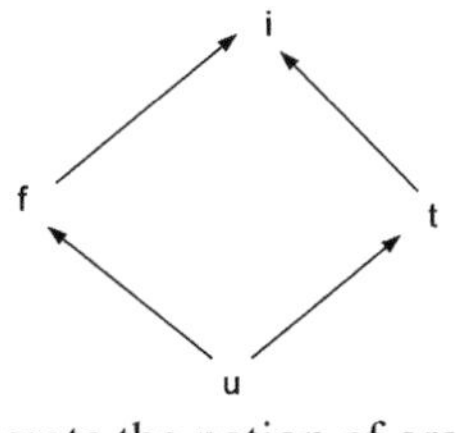

Now as we are going to incorporate the notion of graded consequence in the framework of local logics, an algebraic structure for the meta-language needs to be fixed. We assume the presence of a complete lattice structure over L as per the requirement of the notion of graded consequence in the context of fuzzy sets of premises (cf. Sect. 5.2).

7.2.3 Graded Consequence: Logic for Distributed System

In this section, our attempt is to present the notion of graded consequence in the framework of logic for distributed systems of Barwise and Seligman. In order to achieve this target, we would present the existing notion of graded consequence from a broader perspective, where we consider that each expert, i.e. T_i, is identified with a local logic representing her information and reasoning base.

Let $\{T_i\}_{i\in I}$ be a collection of experts having their respective classifications, denoted as $cl(T_i)$. Let us consider that each T_i is concerned about the same set of tokens Tok and types Typ^*, where Typ^* may be considered as the union of basic atomic predicates (sentences) Typ and their negations. In particular, we may consider that each T_i has opinion regarding which object from Tok satisfies certain properties from Typ^* to what extent, and this information is given by $v(a \models_i \alpha)\ (\in L)$ for $\alpha \in Typ^*$ and $a \in Tok$. So, we consider the classification of T_i, $cl(T_i) = (Tok, Typ^*, \models_i)$, where $\models_i$ is a binary relation on $Tok \times Typ^*$. That is for each $i \in I$, $\models_i$ is a paraconsistent set over $Tok \times Typ$ such that for any $a \in Tok$ and $\alpha \in Typ^*$, $a \models_i \alpha$ can assume any of the four values from $\{u, f, t, i\}$. Specifically, we assign,

$$\begin{aligned} v(a \models_i \alpha) &= t \text{ if } (a, \alpha) \in \models_i \text{ and } (a, \neg\alpha) \notin \models_i, \\ &= f \text{ if } (a, \neg\alpha) \in \models_i \text{ and } (a, \alpha) \notin \models_i, \\ &= u \text{ if } (a, \alpha) \notin \models_i \text{ and } (a, \neg\alpha) \notin \models_i, \\ &= i \text{ if } (a, \alpha) \in \models_i \text{ and } (a, \neg\alpha) \in \models_i. \end{aligned}$$

If we talk in the standard framework of logic with formulas as types and algebraic structures as tokens, then one interpretation of $v(a \models_j \alpha)$ could be according to the expert T_j, with respect to an algebraic structure (model) a the formula α receives the value $v(a \models_j \alpha)$. It is to be noticed that we may relate the existing framework of GCT, where we start with a presumption of a fixed algebraic structure, with that of the new by considering $T_j(\alpha) = v(a \models_j \alpha)$, where Tok can be arbitrarily fixed as a singleton set containing an algebraic structure a for the object language sentences of Typ^*. For the time being, we prefer to choose Typ^* as atomic wffs and their negations, but following (Barwise and Seligman 1997) we may also think of considering other Boolean connectives over Typ.

Now based on $cl(T_i) = (Tok, Typ^*, \models_i)$, the classification of each T_i, below we construct a group-classification combining the information available from different resources.

Definition 7.27 Given $cl(T_j) = (Tok, Typ^*, \models_j)$ for each T_j from a collection $\{T_i\}_{i\in I}$, $cl(Gr) = (Tok_{Gr}, Typ^*, \models_{Gr})$, where $Tok_{Gr} = I \times Tok$. We define an infomorphism $\sigma_{T_j} : cl(Gr) \rightleftarrows cl(T_j)$ such that $\hat{\sigma_{T_j}}(\alpha) = \alpha$, $\check{\sigma_{T_j}}(a) = (j, a)$, and $(j, a) \models_{Gr} \alpha$ iff $a \models_j \alpha$.

As before, also for $cl(Gr)$, the classification of the group, information of the form $(j, a) \models_{Gr} \alpha$ is alternatively expressed as $v((j, a) \models_{Gr} \alpha) \geq t$. This describes that according to expert T_j with respect to $a (\in Tok)$ α is satisfied to some extent $\geq t$.

That is, $cl(Gr)$ works just as a repertoire of the information collected from different T_j's.

We may also consider an invariant $I = (\Sigma, R)$, where $\Sigma \subseteq Typ^*$, on $cl(T_i)$ such that aRb if and only if for any $\alpha \in \Sigma$, $a \models_i \alpha$ iff $b \models_i \alpha$ (i.e. $v(a \models_i \alpha) = v(b \models_i \alpha)$). So, we can construct the quotient classification $cl(T_i)/I$, and hence following (Barwise and Seligman 1997), we have an infomorphism from $cl(T_i)/I$ to $cl(T_i)$. Also by Definition 7.27, we have an infomorphism $\sigma_{T_i} : cl(Gr) \rightleftarrows cl(T_i)$. Hence following Proposition 7.6 the following diagram commutes.

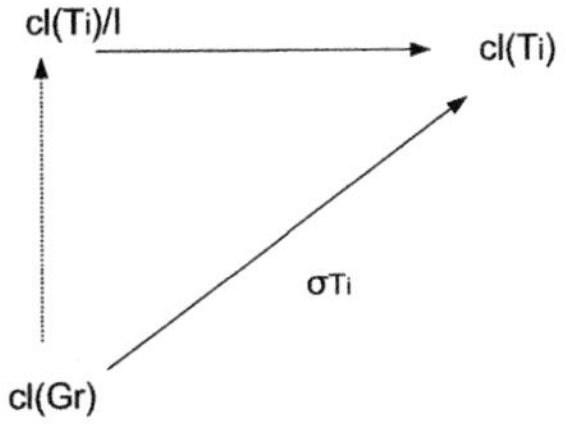

That is, given a pair $(\models_i, a)$ from Tok_{Gr} and $\alpha \in \Sigma$, $(\models_i, a) \models_{Gr} \alpha$ iff $[a]_I \models_{T_i/I} \alpha$. In other words, for certain properties (e.g. element from Σ) if we know from $cl(Gr)$ that $(\models_i, a) \models_{Gr} \alpha$, then we can identify the whole class of structures $[a]_I$ satisfying α. The point of emphasis here is, the way classification for the group, i.e. $cl(Gr)$ has been formed it can preserve information about a particular classification, say $cl(T_i)$, and its structures which are invariant with respect to certain properties of Typ^*.

The classification of each T_i consists of the information about which token (structure) to what extent satisfies some type (property). One simple analogy could be, given a set of objects (Tok) and attributes (Typ^*), $cl(T_i)$ contains the information $a \models_i \alpha$, the degree to which an object a satisfies α. So, $cl(T_i)$ basically a database representing the information system (Pawlak and Skowron 2007a, b, c) of T_i. A database or information system is built on the basis of the knowledge available to an expert. That some token has some property to some extent, varies from expert to expert based on their respective knowledge.

Local Logic for Each T_i's Based on the Respective Belief Profile

Apart from the database, T_j, the expert also has a sense of reasoning based on her belief. This reasoning base reflects the expert's point of view regarding to what extent from a set of properties Γ another set of properties Δ follows. The expert becomes able to consider Δ follows from Γ since according to her prior knowledge there might be some specific cases (i.e. normal tokens) satisfying the relationship, viz. Δ follows from Γ. That is, the reasoning base of T_j is dependent on her information system or database. Following this intuition, we define a local logic (Barwise and Seligman 1997) for each T_j based on its classification $cl(T_j)$ in the sequel below. In this regard, we borrow some notions from Keplicz-Dunin and Szałas (2013) pertaining to belief structure of an individual. In Keplicz-Dunin and Szałas (2013), authors developed an idea of individual belief structure for a group of agents and proposed a notion of group belief structure obtained from that of the individual ones. The work (Keplicz-Dunin and Szałas 2013) is developed in the framework of paraconsistent sets. So, we take up their notion of belief structure for our present purpose of the paper.

In Keplicz-Dunin and Szałas (2013), the notion of belief structure is preceded by some notions. They started with a vocabulary $Const$, a fixed set of constants, Var, a fixed set of variables, and Rel, a fixed set of relation symbols. A literal is an expression of the form $R(\bar{t})$ or $\neg R(\bar{t})$, where $R \in Rel$, and $\bar{t}$ represents a finite sequence of $Const \cup Var$. A ground literal is a literal without any variable, and $\mathcal{G}(Const)$ denotes the set of all ground literals over $Const$. That is, ground literals can be considered as closed atomic formulas and their negations, which in our case is Typ^* without any free variable. The set of all finite subsets of $\mathcal{G}(Const)$ is denoted as $FIN(\mathcal{G}(Const))$.

(a) $v : Var \mapsto Const$ denotes a valuation function. For a literal l, $l(v)$ is the ground literal obtained by replacing each variable x in l by $v(x)$.
(b) Truth value of a literal l with respect to a set of ground literals L and valuation v is given as follows:

$$\begin{aligned} l(L, v) = t, &\text{ if } l(v) \in L \text{ and } \neg l(v) \notin L, \\ i, &\text{ if } l(v) \in L \text{ and } \neg l(v) \in L, \\ u &\text{ if } l(v) \notin L \text{ and } \neg l(v) \notin L, \\ f &\text{ if } l(v) \notin L \text{ and } \neg l(v) \in L. \end{aligned}$$

(c) A constituent C is a member of $\mathcal{C} = FIN(\mathcal{G}(Const))$, and an epistemic profile is a function $\varepsilon : FIN(\mathcal{C}) \mapsto \mathcal{C}$.
(d) By a belief structure over an epistemic profile ε, it is meant $\mathcal{B}^{\varepsilon} = (\mathbb{C}, F)$ such that $F = \varepsilon(\mathbb{C})$, where $\mathbb{C}$ is a non-empty set of finitely many constituents (i.e. $\mathbb{C} \subseteq \mathcal{C}$), and F is the consequent.
(e) Truth value of a formula α with respect to a belief structure $\mathcal{B}^{\varepsilon} = (\mathbb{C}, F)$ and valuation v is given by $\alpha(\mathcal{B}^{\varepsilon}, v) = \alpha(\cup_{C \in \mathbb{C}} C, v)$.
(f) In Keplicz-Dunin and Szałas (2013) they also introduced an operator Bel in the language, and defined the semantics as $Bel(t) = t$, $Bel(i) = i$, and $Bel(u) = Bel(f) = f$.
(g) Then the evaluation of the truth value of a formula with respect to $\mathcal{B}^{\varepsilon} = (\mathbb{C}, F)$ is defined as follows:

$$\begin{aligned} \alpha(\mathcal{B}^{\varepsilon}, v) &= \alpha(\cup_{C \in \mathbb{C}} C, v) \text{ if } \alpha \text{ is not of the form } Bel(\beta), \text{ and} \\ &= Bel(\beta(F, v)) \text{ when } \alpha = Bel(\beta). \end{aligned}$$

Intuitively one can consider that an agent can have a family of sets of information about some facts (ground literals). Each set of information is obtained from a source, and together the family of sets of information forms the set of constituents ($\mathbb{C}$) of the agent. Based on this set of ground literals, the agent can evaluate the degree of truth of a compound formula (cf. (e)). The agent based on the set of constituents ($\mathbb{C}$) can generate a new set of beliefs, a set of consequents F. Deriving a specific set of consequents F from $\mathbb{C}$ depends on the agent's sense of epistemic profile, which is presented by a function ε. With respect to the set of consequents F if agents believe in a formula α, represented as $Bel(\alpha)$, then to what extent *believing* α is true is determined based on the new set of ground literals F.

Below we present how the notion of belief structure (Keplicz-Dunin and Szałas 2013) would be incorporated for our purpose.

Definition 7.28 Let for each T_j there be $C_j, F_j \in FIN(\mathscr{G}(Typ^*))$, which are paraconsistent sets over Typ (i.e. Typ^* contains all atomic formulas and their negations). An epistemic profile for T_j is $(\mathscr{B}_j, \varepsilon_j)$, where $\mathscr{B}_j = (\{C_j\}, F_j)$ and $F_j = \varepsilon_j(\{C_j\})$. Following the terminology of Keplicz-Dunin and Szałas (2013), $\{C_j\}$, or simply C_j, is called the constituent of T_j, and F_j is called the consequent of T_j obtained by applying the epistemic profile function ε_j of T_j on C_j.

Thus, the definition of epistemic profile (Keplicz-Dunin and Szałas 2013) allows an individual to have inconsistent basic beliefs (constituent), and to derive the consequent by applying ε_j on that. As authors in Keplicz-Dunin and Szałas (2013), did not restrict the definition of belief structure by imposing more conditions on $(\mathbb{C}, F)$, we also follow the same line for the time being. So, each expert T_j has its own basic beliefs and a sense of inferencing represented by the epistemic profile function ε_j. Based on the epistemic profile of each T_j, we would define a notion of consequence relation $\vdash_{\varepsilon_j}$ for T_j. Before going into the formal definition let us add an explanation to how Miranda's reasoning worked from the perspective of Dretske's proposal of *legitimately generating new beliefs*, and add a more specific view to the example.

Example 7.1 Miranda has some initial beliefs C_M. Based on those ground literals of C_M Miranda classifies her information that which situation satisfies ($\models_M$) which property to what extent. We present Table 7.1 consisting of some instances $inst_1$, $inst_2$, $inst_3$, ...$inst\text{-}J$ as tokens. Types are some atomic wffs with a free variable x such as x *has the property light in mountain*, represented as *light in mountain*(x), and some more pertaining to Miranda's initial beliefs and knowledge. As for Miranda's knowledge, there are some instances $inst_1, inst_2, inst_3, \ldots inst_n$ which satisfy some of the formulas/properties mentioned in Table 7.1. This is contained in Miranda's initial belief C_M. Now she observes a new situation, viz. $inst\text{-}J$, the situation concerning Judith. This also has some of the properties as $inst_1$, $inst_2$, $inst_3$, ...$inst_n$, and Miranda matches the new case with that of the previous ones. This helps her to generate a new belief regarding whether $inst\text{-}J$ satisfies the formula that *someone is in trouble(x)*. The new observations regarding $inst\text{-}J$ are all part of her new set of beliefs F_M. So, formally the constituent C_M, the consequent F_M, and the normal tokens N_M can be presented as follows.

- $C_M =$ {light in mountain($inst_1$), dark night($inst_1$), troublesome location($inst_1$), someone is in trouble($inst_1$), ...light in mountain($inst_n$), ...¬someone is in trouble($inst_n$)}.
- Miranda may consider $\{inst_1, inst_2, inst_3\}$ as the normal tokens N_M, and these are of special status as any sequent, which Miranda considers as a rule of her theory, is satisfied by all tokens from N_M.

Table 7.1 Table for $\models_M$ evaluated with respect to C_M

	Light in mountain(x)	Dark night(x)	Troublesome location(x)	Someone is in trouble(x)
$inst_1$	t	t	t	t
$inst_2$	t	t	f	t
$inst_3$	t	t	t	t
⋮	⋮	⋮	⋮	⋮
$inst_n$	t	t	t	f
$inst\text{-}J$	t	t	t	?

– So, matching the new situation $inst\text{-}J$ with that of the previous situations Miranda may consider both the possibilities that *someone is in trouble*$(inst\text{-}J)$ and $\neg$*someone is in trouble*$(inst\text{-}J)$. So, her consequent F_M would be such that

$F_M \supseteq$ {light in mountain($inst\text{-}J$), dark night($inst\text{-}J$), troublesome location($inst\text{-}J$), someone is in trouble($inst\text{-}J$), $\neg$someone is in trouble($inst\text{-}J$)}.

So, according to Miranda to check whether for $inst\text{-}J$, $\alpha =$ someone is in trouble(x) follows legitimately from $\Gamma =$ {light in mountain(x), dark night(x)}, the following needs to be satisfied.

$$\Gamma_{x|inst-J} \vdash_M \alpha_{x|inst-J} \text{ iff}$$

$(\Gamma_{x|inst-J} \subseteq F_M$ implies $\alpha_{x|inst-J} \in F_M)$ and $\forall_{a \in N_M}(a \models_M \beta$ for all $\beta \in \Gamma$ implies $a \models_M \alpha)$.

Here $\Gamma_{x|inst-J}$ denotes that all occurrences of the variable x in all formulas in Γ are replaced by $inst\text{-}J$. The satisfaction relation $\models_M$ is a four-valued relation, and $a \models_M \gamma$ means for a token a, $v(a \models_M \gamma) = \gamma(C_M, x|a) \geq t$ for a wff γ (cf. item (b) above Definition 7.28). In other words, when the variable x in γ is replaced by a, $\gamma_{x|a} \in C_M$. So, in the first part, Miranda checks if premises (Γ), considering with respect to $inst\text{-}J$, belong to her new beliefs F_M, then conclusion α with respect to $inst\text{-}J$ also belongs to F_M. Moreover, it is checked if every instance from her normal tokens N_M too satisfies the sequent $\langle \Gamma, \alpha \rangle$. In other words, in order to derive $\Gamma_{x|inst-J} \vdash_M \alpha_{x|inst-J}$ Miranda checks both in terms of her newly generated belief as well as previously available situations which she considers as normal tokens.

□

The above exemplification of Miranda's story leads us towards the following definition.

Definition 7.29 (i) Each T_i is associated with a classification $cl(T_i) = (Tok, Typ^*, \models_i)$ and a belief structure $\mathscr{B}^{\varepsilon_i} = (C_i, F_i)$ such that at a point of time, say t, F_i is obtained from C_i using the epistemic profile function ε_i, and $C_i \subseteq F_i$. For any $\alpha \in Typ^*$ possibly with a free variable x and $a \in Tok$, $v(a \models_i \alpha) = \alpha(C_i, x|a)$.

(ii) $N_i \neq \phi$ is a subset of Tok.
(iii) For some $\Gamma, \Delta \subseteq Typ^*$ such that Γ, Δ share the variable x and $a \in Tok$, $\Gamma_{(x|a)} \vdash_{\varepsilon_i} \Delta_{(x|a)}$ iff ($\Gamma_{(x|a)} \subseteq F_i$ implies $\Delta_{(x|a)} \cap F_i \neq \phi$), and $\forall_{b \in N_i}(b \models_i \beta$ for all $\beta \in \Gamma$ implies $b \models_i \alpha$ for some $\alpha \in \Delta$).
(iv) The sequents of the form $\langle \Gamma, \Delta \rangle$ are considered to be a constraint of $Th(T_i) = (Typ^*, \vdash_{\varepsilon_i})$ if for all $a \in Tok$ - N_i, $\Gamma_{(x|a)} \vdash_{\varepsilon_i} \Delta_{(x|a)}$.

It is to be noticed, that in the item (iii) of Definition 7.29, the expressions such as $b \models_i \beta$ stands for $v(b \models_i \beta) = \beta(C_i, x|b) \geq t$ (cf. (i) of Definition 7.29). More specifically, $b \models_i \beta$ represents that the formula $\beta(x|b) \in C_i$, where the free variable x in β is replaced by a token (i.e. situation) b. Also, it is to be noted that the language of Typ^* contains formulas with one free variable x. This is basically intended to accommodate statements about different situations, where x is a variable ranging over situations.

Theorem 7.8 *If $\langle \Gamma, \Delta \rangle$ is a constraint of $Th(T_i)$, then for each $b \in N_i$, $\Gamma_{(x|b)} \vdash_{\varepsilon_i} \Delta_{(x|b)}$.*

Proof Let $\langle \Gamma, \Delta \rangle$ be a constraint of $Th(T_i)$ and $b \in N_i$.

So, $b \models_i \beta$ for all $\beta \in \Gamma$ implies $b \models_i \alpha$ for some $\alpha \in \Delta$. That is, we have the following.

Whenever $v(b \models_i \beta) = \beta(C_i, x|b) \geq t$ for all $\beta \in \Gamma, v(b \models_i \alpha) = \alpha(C_i, x|b) \geq t$ for some $\alpha \in \Delta$, i.e. whenever $\beta_{(x|b)} \in C_i \subseteq F_i$ for all $\beta \in \Gamma$, $\alpha_{(x|b)} \in C_i \subseteq F_i$ for some $\alpha \in \Delta$.

That is, $\Delta_{(x|b)} \cap F_i \neq \phi$ provided $\Gamma_{(x|b)} \subseteq F_i$. Hence, we can write $\Gamma_{(x|b)} \vdash_{\varepsilon_i} \Delta_{(x|b)}$. □

It is to be noted that for $a \in Tok$, $\Gamma_{(x|a)} \vdash_{\varepsilon_i} \Delta_{(x|a)}$ does not necessarily imply that $a \in N_i$.

Theorem 7.9 *$(Typ^*, \vdash_{\varepsilon_i})$ is a regular theory.*

Proof Proofs of identity and weakening are straightforward.
To prove global cut, let for any partition (Σ_0, Σ_1) of $\Sigma \subseteq Typ^*$, $\langle \Gamma \cup \Sigma_0, \Delta \cup \Sigma_1 \rangle$ be a constraint of $Th(T_i)$. That is, for any (Σ_0, Σ_1) of Σ and $a \in Tok - N_i$ if $\Gamma \cup \Sigma_{0(x|a)} \subseteq F_i$ then $(\Delta \cup \Sigma_1)_{(x|a)} \cap F_i \neq \phi$. We want to prove that $\langle \Gamma, \Delta \rangle$ is a constraint of $Th(T_i)$. Let us consider an $a \in Tok$ - N_i such that $\Gamma_{(x|a)} \subseteq F_i$. We need to prove $\Delta_{(x|a)} \cap F_i \neq \phi$. Now for $a \in Tok - N_i$, there are two cases. (1) $\Sigma_{(x|a)} \cap F_i = \phi$ and (2) $\Sigma_{(x|a)} \cap F_i \neq \phi$. For (1), we can choose the partition (ϕ, Σ). So, $\langle \Gamma, \Delta \cup \Sigma \rangle$ being a constraint of $Th(T_i)$, we have $\Delta_{(x|a)} \cap F_i \neq \phi$. For (2), let us consider the partition (Σ_0, Σ_1) of Σ such that $\Sigma_{0(x|a)} = \Sigma_{(x|a)} \cap F_i$ and $\Sigma_{1(x|a)} = \Sigma_{(x|a)} - \Sigma_{0(x|a)}$. So, now $\langle \Gamma \cup \Sigma_0, \Delta \cup \Sigma_1 \rangle$ being a constraint of $Th(T_i)$, and $\Gamma \cup \Sigma_{0(x|a)} \subseteq F_i$, we have $\Delta_{(x|a)} \cap F_i \neq \phi$. ...(A)

Now for the second part, let for an arbitrary $b \in N_i, b \models_i \beta$ for all $\beta \in \Gamma$. We want to prove $b \models_i \alpha$ for some $\alpha \in \Delta$. As $\langle \Gamma \cup \Sigma_0, \Delta \cup \Sigma_1 \rangle$ is a constraint of $Th(T_i)$ for any partition (Σ_0, Σ_1) of Σ, with respect to the partitions (Σ, ϕ) and (ϕ, Σ) we have, respectively, the following constraints $\langle \Gamma \cup \Sigma, \Delta \rangle$ and $\langle \Gamma, \Delta \cup \Sigma \rangle$ in $Th(T_i)$.

Now two cases arise. (1) $b \models_i \beta$ for all $\beta \in \Sigma$ and (2) it is not that $b \models_i \beta$ for all $\beta \in \Sigma$.

For (1), we choose the constraint $\langle \Gamma \cup \Sigma, \Delta \rangle$. Then as $b \models_i \beta$ for all $\beta \in \Gamma \cup \Sigma$, $b \models_i \alpha$ for some $\alpha \in \Delta$. For (2) we again have two subcases. Either $b \not\models_i \beta$ for all $\beta \in \Sigma$ or $b \models_i \beta$ for some $\beta \in \Sigma$. For the first subcase, we choose the constraint $\langle \Gamma, \Delta \cup \Sigma \rangle$ of $Th(T_i)$. Then as $b \models_i \beta$ for all $\beta \in \Gamma$ and $b \not\models_i \beta$ for all $\beta \in \Sigma$, $b \models_i \alpha$ for some $\alpha \in \Delta$.

For the second subcase, let $\Sigma_0 = \{\beta \in \Sigma : b \models_i \beta\}$ and $\Sigma_1 = \Sigma - \Sigma_0$. Then for the constraint $\langle \Gamma \cup \Sigma_0, \Delta \cup \Sigma_1 \rangle$, as $b \models_i \beta$ for all $\beta \in \Gamma \cup \Sigma_0$ and $b \not\models_i \beta$ for all $\beta \in \Sigma_1$, $b \models_i \alpha$ for some $\alpha \in \Delta$. So, combining all the cases we have $b \models_i \alpha$ for some $\alpha \in \Delta$. ...(B)

Hence combining (A) and (B) we can conclude that $\langle \Gamma, \Delta \rangle$ is a constraint of $Th(T_i)$.
□

Definition 7.30 For each T_i, the local logic of T_i is $LL(T_i) = (Tok, Typ^*, \models_i, \vdash_{\varepsilon_i}, N_i)$ such that (i) $cl(T_i) = (Tok, Typ^*, \models_i)$ is the classification of T_i, (ii) $Th(T_i) = (Typ^*, \vdash_{\varepsilon_i})$ is a regular theory, and (iii) $N_i \subseteq Tok$ is the set of normal tokens.

Deterministic Local Logic for Each Individual T_i

So, we have constructed a local logic $LL(T_i) = (Tok, Typ^*, \models_i, \vdash_{\varepsilon_i}, N_i)$ for each T_i based on the classification $cl(T_i) = (Tok, Typ^*, \models_i)$ such that $(Typ^*, \vdash_{\varepsilon_i})$ is a regular theory. From now onwards we shall denote the set of all constraints of $Th(T_i)$ by $S_{\vdash_{\varepsilon_i}}$. We now want to construct a deterministic consequence relation $\vdash^D_{\varepsilon_i}$ out of $\vdash_{\varepsilon_i}$ as in the context of graded consequence we consider *following of a single formula from a set of formulae* and uncertainty about derivation of a formula is captured by allowing a graded (i.e. non-crisp) notion of derivation. So, from $\vdash_{\varepsilon_i}$, a two-valued non-deterministic relation we propose to define a many-valued deterministic relation $\vdash^D_{\varepsilon_i}$.

Definition 7.31 Given $Th(T_i) = (Typ^*, \vdash_{\varepsilon_i})$, $Th_D(T_i) = (Typ^*, \vdash^D_{\varepsilon_i})$ is defined as follows:
(i) $S_{\vdash^D_{\varepsilon_i}} = \cup_\Gamma \{\langle \Gamma, \alpha \rangle$: for some Δ, $\langle \Gamma, \Delta \rangle \in S_{\vdash_{\varepsilon_i}}$ and for some $a \in Tok$, $\alpha_{(x|a)} \in \Delta_{(x|a)} \cap F_i\}$.
(ii) $Th_D(T_i)$ consists of all sequents of $S_{\vdash^D_{\varepsilon_i}}$.

The definition of $\vdash_{\varepsilon_i}$, proposed in Definition 7.29, is a two-valued relation between two subsets of formulae. If we restrict the right-hand side of $\vdash_{\varepsilon_i}$ to a single formula even then it remains a two-valued relation. Here a sense of four-valuedness is hidden in the set F_i, which is a paraconsistent set. While defining $\vdash^D_{\varepsilon_i}$ we have chosen those sequents $\langle \Gamma, \Delta \rangle$ such that $\langle \Gamma, \Delta \rangle \in S_{\vdash_{\varepsilon_i}}$. Now it may happen that though for all $o \in Tok$, $\Delta_{(x|o)} \cap F_i \neq \phi$, for none of the formula $\beta \in \Delta$, $\beta_{(x|o)} \in F_i$ holds for all $o \in Tok$. In particular, there may be a case where for some α, $\alpha_{(x|a)} \in \Delta_{(x|a)} \cap F_i$ holds for only a single $a \in Tok$. We attempt to address this as *there is an evidence* *that α follows from Γ*. Consequently, the possibility of a four-valued relation comes up, and this is what we can see in Definition 7.31.

The above defined notion of constraint of $Th_D(T_i)$ can be presented as ordinary set as well as a paraconsistent relation in the following way.
For $\langle \Gamma, \alpha \rangle \in S_{\vdash^D_{\varepsilon_i}}$ and $\langle \Gamma, \neg\alpha \rangle \notin S_{\vdash^D_{\varepsilon_i}}$, we write $v(\Gamma \vdash^D_{\varepsilon_i} \alpha) = $ t,
For $\langle \Gamma, \alpha \rangle \notin S_{\vdash^D_{\varepsilon_i}}$ and $\langle \Gamma, \neg\alpha \rangle \in S_{\vdash^D_{\varepsilon_i}}$, we write $v(\Gamma \vdash^D_{\varepsilon_i} \alpha) = $ f,
For $\langle \Gamma, \alpha \rangle \notin S_{\vdash^D_{\varepsilon_i}}$ and $\langle \Gamma, \neg\alpha \rangle \notin S_{\vdash^D_{\varepsilon_i}}$, we write $v(\Gamma \vdash^D_{\varepsilon_i} \alpha) = $ u, and
For $\langle \Gamma, \alpha \rangle \in S_{\vdash^D_{\varepsilon_i}}$ and $(\Gamma, \neg\alpha) \in S_{\vdash^D_{\varepsilon_i}}$, we write $v(\Gamma \vdash^D_{\varepsilon_i} \alpha) = $ i.

From now onwards, we shall interchangeably write $\Gamma \vdash^D_{\varepsilon_i} \alpha$ and $v(\Gamma \vdash^D_{\varepsilon_i} \alpha) \geq t$. If we write $\Gamma \nvdash^D_{\varepsilon_i} \alpha$, that means $v(\Gamma \vdash^D_{\varepsilon_i} \alpha) \leq f$. Also one should note that whenever $\langle \Gamma, \alpha \rangle \in S^D_{\vdash_{\varepsilon_i}}$, $v(\Gamma \vdash^D_{\varepsilon_i} \alpha) \geq t$, and whenever $\langle \Gamma, \alpha \rangle \notin S^D_{\vdash_{\varepsilon_i}}$, $v(\Gamma \vdash^D_{\varepsilon_i} \alpha) \leq f$. Below, we present the counterpart of regular theory for a deterministic notion of consequence. That means, we need to show that the deterministic notion of consequence satisfies identity/overlap, dilution/weakening, and (finite) cut.

Theorem 7.10 *The theory $Th_D(T_i) = (Typ^*, \vdash^D_{\varepsilon_i})$ satisfies overlap, dilution and cut.*

Proof Proof of overlap is straightforward. In order to prove dilution, let $\langle \Gamma, \alpha \rangle \in S_{\vdash^D_{\varepsilon_i}}$ and $\Gamma \subseteq \Gamma'$. Then for some $\Delta \subseteq Typ^*$, $\langle \Gamma, \Delta \rangle \in S_{\vdash_{\varepsilon_i}}$ and for some $a \in Tok$, $\alpha_{(x|a)} \in \Delta_{(x|a)} \cap F_i$. Now as $Th(T_i)$ satisfies weakening, $\langle \Gamma', \Delta \rangle \in S_{\vdash_{\varepsilon_i}}$, and we already have that for some $a \in Tok$, $\alpha_{(x|a)} \in \Delta_{(x|a)} \cap F_i$. So, $\langle \Gamma', \alpha \rangle \in S_{\vdash^D_{\varepsilon_i}}$.

To prove cut, let us assume $\langle \Gamma \cup \{\alpha\}, \beta \rangle \in S_{\vdash^D_{\varepsilon_i}}$ and $\langle \Delta, \alpha \rangle \in S_{\vdash^D_{\varepsilon_i}}$. So, there are $\Delta_1, \Delta_2 \subseteq Typ^*$ such that $\langle \Gamma \cup \{\alpha\}, \Delta_1 \rangle \in S_{\vdash_{\varepsilon_i}}$ and $\langle \Delta, \Delta_2 \rangle \in S_{\vdash_{\varepsilon_i}}$, and for some $a, b \in Tok$, $\beta_{(x|a)} \in \Delta_{1(x|a)} \cap F_i$, and $\alpha_{(x|b)} \in \Delta_{2(x|b)} \cap F_i$. So, $\alpha \in \Delta_2$ and $\beta \in \Delta_1$. Let $\Delta' = \Delta_2 - \{\alpha\}$. So, by weakening $\langle \Gamma \cup \Delta \cup \{\alpha\}, \Delta_1 \cup \Delta' \rangle \in S_{\vdash_{\varepsilon_i}}$ and $\langle \Gamma \cup \Delta, \Delta_1 \cup \Delta' \cup \{\alpha\} \rangle \in S_{\vdash_{\varepsilon_i}}$.

Hence by finite cut, $\langle \Gamma \cup \Delta, \Delta_1 \cup \Delta' \rangle \in S_{\vdash_{\varepsilon_i}}$. Now as for some $a \in Tok$, $\beta_{(x|a)} \in \Delta_{1(x|a)} \cap F_i$, $\beta_{(x|a)} \in \Delta_1 \cup \Delta'_{(x|a)} \cap F_i$. Hence, $\langle \Gamma \cup \Delta, \beta \rangle \in S_{\vdash^D_{\varepsilon_i}}$. □

Thus, from the theory $Th(T_i) = (Typ^*, \vdash_{\varepsilon_i})$, we obtained a deterministic theory $(Typ^*, \vdash^D_{\varepsilon_i})$. Let $N^D_{\varepsilon_i}$ be the normal tokens satisfying all constraints of $S_{\vdash^D_{\varepsilon_i}}$. For a token $a \in Tok$, if a satisfies a constraint, say $\Gamma \vdash^D_{\varepsilon_i} \alpha \in S_{\vdash^D_{\varepsilon_i}}$, then clearly a satisfies $\Gamma \vdash_{\varepsilon_i} \Delta \in S_{\vdash_{\varepsilon_i}}$ where $\alpha \in \Delta$. Let us call the set of all tokens satisfying a constraint, say $\Gamma \vdash^D_{\varepsilon_i} \alpha$, as *realization* of $\Gamma \vdash^D_{\varepsilon_i} \alpha$, and denote it as $Rlz(\Gamma \vdash^D_{\varepsilon_i} \alpha)$. Then by the above observation $Rlz(\Gamma \vdash^D_{\varepsilon_i} \alpha) \subseteq Rlz(\Gamma \vdash_{\varepsilon_i} \Delta)$ where $\alpha \in \Delta$.

Proposition 7.9 *There is an embedding $emb : S_{\vdash^D_{\varepsilon_i}} \mapsto S_{\vdash_{\varepsilon_i}}$.*

Proof Let us enumerate all subsets of Typ^* as $\Delta_1, \Delta_2, \Delta_3, \ldots$, and all constraints of $S_{\vdash^D_{\varepsilon_i}}$ as: $\Gamma_1 \vdash^D_{\varepsilon_i} \alpha_{11}, \Gamma_1 \vdash^D_{\varepsilon_i} \alpha_{12}, \Gamma_1 \vdash^D_{\varepsilon_i} \alpha_{13} \ldots, \Gamma_2 \vdash^D_{\varepsilon_i} \alpha_{21}, \Gamma_2 \vdash^D_{\varepsilon_i} \alpha_{22}, \Gamma_2 \vdash^D_{\varepsilon_i} \alpha_{23} \ldots$, where $\Gamma_1 \neq \Gamma_2 \neq \Gamma_3 \ldots$ and $\alpha_{i1} \neq \alpha_{i2} \neq \alpha_{i3} \ldots$ for each i.

For $\Gamma_1 \vdash^D_{\varepsilon_i} \alpha_{11}$, there must be some Δ_j such that $\alpha_{11} \in \Delta_j$ and $\Gamma_1 \vdash_{\varepsilon_i} \Delta_j$. We assign $emb(\Gamma_1 \vdash^D_{\varepsilon_i} \alpha_{11}) = \Gamma_1 \vdash_{\varepsilon_i} \Delta_k$ such that Δ_k is the first such Δ_j in the list of enumeration. Then for $\Gamma_1 \vdash^D_{\varepsilon_i} \alpha_{12}$, we must have some Δ_l, appeared first in the enumeration, such that $\Gamma_1 \vdash_{\varepsilon_i} \Delta_l$ and $\alpha_{12} \in \Delta_l$. So, $emb(\Gamma_1 \vdash^D_{\varepsilon_i} \alpha_{12}) = \Gamma_1 \vdash_{\varepsilon_i} \Delta_l$. Thus, continuing in this way, it can be shown that emb is a function from $S_{\vdash^D_{\varepsilon_i}}$ to $S_{\vdash_{\varepsilon_i}}$. □

Let us denote $N_{emb(S_{\vdash^D_{\varepsilon_i}})} = \{a \in Tok : a$ satisfies every constraint of $emb(S_{\vdash^D_{\varepsilon_i}}) \subseteq S_{\vdash_{\varepsilon_i}}\}$. So, from the above proposition $N_i \subseteq N_{emb(S_{\vdash^D_{\varepsilon_i}})}$. Also, for any $a \in N^D_{\varepsilon_i}$, $a \in Rlz(\Gamma \vdash^D_{\varepsilon_i} \alpha)$ for any $\Gamma \vdash^D_{\varepsilon_i} \alpha \in S_{\vdash^D_{\varepsilon_i}}$. Therefore, $a \in Rlz(\Gamma \vdash_{\varepsilon_i} \Delta)$ for all $\Gamma \vdash_{\varepsilon_i} \Delta \in emb(S_{\vdash^D_{\varepsilon_i}}) \subseteq S_{\vdash_{\varepsilon_i}}$. Hence, $N^D_{\varepsilon_i} \subseteq N_{emb(S_{\vdash^D_{\varepsilon_i}})}$.

Definition 7.32 (i) Let $(Typ^*, \vdash)$ be a regular theory, and $\vdash^D$ be a deterministic consequence relation on Typ^*, obtained from $\vdash$, such that $emb(S_{\vdash^D}) \subseteq S_{\vdash}$. Then $\vdash^D$ is called a deterministic restriction of $\vdash$ on Typ^*. $(Typ^*, \vdash^D)$ is called the deterministic restriction of the theory $(Typ^*, \vdash)$.
(ii) If a deterministic restriction of a regular theory satisfies overlap, dilution and cut, then the theory is called a deterministic regular theory.

From Theorem 7.10, we can see $(Typ^*, \vdash^D_{\varepsilon_i})$ is a deterministic restriction of the regular theory $(Typ^*, \vdash_{\varepsilon_i})$, and thus $(Typ^*, \vdash^D_{\varepsilon_i})$ is a deterministic regular theory.

Definition 7.33 A deterministic local logic for T_i is $LL_D(T_i) = (Tok, Typ^*, \models_i, \vdash^D_{\varepsilon_i}, N^D_{\varepsilon_i})$, with $N^D_{\varepsilon_i} (\subseteq Tok)$ as the set of normal tokens.

After setting up the ground regarding specification of each expert T_i, given in terms of the respective deterministic local logic of T_i, we need to show that $(Typ^*, \vdash^D_{\varepsilon_i})$, the deterministic restriction of the regular theory $(Typ^*, \vdash_{\varepsilon_i})$ is nothing but a translation (interpretation) of the set of constraints of $(Typ^*, \vdash_{\varepsilon_i})$ to the set of constraints of $(Typ^*, \vdash^D_{\varepsilon_i})$.

Proposition 7.10 *Let $L_{\vdash_{\varepsilon_i}}$ and $L_{\vdash^D_{\varepsilon_i}}$ be the languages with respective classifications, viz. $(S_{\vdash_{\varepsilon_i}}, N_i, \models_{\varepsilon_i})$ and $(S_{\vdash^D_{\varepsilon_i}}, N^D_{\varepsilon_i}, \models^D_{\varepsilon_i})$, where $\Gamma \vdash_{\varepsilon_i} \Delta \models_{\varepsilon_i} a$ means if for all $x \in \Gamma$, $a \models_i x$ then $a \models_i \beta$ for some $\beta \in \Delta$, and similarly, $\Gamma \vdash^D_{\varepsilon_i} \alpha \models^D_{\varepsilon_i} a$ means if for all $x \in \Gamma$, $a \models_i x$ then $a \models_i \alpha$. Then $\mathscr{I} : (S_{\vdash_{\varepsilon_i}}, N_i, \models_{\varepsilon_i}) \rightleftarrows (S_{\vdash^D_{\varepsilon_i}}, N^D_{\varepsilon_i}, \models^D_{\varepsilon_i})$ is an interpretation function such that $\check{\mathscr{I}}(\Gamma \vdash^D_{\varepsilon_i} \alpha) = emb(\Gamma \vdash^D_{\varepsilon_i} \alpha) \in S_{\vdash_{\varepsilon_i}}$ and for each $a \in N_i$, $\hat{\mathscr{I}}(a) = a_{\mathscr{I}}$ such that $\check{\mathscr{I}}(\Gamma \vdash^D_{\varepsilon_i} \alpha) \models_{\varepsilon_i} a$ iff $\Gamma \vdash^D_{\varepsilon_i} \alpha \models^D_{\varepsilon_i} a_{\mathscr{I}}$.*

In the above proposition, the classification of the languages $L_{\vdash_{\varepsilon_i}}$ and $L_{\vdash^D_{\varepsilon_i}}$, respectively, contain $S_{\vdash_{\varepsilon_i}}$ and $S_{\vdash^D_{\varepsilon_i}}$ as their tokens, whereas the types are, respectively, the normal tokens considered in $LL(T_i)$ and $LL_D(T_i)$. In the definition of interpretation the assignment of $\hat{\mathscr{I}}(a) = a_{\mathscr{I}}$ such that $\check{\mathscr{I}}(\Gamma \vdash^D_{\varepsilon_i} \alpha) \models_{\varepsilon_i} a$ iff $\Gamma \vdash^D_{\varepsilon_i} \alpha \models^D_{\varepsilon_i} a_{\mathscr{I}}$ is possible as $N_i \subseteq N_{emb(S_{\vdash^D_{\varepsilon_i}})}$ and $N^D_{\varepsilon_i} \subseteq N_{emb(S_{\vdash^D_{\varepsilon_i}})}$, and so for each $a \in N_i$, we can have a unique $a_I \in N^D_{\varepsilon_i}$ such that $\Gamma \vdash^D_{\varepsilon_i} \alpha \models^D_{\varepsilon_i} a_{\mathscr{I}}$ iff $emb(\Gamma \vdash^D_{\varepsilon_i} \alpha) \models_{\varepsilon_i} a$.

So, we have a translation of the regular theory $(Typ^*, \vdash_{\varepsilon_i})$ to its deterministic counterpart $(Typ^*, \vdash^D_{\varepsilon_i})$. Now we shall construct a local logic for the group based on the classification $cl(Gr)$ (Definition 7.27), and the local logics $LL_D(T_i)$ for each T_i.

Deterministic Local Logic for a Group of T_i's

As the group only works as a repertoire of the information available to the T_i's, it does not have any distinctive role of using epistemic profile in order to derive a set of consequents from a set of constituents. So, we impose the following definition for the local logic of the group.

Definition 7.34 $LL_D(Gr) = (I \times Tok, Typ^*, \models_{Gr}, \vdash_{Gr}, N_{Gr})$ where
(i) $cl(Gr) = (I \times Tok, Typ^*, \models_{Gr})$,
(ii) $Th(Gr) = (Typ^*, \vdash_{Gr})$ where $v(\Gamma \vdash_{Gr} \alpha) = \inf_{i \in I} v(\Gamma \vdash^D_{\varepsilon_i} \alpha)$, and
(iii) $N_{Gr} = \cup_{i \in I}(\{i\} \times N^D_{\varepsilon_i})$.

So, $\vdash_{Gr}$ basically specifies $\Gamma \vdash_{Gr} \alpha$ iff for all $i \in I$, $\Gamma \vdash^D_{\varepsilon_i} \alpha$. It can be shown that $(Typ^*, \vdash_{Gr})$ satisfies the criterion of a (deterministic) regular theory as intersection of a set of regular theories also satisfies the conditions of a regular theory. We can also check that N_{Gr}, defined above, turns out to be the set of normal tokens for $Th(Gr)$. Let us consider any constraint $\Gamma \vdash_{Gr} \alpha \in S_{\vdash_{Gr}}$ (i.e. $v(\Gamma \vdash_{Gr} \alpha) \geq t$). Then by definition of $\vdash_{Gr}$, we can obtain $\Gamma \vdash^D_{\varepsilon_i} \alpha \in S_{\vdash^D_{\varepsilon_i}}$ for each $i \in I$. That is, for any $i \in I$, for any $a \in N^D_{\varepsilon_i}$, a satisfies the constraint $\langle \Gamma, \alpha \rangle$. That is, for any $a \in \cup_{i \in I} N^D_{\varepsilon_i}$, a satisfies $\langle \Gamma, \alpha \rangle$. That is, for any $i \in I$, $a\,(\in N^D_{\varepsilon_i})$ satisfies $\langle \Gamma, \alpha \rangle$ iff $(i, a)\,(\in Tok_{Gr})$ satisfies $\langle \Gamma, \alpha \rangle$. That is, for any $(i, a) \in \cup_{i \in I}(\{i\} \times N^D_{\varepsilon_i})$, $(i, a) \in N_{Gr}$. On the other hand, let $(i, a) \in N_{Gr}$. Then for any constraint $\Gamma \vdash_{Gr} \alpha$ of $Th(Gr)$, (i, a) satisfies the sequent $\langle \Gamma, \alpha \rangle$. That is, $a \in N^D_{\varepsilon_i}$ satisfies the sequent. So, $a \in \cup_{i \in I} N^D_{\varepsilon_i}$. Hence $(i, a) \in \cup_{i \in I}(\{i\} \times N^D_{\varepsilon_i})$. So, $N_{Gr} \subseteq \cup_{i \in I}(\{i\} \times N^D_{\varepsilon_i})$.

Following the notion of logic infomorphism in Barwise and Seligman (1997), we now show that there is a logic infomorphism from $LL_D(Gr)$ to $LL_D(T_i)$ for each T_i.

Proposition 7.11 *Given the local logics $LL_D(T_i)$ for T_i and $LL_D(Gr)$, let the function $f_i : (I \times Tok, Typ^*, \models_{Gr}, \vdash_{Gr}, N_{Gr}) \rightleftarrows (Tok, Typ^*, \models_i, \vdash^D_{\varepsilon_i}, N^D_{\varepsilon_i})$ be defined as follows.*
(i) $cl(Gr) \rightleftarrows cl(T_i)$ is an infomorphism (as defined in Definition 7.27)
(ii) $f_i : Th(Gr) \to Th(T_i)$, where $\hat{f}_i(\alpha) = \alpha$, and
(iii) for any $a \in N^D_{\varepsilon_i}$, $\check{f}_i(a) = (i, a)$.
Then $\check{f}_i(N^D_{\varepsilon_i}) \subseteq N_{Gr}$, f_i is a theory interpretation from $Th(Gr)$ to $Th(T_i)$, and hence $f_i : LL_D(Gr) \rightleftarrows LL_D(T_i)$ is a logic infomorphism.

Proof The proof of $\check{f}_i(N^D_{\varepsilon_i}) \subseteq N_{Gr}$ is straightforward from the definition of $\check{f}_i$ and N_{Gr}. Also, from the definitions of $\hat{f}_i$ and $\vdash_{Gr}$ it is immediate that if $\Gamma \vdash_{Gr} \alpha$, then $\hat{f}_i(\Gamma) \vdash^D_{\varepsilon_i} \hat{f}_i(\alpha)$. That is, $\hat{f}_i$ is a theory interpretation. Hence, that $f_i : LL_D(Gr) \rightleftarrows LL_D(T_i)$ is a logic informorphism follows directly from the definitions of $LL_D(Gr)$ and f_i. □

We now construct the logic for decision-maker, denoted as $LL_D(DM)$. In the construction of $LL_D(Gr)$, neither the semantic satisfiability relation $\models_{Gr}$ nor $\vdash_{Gr}$ play any distinctive role other than accumulating information from $LL_D(T_i)$'s. In contrast, $LL_D(DM)$, the logic for decision-maker is expected to play a role more than that of accumulation of information. Thus, we propose $LL_D(DM)$ to be such that it would base on $LL_D(Gr)$, and moreover would have its own local logic.

Deterministic Local Logic for Decision-Maker

The purpose of this section is to relate the notion of logic for a distributed network (of agents) (Barwise and Seligman 1997), and the theory of graded consequence, which endorses to have, generally, a different reasoning base for decision-making than that of the subject/language of concern. Keeping this in mind, after constructing the logics for individual experts following Barwise and Seligman's local logic, below we are going to design $LL_D(DM)$, the local logic for the decision-maker, in accordance with the development of GCT. Throughout this work, we stick to defining the theory of a local logic of each expert, i.e. $(Th(T_i), \vdash_{\varepsilon_i})$, based on the respective belief profiles. In the context of decision-maker, the same will be additionally constrained with the aggregation scheme of the semantic definition of graded consequence relation.

While designing the local logic for a group of experts $\{T_i\}_{i \in I}$ we have observed that the set of tokens for the group is nothing but the disjoint union of the copies of the tokens of each expert, and the theory consists of those sequents which are common to all the experts. In contrast to the local logic for the group, local logic for the decision-maker may contain a different set of tokens than that of the group as decision-maker is considered to be a different agent from those who belong to the group. In the context of Definition 7.35, for simplicity, decision-maker is considered to have all the tokens of the group and some more. So, if the local logic for the decision-maker is denoted by $LL_D(DM) = (Tok_{DM}, Typ^*, \models_{DM}, \vdash_{DM}, N_{DM})$, then the first condition we assume

$$\text{(i) } Tok_{Gr} \subseteq Tok_{DM}.$$

As for the information the decision-maker depends on the information gathered from the group, when restricted to the tokens of the group, i.e. Tok_{Gr}, the satisfiability relation of the decision-maker does not differ from that of the group. So we impose

$$\text{(ii) } \models_{DM} = \models_{Gr} \text{ when restricted to the set } Tok_{Gr}.$$

Moreover, the decision-maker can select some cases from N_{Gr} as her normal tokens. So, we consider

$$\text{(iii) } N_{DM} \subseteq N_{Gr}.$$

The initial set of constituents (C_{DM}), that is basic beliefs, of the decision-maker contains set of consequents F_i (set of final beliefs) of each expert T_i as well as some more from her own experience. Applying her own epistemic profile ε^{DM} on C_{DM} the set of consequents F_{DM} can be obtained. In the present context, F_{DM} is supposed to contain the union of all beliefs that are collected from each source/expert. So, we impose

(iv) The epistemic profile for the decision-maker is $\mathscr{B}_{DM} = (C_{DM}, F_{DM})$ where $\{F_i\}_{i\in I} \subseteq C_{DM}, F_{DM} = \varepsilon^{DM}(C_{DM}) \supseteq \cup_{C\in C_{DM}} C$.

There can be different ways of aggregating sets of beliefs of experts. So, it is quite evident that the status of a group of agent is not in general the same as that of the decision-maker. Below, we present one such possible way of designing the local logic of a decision-maker based on that of the experts.

Finally, it is the turn for designing $\vdash_{DM}$. To check to what extent a formula (type) α follows from a set Γ of formulae (types) the decision-maker aggregates the information collected from different T_i's. The definition here is based on the basic idea of semantic consequence.

In the present context, both Γ and F_i's are represented by four-valued membership functions, and so the notion of semantic consequence is in accordance with the notion of graded semantic consequence given in Definition 5.4. Besides, two more additional aspects have been considered here. One is that T_i's here have more concrete structures for their information systems, given in terms of $cl(T_i)$'s, and logics based on $cl(T_i)$'s and their epistemic profiles $(\mathscr{B}_i, \varepsilon_i)$'s. The other is that the concerned set of sentences (i.e. Typ^*) accommodates formulas with a single variable, representing situations, as well. So, in the present context $v(\Gamma \vdash_{DM} \alpha)$ is proposed with a modification of Definition 5.4. In order to aggregate the opinions of the experts T_i's, for checking if $\Gamma \vdash_{DM} \alpha$, the decision-maker looks at each individual expert's consequent F_i by applying the respective projection function Π_i on her consequent set F_{DM}. So, we need

(v) for each $i \in I$, a projection function $\Pi_i : F_{DM} \mapsto \{F_i\}_{i\in I}$ such that $\Pi_i(\alpha) = F_i$.

Now, for each token $b \in N_{DM}$, there must be a member of $\cup_{i\in I} N^D_{\varepsilon_i}$; i.e. it must be a normal token of some of the experts. In other words, as normal tokens of the decision-maker are some of those which are the collective normal tokens of the group of agents, each normal token of the decision-maker has a corresponding counterpart in the sets of normal tokens of some of the experts. So, we need

(vi) a projection function $\Pi_{Tok} : N_{Gr} \mapsto \cup_{i\in I} N^D_{\varepsilon_i}$ such that $\Pi_{Tok}(i, a) = a$.

In order to decide whether $v(\Gamma \vdash_{DM} \alpha) \geq t$ (i.e. $\Gamma \vdash_{DM} \alpha$), the decision-maker needs to check, for each of her normal token b and $i \in I$, whenever $\Gamma_{(x|\Pi_{Tok}(b))} \subseteq \Pi_i(F_{DM})$, $\alpha_{(x|\Pi_{Tok}(b))} \in \Pi_i(F_{DM})$ (i.e. $F_i(\alpha_{(x|\Pi_{Tok}(b))}) \geq t$). Hence finally we need

$$\text{(vii) } v(\Gamma \vdash_{DM} \alpha) = \inf_{i\in I} \inf_{b\in N_{DM}} \{\Pi_i(F_{DM})(\alpha_{(x|\Pi_{Tok}(b))}) : \Gamma_{(x|\Pi_{Tok}(b))} \subseteq \Pi_i(F_{DM})\}$$
$$= \inf_{i\in I} \inf_{b\in N_{DM}} \{F_i(\alpha_{(x|\Pi_{Tok}(b))}) : \Gamma_{(x|\Pi_{Tok}(b))} \subseteq F_i\}.$$

Thus, the local logic for decision-maker is given as follows.

Definition 7.35 $LL_D(DM) = (Tok_{DM}, Typ^*, \models_{DM}, \vdash_{DM}, N_{DM})$ is said to be the local logic for the decision-maker where the components of $LL_D(DM)$ follow the conditions (i) to (vii), mentioned above.

We have seen that $v(\Gamma \vdash_{DM} \alpha) = \inf_{i \in I} \inf_{b \in N_{DM}} \{F_i(\alpha_{(x|\Pi_{Tok}(b))}) : \Gamma_{(x|\Pi_{Tok}(b))} \subseteq F_i\}$ is designed in the line of Definition 5.4. Moreover, the axioms $(GC_f 1)$ to $(GC_f 3)$ of Sect. 5.2 are the many-valued counterparts of the conditions of a deterministic regular theory. So, following the Representation Theorem (Dutta and Chakraborty 2009), we can show that $(Typ^*, \vdash_{DM})$ forms a deterministic regular theory (in four-valued context).

Our next target is to establish a logic infomorphism from $LL_D(DM)$ to $LL_D(Gr)$. In order to do so, we shall impose some conditions on the relationship between $LL_D(Gr)$ and $LL_D(DM)$ bringing in some analogy from the decision-making story of Judith and Miranda. Example 7.1, thus, is extended as follows.

Example 7.2 The communication of some message from Judith to Miranda has taken place in a context where Miranda is not aware of every possible instances of Judith's information system and properties satisfied by them. Though Miranda and Judith do not have any access to their information systems, some message is communicated between them. In order to analyse the flow of information, let us consider a set of instances say, $inst_1, inst_2, inst_3, \ldots, inst_n$ as already available tokens at time t, before the occurrence of Judith's event. Let us call the specific event of Judith as $inst\text{-}J$. Miranda is not sure whether the situation of $inst\text{-}J$ satisfies *reflection of moonlight from water(x)* or not. So, to her *reflection of moonlight from water(*$inst\text{-}J$*)* does have the value u. But to Judith *reflection of moonlight from water(*$inst\text{-}J$*)* must have the value 0, as she knows that this is not the case. Despite that there are certain types like *light in mountain(x), dark night(x), troublesome location(x)* with respect to which some tokens from $inst_1, inst_2, \ldots, inst_n$ have the same behaviour as $inst\text{-}J$. That is, observationally $inst\text{-}J$ is equivalent to some instances from $inst_1, inst_2, \ldots, inst_n$ where *light in mountain(x), dark night(x), troublesome location(x)* have the same values. This brings about to draw Miranda that $inst\text{-}J$ is similar to those situations where someone was in trouble. So, two different tokens for two different agents may satisfy the same set of types (properties), and this in turn helps two agents, even without direct communication, to relate each others' mind. □

In the light of Example 7.2, below we are going to introduce an assumption in order to tie the classification of the group and that of the decision-maker in a thread.

(A_1) For a non-empty set Σ $(\subseteq Typ^*)$, there are $(i, a) \in Tok_{Gr}$ and $b \in Tok_{DM}$ such that $(i, a) \models_{Gr} \alpha$ iff $b \models_{DM} \alpha$ for $\alpha \in \Sigma$.

In such case, as $Tok_{Gr} \subseteq Tok_{DM}$, we say (i, a) and b are related by some invariant relation, denoted as $(i, a)\ R\ b$, and the context of the invariant relation is denoted as $I = (\Sigma, R)$. For instance, in the case of the above illustration one can choose $\Sigma = \{$light in mountain(x), dark night(x), troublesome location(x)$\}$. We are now going to consider the quotient local logic of the decision-maker.

Definition 7.36 Given $LL_D(DM)$ and (A_1), we define the quotient local logic of $LL_D(DM)$ as $LL_D(DM)_{|I} = (Tok_{DM|_I}, \Sigma, \models_{DM|I}, \vdash_{DM_{|\Sigma}}, N_{DM|I})$ where $Tok_{DM|I} = \{[a]_I : a \in Tok_{DM}\}$, $a \models_{DM} \alpha$ iff $[a]_I \models_{DM|I} \alpha$, $\vdash_{DM_{|\Sigma}}$ is the restriction of $\vdash_{DM}$ on Σ, and $N_{DM|I} = \{[a]_I : a \in N_{DM}\}$.

Proposition 7.12 *Given the local logics $LL_D(DM)_{|I}$ and $LL_D(Gr)$, we define the function*
$r : (Tok_{DM|_I}, \Sigma, \models_{DM|I}, \vdash_{DM} |_{\Sigma}, N_{DM|I}) \rightleftarrows (Tok_{Gr}, Typ^*, \models_{Gr}, \vdash_{Gr}, N_{Gr})$ *as follows:*
(i) $cl(DM)_{|I} \rightleftarrows cl(Gr)$ is an infomorphism where each $\alpha \in \Sigma$ from $cl(DM)_{|I}$ is mapped to itself, and each $a \in Tok_{Gr}$ is mapped to $[a]_I$ of $Tok_{DM|I}$ such that $[a]_I \models_{DM|I} \alpha$ iff $a \models_{Gr} \hat{r}(\alpha)$,
(ii) $r : Th(DM)_{|I} \to Th(Gr)$, where $\hat{r}(\alpha) = \alpha$, and
(iii) for any $a \in N_{Gr}$, $\check{r}(a) = [a]_I$.
Then $\check{r}(N_{Gr}) \subseteq N_{DM}$, $r : Th(DM)_{|I} \to Th(Gr)$ is a theory interpretation, and hence $r : LL_D(DM)_{|I} \rightleftarrows LL_D(Gr)$ is a logic infomorphism.

Proof The infomorphism $cl(DM)_{|I} \rightleftarrows cl(Gr)$ is clear from the construction of $LL_D(DM)_{|I}$. We first show that $r : Th(DM)_{|I} \to Th(Gr)$ is a theory interpretation.

Let $\Gamma \vdash_{DM_{|\Sigma}} \alpha$, i.e. $v(\Gamma \vdash_{DM} |_{\Sigma} \alpha)$
$= \inf_i \inf_{b \in N_{DM}} \{F_i(\alpha_{(x|\Pi_{Tok}(b))}) : \Gamma_{(x|\Pi_{Tok}(b))} \subseteq F_i\} \geq t$.
That is, for each $i \in I$, and $b \in N_{DM}$, $\Gamma_{(x|\Pi_{Tok}(b))} \subseteq F_i$ implies $F_i(\alpha_{(x|\Pi_{Tok}(b))}) \geq t$ (or in other words $\alpha_{(x|\Pi_{Tok}(b))} \in F_i$).

As $\hat{r}$ is an identity function, we need to show that $\Gamma \vdash_{Gr} \alpha$, i.e. $\Gamma \vdash^D_{\varepsilon_i} \alpha$ for each $i \in I$. If not, then for some $j \in I$, $\Gamma \nvdash^D_{\varepsilon_j} \alpha$. That is, for no $a \in Tok$, if $\Gamma_{(x|a)} \subseteq F_j$ then $\alpha_{(x|a)} \in F_j$.
This contradicts the assumption that for any $i \in I$, and $b \in N_{DM}$ (i.e. $b \in N_{Gr}$ and $\Pi_{Tok}(b) \in Tok$), $\Gamma_{(x|\Pi_{Tok}(b))} \subseteq F_i$ implies $\alpha_{(x|\Pi_{Tok}(b))} \in F_i$. Hence $\hat{r}(\Gamma) \vdash_{Gr} \hat{r}(\alpha)$.
Now, we have $S_{\vdash_{DM|\Sigma}} \subseteq S_{\vdash_{Gr}}$. That is, $\langle \Gamma, \alpha \rangle \in S_{\vdash_{DM|\Sigma}}$ implies $\langle \Gamma, \alpha \rangle \in S_{\vdash_{Gr}}$.
Now for $a \in N_{Gr}$, $a \in Rlz(\Gamma \vdash_{Gr} \alpha)$ as $\Gamma \vdash_{Gr} \alpha$. That is, $a \models_{Gr} x$ for all $x \in \Gamma$ implies $a \models_{Gr} \alpha$ iff $[a]_I \models_{DM|I} x$ for all $x \in \Gamma$ implies $[a]_I \models_{DM|I} \alpha$. So, $[a]_I \in Rlz(\Gamma \vdash_{DM} |_{\Sigma} \alpha)$. That is, for each $a \in N_{Gr}$, $\check{r}(a) = [a]_I \in N_{DM|I}$. Hence, $\check{r}(N_{Gr}) \subseteq N_{DM|I}$. Thus we have $r : LL_D(DM)_{|I} \rightleftarrows LL_D(Gr)$ is a logic infomorphism. □

The constructions made in this section, thus, shows that the way information collected from different sources ($\{T_i\}_{i \in I}$) is aggregated in the framework of the theory of graded consequence can be viewed as a flow of information, among different sources/experts and the decision-maker, as given in the following Fig. 7.1. The direction of the logic infomorphism is from the quotient local logic of the decision-maker to the local logic of $LL_D(Gr)$, which is basically a repertoire of information collected from different experts. The direction of logic infomorphism may be interpreted as the direction of *accessibility of information content*; that is we may read, $LL_D(DM)_{|I}$ can *access* the information collected in $LL_D(Gr)$, and $LL_D(Gr)$ can *access* the information available at each of $LL_D(T_i)$.

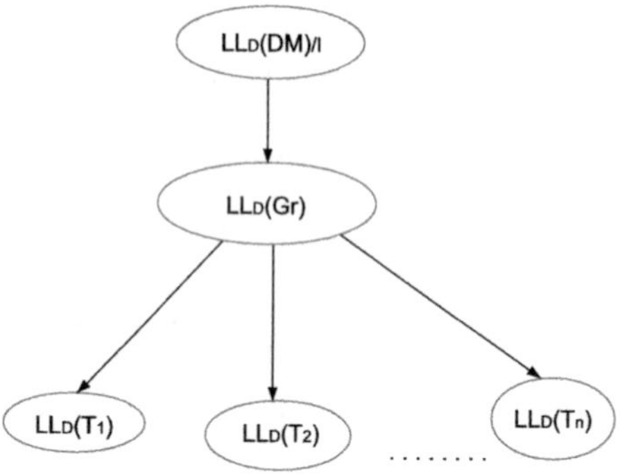

Fig. 7.1 Information flow in GCT

From the definitions of $(Typ^*, \vdash_{Gr})$ of $LL_D(Gr)$ and $(Typ^*, \vdash_{DM})$ of $LL_D(DM)$, one may think why at all we need to incorporate two different levels, one for the *group* and the other for the *decision-maker*. In this regard, some explanation is needed to make our purpose behind designing the *local logic for the group* and the *local logic for the decision-maker* clear.

(i) Though for determining α follows from Γ, both DM (decision-maker) and Gr (the group) are basing on each expert T_i, DM is equipped with more tokens and an extended set of initial beliefs, whereas that of Gr are just the accumulation of what is available at each expert T_i. Moreover, for Gr, there is no belief structure assumed; F_{DM} on the other hand, contains a belief structure and projection functions to access respective consequent beliefs F_i for each expert T_i. Moreover, N_{Gr}, the set of normal tokens of the group is just the disjoint union of the copies of the normal tokens of each expert. Whereas DM can be selective in choosing a subclass of N_{Gr} as her normal tokens. Advantage of DM over Gr may become more clear from the following example.

(ii) Let there be a collection of experts $\{T_i\}_{i \in I}$. Let for some α, $F_j(\alpha) = u$ for some $j \in I$, and for the rest i $(\neq j)$, $F_i(\alpha) > u$. If we consider Gr, then for some $\Gamma \subseteq Typ^*$, $v(\Gamma \vdash_{Gr} \alpha) = u$, i.e. $\Gamma \nvdash_{Gr} \alpha$. But in case of DM, as $F_{DM} \supseteq F_i$ for each i, DM might have the choice to eliminate F_j and consider the collection $\{T_i\}_{i \in I, i \neq j}$. In that case, $\vdash_{DM}$ can be refined to a new consequence relation $\vdash^r_{DM}$ such that $v(\Gamma \vdash^r_{DM} \alpha) = \inf_{i \neq j} \inf_{a \in N_{DM}} \{F_i(\alpha_{(x|\Pi_{Tok}(a))}) : \Gamma_{(x|\Pi_{Tok}(a))} \subseteq F_i\}$. The obtained theory $(Typ^*, \vdash^r_{DM})$ based on the rest of the experts of the collection still satisfies the criterion of regular theory. Moreover as the previous theory $(Typ^*, \vdash_{DM})$ can be interpreted in the new theory $(Typ^*, \vdash^r_{DM})$ we can establish a refinement, basically a logic infomorphism, from $(Tok_{DM}, Typ^*, \models_{DM}, \vdash_{DM}, N_{DM})$ to $(Tok_{DM}, Typ^*, \models_{DM}, \vdash^r_{DM}, N_{DM})$.

7.3 Sorites Paradox in the Light of GCT

Any discourse on vagueness invariably leads to Sorites paradox. As a 'proof' of this statement, we would suggest the readers to browse through the index of any book on vagueness, for example, the book 'Vagueness in Context' by Shapiro (2006), or the book entitled 'Theories of Vagueness' by Keefe (2000). In this section, we shall establish a connection of GCT with this paradox as GCT also deals with vagueness. Let us first describe (although it is a very well-known topic) what does this mean.

The paradox of the heap or 'Sorites paradox', as this and similar other paradoxes are called, has its origin in the Greek antiquity. Centuries back, Eubulides, the Megarian, observed that a chain of apparently valid arguments setting off with evidently true premises ultimately leads to a patently false conclusion. Eubulides' version of the paradox is as follows. Let there be a heap of sand. If one grain of sand is taken away it remains a heap. If another grain is taken away it still remains a heap. In fact, if n grains make a heap so do $n - 1$ grains, written a bit formally

$$\forall n(\text{Heap}(n) \rightarrow \text{Heap}(n-1)).$$

So, by repeated use of the logical rule of Universal Instantiation (UI) and Modus Ponens (MP), one arrives at the conclusion that when all the grains are taken away even then the heap is there—an obvious paradox. To be noted that the predicate 'is heap' is a vague one in the sense that after a certain stage, which is not also determinate, the collection of grains turn into borderline instance of the predicate. Secondly, to generate the paradoxical conclusion the number of steps in the above argumentation has to be large and finite. Thus, three things are essential—vagueness of a predicate P involved, a finite chain of objects a_i, $i = 1, 2, \ldots, n$, to which the predicate is being applied, and indistinguishability (with respect to P) of the successive objects a_i and a_{i+1}. In other words, the quantified inductive rule is not an essentiality in the premises. So, the Sorites paradox involving a vague predicate P can be stated as follows.

One starts with Pa_1 as true, and a collection of conditional premises of the form 'if Pa_i then Pa_{i+1}' for $1 \leq i \leq n$, which are also true. Then by repetitive application of MP (mark that UI is not required) one arrives at the obviously false conclusion Pa_n, for some suitably large n.

Other predicates usually referred to in the discussion of Sorites paradox are predicates like 'bald' or colour terms like 'red', etc.

Philosophers, semanticists and logicians have addressed the paradox from various (meta-theoretic) angles, and there are various accounts of that Fine (1975), Keefe (2000), Shapiro (2006), Parikh (1983). Of these efforts, one broad divide rests upon degree-theoretic and non-degree-theoretic approaches. Among the second group, some important approaches are the supervaluationist (Fine 1975; Keefe 2000), intuitionist (Dummett 1975), and contextualist (Graff 2000; Raffman 1996). We shall however present here a few approaches belonging to the first group, and within this group present how GCT may be used to offer a solution. This is to note that here we shall not make any kind of judgement between the solutions offered. We only want to emphasize upon that GCT presents an alternative approach to the problem.

7.3.1 *Different Approaches Towards Sorites Paradox*

In this subsection, we shall discuss a few approaches to deal with sorites paradox.

Tye's Solution for Sorites Paradox with a Three-Valued Logic

Tye (1994) adopts a three-valued semantics framed after Kleene's three-valued logic. The third value is called 'indefinite'. Tye observed that there are borderline bald men with say, n hairs, who would not cease to be bald by addition of one hair on his head. Thus, there is some n, for which both the statements 'a man with n hairs on his head is bald' and 'a man with $n + 1$ hairs on his head is bald' are indefinite. Hence according to Kleene's three-valued matrix, of one kind (Katsura 2008), for computing 'if-then' ('$\supset$'), with respect to such n, the conditional statement 'if a man with n hairs on his head is bald then a man with $n + 1$ hairs on his head is bald' will be indefinite. According to Tye, the initial few statements in the array, having the form 'a man with n hairs on his head is bald', where n ranges from 0 to 1,000,000 perhaps, are true, also the last few statements of the same form are false, and in between somewhere in the array, there are statements that are indefinite. But one could never say where in the array, the statements of the form 'a man with n hairs on his head is bald' cease to be true and become indefinite, and also at which point in the array indefinite statements end and false statements begin. Tye's approach respects tolerance of a vague predicate to minute changes. But by admitting one of the premises to be non-true he dissolves the paradox instead of giving any solution to the paradoxical situation where starting from true premises, following a valid rule(s) of inference, one arrives at a false conclusion.

Sorites Paradox from the Angle of Goguen's Fuzzy Logic

As an example of fuzzy logical approach to the sorites paradox, we present Goguen's (1968) approach based on fuzzy set theory. According to Goguen, the conditional statements of the form 'if a man with i hairs on his head is bald then a man with j hairs on his head is bald' should provide a way so that from the truth value of 'a man with i hairs on his head is bald' one can derive the truth value of the statement 'a man with j hairs on his head is bald'. He suggested to represent the conditional premise by a fuzzy relation $H(i, j)$, read as, 'the relative baldness of a man with j hairs on his head with respect to the baldness of a man with i hairs on his head' so that $H(i, j)$ satisfies the following equation.

$$B(j) = H(i, j).B(i),$$

where 'B' denotes a fuzzy set corresponding to 'bald'. That is, $H(i, j) = \frac{B(j)}{B(i)}$. Now, as the fuzzy set B, representing 'bald' is continuous and $B(i)$ monotone decreasing in nature, for some k, $H(k-1,k)$ is non-unit. Hence, if $B(0)$ is 1, then for the series of natural numbers, from 0 to 1,000,000, $B(1,000,000) = \Pi_{i=1}^{1,000,000} H(i-1,i)$, which is a result of repetitive product of non-unit numbers, and that might be very close to zero as the number of steps increases. This explains why the conclusion of the sorites appears to be false.

There are two problems in this solution. The first is the same as in Tye's case, where one of the premises is admitted to be non-true. The second is the case where it is admitted that for some k, $H(k-1,k)$ is non-unit, i.e. $B(k-1) > B(k)$. This goes against the idea that a vague predicate is tolerant to mintue changes.

Gaines Proposal for Sorites Paradox

Similar to Goguen, Gaines (1977) interprets the implication in such a way that the value of the conditional premise is not necessarily 1 (true). He suggests to choose the value of the conditional from among the following:

$$a \to b = \min(1, 1 - a + b)$$
$$a \to b = \min(1, \frac{b}{a}).$$

It is known at present that the first is Łukasiewicz's implication and the second is Goguen's implication (both residua). After that, as in Goguen's approach, he takes the product $P(a_{i-1}.(P(a_{i-1}) \to P(a_i))$ to obtain the value of $P(a_i)$. The predicate P giving monotonically decreasing values $(P(a_{i-1}) \to P(a_i)) < 1$, and hence $P(a_i) < P(a_{i-1})$. As the number of application of the MP rule increases the value of $P(a_i)$ gradually diminishes and approach to 0. Gaines' representation in Gaines (1977) is slightly different but the above is the essence.

Edgington's Many-Valued Approach to Sorites Paradox Based on Probability

Edgington (1997) also embraces a degree-theoretic approach to give an account of reasoning in vague context. What distinguishes her approach from other degree theories, including GCT, is that it uses probability theory as providing a general structure for calculating logical compositions (not necessarily truth functional) of different degrees of verity of sentences. In her approach, each conditional premise of the form '$Pa_n \supset Pa_{n+1}$ has a degree of verity slightly less than clearly true (1). However, as the deduction proceeds small unverities (1 - degree of verity) of the premises mount up to yield a conclusion which is clearly false (0). Yet each step of the argument is valid; because, the unverity of the conclusion never exceeds the sum of the unverities of the premises. This is how Edgington distinguishes arguments, where fall in the values of the conclusion is constrained by the values of the premises, from 'genuinely invalid' arguments, where such constraint does not work.

7.3.2 Approach to Sorites Paradox in the Context of GCT

Theory of graded consequence has an advantageous point over the other theories in that it admits a degree of an inference rule other than 1; that is, it has a formal mechanism to admit weak inference rules. While in other approaches the conditional used has been accepted to be weak, in GCT the rule (in this case MP) is taken to be of strength less than the full (1). Let us take B to denote the predicate 'bald', and the sequence $a_1, a_2, \ldots, a_n$, as before as B-applicable objects, n is a suitably large number. The number of hairs on the head a_{k+1} is only a few more than a_k so that in ordinary language if a person with a_k number of hairs is considered to be bald so is considered to be a person with a_{k+1} number of hairs. Also, a person with a_1 number of hairs is assumed to be definitely bald. In a typical case of vague predicate, represented by fuzzy set, a few a_k's towards the beginning have value $B(a_k) = 1$, a few towards the end have $B(a_k) = 0$, and for some in the middle $B(a_k)$ is around .5. These matters are only definite. There cannot be in principle a pair (a_k, a_{k+1}) such that $B(a_k) = 1$ and $B(a_{k+1}) < 1$ or similarly $B(a_k) > 0$ and $B(a_{k+1}) = 0$. GCT represents this situation in the following way (Dutta et al. 2013).

Let us reframe the paradox. In the context of classical logic if $\alpha_1, \alpha_2, \ldots, \alpha_n$ are formulas then $\{\alpha_1, \alpha_1 \supset \alpha_2, \alpha_2 \supset \alpha_3, \ldots, \alpha_{n-1} \supset \alpha_n\} \vdash \alpha_n$ (syntactically). Because of soundness, $\{\alpha_1, \alpha_1 \supset \alpha_2, \alpha_2 \supset \alpha_3, \ldots, \alpha_{n-1} \supset \alpha_n\} \models \alpha_n$ holds. That is, for all states of affair T, if $T(\alpha_1) = T(\alpha_1 \supset \alpha_2) = \ldots = T(\alpha_{n-1} \supset \alpha_n) = 1$, then $T(\alpha_n) = 1$ also. Now when α_i represents $B(a_i)$, a_i *is bald*, for $i = 1, 2, \ldots, n$, $T(\alpha_1) = T(B(a_1)) = B(a_1)$ following the Tarskian paradigm a_1 *is bald is true if and only if* a_1 *is bald*. That is, here we have an actual state of affairs T such that, $T(\alpha_1) = B(a_1)$, $T(\alpha_1 \supset \alpha_2) = B(a_1) \rightarrow_o B(a_2)$, and so on, where $\rightarrow_o$ is an operator for computing object language implication $\supset$. Now the paradox arises because $T(\alpha_1) = B(a_1) = 1$, and $T(\alpha_k \supset \alpha_{k+1}) = B(a_k) \rightarrow B(a_{k+1}) = 1$ for $1 \leq k \leq n-1$, but $T(\alpha_n) = B(a_n) = 0$. So, where does the problem lie? GCT provides an answer to the problem by accommodating that the rule MP in this context is not of full strength; it is rather a weak rule with some grade $|MP| < 1$. Why is that, and how would the value $|MP|$ be computed will be discussed afterwards. Assuming $|MP| < 1$, let us recall the derivation procedure in GCT (cf. Chap. 2). We get the following chain of an argumentation.

It should be noted that at stage 3, the value $1 *_m 1 *_m |MP|$ is not the actual value of the sentence a_2 is bald, but the value of the derivability degree of this sentence from the two premises. $B(a_2)$ is quite likely to be equal to 1, but what $1 *_m 1 *_m |MP|$ is showing is the value transmitted in the sentence by the rule whose strength is $|MP| < 1$. What is certain that because of soundness the value $1 *_m 1 *_m |MP| \leq B(a_2)$.

Thus, the value of the derivation that a_n is bald would be $|MP|^{n-1}$, and this value gets transmitted to the conclusion as well. If $|MP| < 1$ and n is sufficiently large, $|MP|^{n-1}$ would be close to 0. Since the MP rule is weak the value transmitted to α_n through the syntactic derivation process can be made as small as we want. So, according to this framework of syntactic derivation, it is no longer a necessity that

Table 7.2 Derivation of Sorites chain in GCT

			Step value	Value transmitted (value of derivation)
1.	$\alpha_1\ (\equiv B(a_1))$	(Pr)	$B(a_1) = 1$	1
2.	$\alpha_1 \supset \alpha_2\ (\equiv B(a_1) \supset B(a_2))$	(Pr)	$B(a_1) \to_o B(a_2) = 1$	$1 *_m 1$
3.	$\alpha_2\ (\equiv B(a_2))$	(MP)	$\|MP\|$	$1 *_m 1 *_m \|MP\|$
4.	$\alpha_2 \supset \alpha_3\ (\equiv B(a_2) \supset B(a_3))$	(Pr)	$B(a_2) \to_o B(a_3) = 1$	$1 *_m 1 *_m \|MP\| *_m 1$
5.	$\alpha_3\ (\equiv B(a_3))$	(MP)	$\|MP\|$	$1 *_m 1 *_m \|MP\| *_m 1 *_m \|MP\|$
	$\vdots$		$\vdots$	$\vdots$

starting from premises with step value 1 one has to reach a conclusion with step value 1 as well.

GCT gives a measure of the strength of derivability of the sentence 'a_n is bald' from the sentences 'a_1 is bald' and 'if a_k is bald then a_{k+1} is bald', $k = 1, 2, \ldots n-1$, all of them being true (1). It also says that the actual value of the conclusion 'a_n is bald' does not fall below $|MP|^{n-1}$. Here is the advantage of this model that it keeps the paradoxical nature of the argument alive since $B(a_n)$ is in fact 0. But this is only mathematical technicality, if the value $B(a_n) = 10^{-1000}$, it is greater than 0 but the sentence 'a_n is bald' is recognized as false by the language users. Given any small ε ($>$0) it is always possible to have n such that $|MP|^{n-1} < \varepsilon$.

The crucial point is that $0 < |MP| < 1$. How could this value be determined? Does GCT offer some method or is it totally ad hoc? To have this it is necessary to depend on the semantic base $\{T_i\}_{i\in I}$, a collection of T_i's assigning values to the atomic sentences, namely 'a_k is bald'. As formulated in Chap. 2, we have

$$\begin{aligned}|MP| &= \inf_{\alpha,\beta} gr(\{\alpha, \alpha \supset \beta\} \approx\!\!\!\!|_{\{T_i\}_{i\in I}}\ \beta).\\ &= \inf_{\alpha,\beta} \inf_i [\{T_i(\alpha) \wedge (T_i(\alpha) \to_o T_i(\beta))\} \to_m T_i(\beta)] \qquad \ldots(A).\end{aligned}$$

Since $a \to_m b = 1$ iff $a \leq b$, we need to consider only those α, β and T_i's such that

$$T_i(\alpha) \wedge (T_i(\alpha) \to_o T_i(\beta)) > T_i(\beta) \qquad \ldots(A1).$$

Are there plenty of such α, β and T_i's in this context of Sorites like situation? The above inequality also implies that

$$T_i(\alpha) \geq T_i(\alpha) \wedge (T_i(\alpha) \to_o T_i(\beta)) > T_i(\beta).$$

So, we look for α, β such that $T_i(\alpha) > T_i(\beta)$. We focus on the atomic formulae only, i.e. sentences appeared at steps 1, 3, 5, (Table 7.2) and so on up to n. Since B is monotonic decreasing function there are recognizable cases like a_k and a_l with $k < l$, such that $B(a_k) > B(a_l)$. This function B is supposed to represent the actual membership function for baldness, which we assume to exist but not always knowable. It is not expected that $T_i(a_k \text{ is bald}) = T_i(\alpha_k)$ is to be exactly $B(a_k)$ and $T_i(a_l \text{ is bald}) = T_i(\alpha_l)$ is to be exactly $B(a_l)$, but these valuations $\{T_i\}_{i\in I}$ should comply with the fact that a_k is more bald than a_l, i.e. $T_i(\alpha_k) > T_i(\alpha_l)$. There is further restriction to be followed.

(A1) contains a component $T_i(\alpha) \rightarrow_o T_i(\beta)$ in the left-hand side of $>$. Since we have not yet imposed any restriction on the object-level $\rightarrow_o$, $T_i(\alpha) > T_i(\beta)$ does not automatically imply $T_i(\alpha) \rightarrow_o T_i(\beta) \neq 1$. So, there may be two possibilities, viz. $T_i(\alpha) \rightarrow_o T_i(\beta) = 1$ and $T_i(\alpha) \rightarrow_o T_i(\beta) < 1$. In the first case, the first restriction $T_i(\alpha) > T_i(\beta)$ becomes compatible with (A1) automatically. In the second case, one needs to choose α, β such that (A1) holds. Generally, the many-valued implications, which can be chosen for $\rightarrow_o$, obey the property that $T_i(\alpha) \rightarrow_o T_i(\beta) \geq T_i(\beta)$. An implication that gives $T_i(\alpha) \rightarrow_o T_i(\beta) = T_i(\beta)$ cannot be chosen since in that case (A1) will fail. Hence taking into consideration the above constraints, α, β and T_i are to be properly chosen.

What are these T_i's? Technically these are mapping from the set of atomic sentences to the truth-value set [0, 1]. These will be mathematically infinite in number. But we pick a handful of them, the 'experts', from within the community of language users who will be asked to assign the degree of baldness according to his/her perception and experience. Thus, for the sentence α_k, which is 'a_k is bald', the expert T_i will assign a value $T_i(\alpha_k)$ as the degree of baldness of a person with a_k number of hairs. A set of such experts would constitute the set $\{T_i\}_{i\in I}$, and based on their opinion the strength of the rule MP will be calculated by (A). Expectedly, the language user community will accept the strength of the rule since T_i's are experts in the use of the predicate 'bald'.

It may not be necessary to range α, β over the whole set of atomic sentences. A random sampling on the atomic sentences may be sufficient for a particular case. Of course, for an efficient use of the rule and for better understanding a particular Sorites like paradoxical situation, it is good to include as many atomic sentences as possible.

Hence, in opposition to Tye's many-valued approach, Goguen's fuzzy approach, and Edgington's degree-theoretic approach, GCT neither needs to assume one of the premises to be non-true nor needs to assume existence of a cut off point violating the linguistic character of vague predicates. However, GCT banks on the weakness of the MP rule.

References

Alcalde, C., Burusco, A., Fuentes-Gonzlez, R.: A constructive method for the definition of interval-valued fuzzy implication operators. Fuzzy Sets Syst. **153**(2), 211–227 (2005)

Barwise, J., Seligman, J.: *Information Flow: The Logic of Distributed Systems*. Cambridge University Press (1997)

Bedregal, B.R.C., et al.: On interval fuzzy S-implications. Inf. Sci. **180**, 1373–1389 (2010)

Bedregal, B.R.C., Santiago, R.H.N.: Interval representations, Łukasiewicz implicators and Smets-Magrez axioms. Inf. Sci. **221**, 192–200 (2013)

Bedregal, B.R.C., Takahashi, A.: The best interval representations of t-norms and automorphisms. Fuzzy Sets Syst. **157**, 3220–3230 (2006)
Chakraborty, M.K.: Graded consequence: further studies. J. Appl. Non-Classical Log. **5**, 227–237 (1995)
Chakraborty, M.K., Basu, S.: Graded consequence and some metalogical notions generalized. Fundam. Inform. **32**, 299–311 (1997)
Chakraborty, M.K., Dutta, S.: Graded consequence revisited. Fuzzy Sets Syst. **161**, 1885–1905 (2010)
Cornelis, C., Deschrijver, G., Kerre, E.E.: Implication in intuitionistic fuzzy and interval-valued fuzzy set theory: construction, classification, application. Int. J. Approx. Reason. **35**, 55–95 (2004)
Deschrijver, G., Kerre, E.E.: Classes of intuitionistic fuzzy t-norms satisfying the residuation principle, Int. J. Uncertain. Fuzziness Knowl.-Based Syst. **11**, 691–709 (2003)
Deschrijver, G., Cornelis, C.: Representability in interval-valued fuzzy set theory. Int. J. Uncertain. Fuzziness Knowl.-Based Syst. **15**(3), 345–361 (2007)
Dretske, Fred I.: Knowledge and the Flow of Information. The MIT Press, Cambridge, Massachusetts (1981)
Dummett, M.: Wang's paradox. Synthese **30**, 301–324 (1975)
Dutta, S., Basu, S., Chakraborty, M.K.: Many-valued logics, fuzzy logics and graded consequence: a comparative appraisal. In: Lodaya, K. (ed.) Proceedings of ICLA 2013, LNCS 7750, pp. 197–209. Springer, Heidelberg (2013)
Dutta, S., Bedregal, B.R.C., Chakraborty, M.K.: Some instances of graded consequence in the context of interval-valued semantics. In: Banerjee, M., Krishna, S. (eds.) ICLA 2015, LNCS 8923, pp. 74–87 (2015)
Dutta, S., Chakraborty, M.K.: Rule modus ponens vis-á-vis explosiveness condition in graded perspective. In: Lowen and Roubens (eds.) Proceedings of the International Conference on Rough Sets, Fuzzy Sets and Soft Computing, held on 5–7 November 2009 at Tripura University, pp. 271–284. SERIALS (2009)
Dutta, S., Skowron, A., Chakraborty, M.K.: Information flow in logic for distributed systems: extending graded consequence. Inf. Sci. **491**(2019), 232–250 (2019)
Edgington, D.: Vagueness by degrees. In: Keefe, R., Smith, P. (eds.) Vagueness: A Reader, pp. 294–316. MIT Press, Massachusetts (1997)
Fine, K.: Vagueness, truth and logic. Synthese **30**, 265–300 (1975)
Gaines, B.R.: Foundations of fuzzy reasoning. In: Gupta, M.M., Saridis, G.N., Gaines, B.R (eds.) Fuzzy Automata and Decision Processes, pp. 19–75. Elsevier, North Holland Inc., New York (1977)
Gasse B.V., et al.: On the properties of a generalized class of t-norms in interval-valued fuzzy logics. New Math. Nat. Comput. **2**(1), 29–41, World Scientific Publishing Company (2006)
Goguen, J.A.: The logic of inexact concept. Synthese **19**, 325–373 (1968)
Graff, D.: Shifting sands: an interest-relative theory of vagueness. Philos. Topics **28**, 45–81 (2000)
Katsura, S.: Nagarjuna and the tetralemma. In: Silk, J.A (ed.) Buddhist Studies: The Legacy of Godjin M. Nagas. Motilal Banarsidass Pvt. Ltd., Delhi (2008)
Keefe, R.: *Theories of Vagueness*. Cambridge University Press (2000)
Keplicz-Dunin, B., Szałas, A.: Taming complex belief. In: Nguyen, N.T. (ed.) Transactions on CCI XI, LNCS 8065, pp. 1–21 (2013)
Li, D., Li, Y.: Algebraic structures of interval-valued fuzzy (S, N)-implications. Int. J. Approx. Reason. **53**(6), 892–900 (2012)
Parikh, R.: The problem of vague predicates. In: Cohen, R.S., Wartofsky, M. (eds.) Language, Logic, and Method, pp. 241–261. D. Ridel Publishing Company (1983)
Pawlak, Z., Skowron, A.: Rough sets and Boolean reasoning. Inf. Sci. **177**(1), 41–73 (2007a)
Pawlak, Z., Skowron, A.: Rough sets: some extensions. Inf. Sci. **177**(1), 28–40 (2007b)
Pawlak, Z., Skowron, A.: Rudiments of rough sets. Inf. Sci. **177**(1), 3–27 (2007c)
Pawlak, Z.: Rough sets. Int. J. Comp. Inf. Sci. **11**, 341–356 (1982)

Raffman, D.: Vagueness and context relativity. Philos. Stud. **81**, 175–192 (1996)

Shapiro, S.: *Vagueness in Context*. Clarendon Press (2006)

Shoesmith, D.J., Smiley, T.J.: Multiple Conclusion Logic. Cambridge University Press, Cambridge (1978)

Tye, M.: Sorites paradoxes and the semantics of vagueness. In: Tomberlin, J.E. (ed.) Philosophical Perspectives: Logic and Language, vol. 8, pp. 189–206. Ridgeview Publishing Co., Atascadero (1994)

Vitória, A., Małuszyński, J., Szałas, A.: Modeling and reasoning with paraconsistent rough sets. Fundam. Inform. **97**, 405–438 (2009)

Index

M. K. Chakraborty and S. Dutta, *Theory of Graded Consequence*, Logic in Asia: Studia Logica Library, https://doi.org/10.1007/978-981-13-8896-5